AIRLINE DEREGULATION AND LAISSEZ-FAIRE MYTHOLOGY

Airline Deregulation and Laissez-Faire Mythology

PAUL STEPHEN DEMPSEY
& ANDREW R. GOETZ

Q

QUORUM BOOKS
WESTPORT, CONNECTICUT · LONDON

387.71
D389a

Library of Congress Cataloging-in-Publication Data

Dempsey, Paul Stephen.
    Airline deregulation and laissez-faire mythology / Paul Stephen
Dempsey and Andrew R. Goetz.
        p.   cm.
    Includes index.
    ISBN 0–89930–693–4 (alk. paper)
    1. Airlines—United States—Deregulation.   2. Aeronautics,
Commercial—United States—Deregulation.   3. Aeronautics and state—
United States.   I. Goetz, Andrew R.   II. Title.
HE9803.A4D46   1992
387.7′1—dc20   91–35688

British Library Cataloguing in Publication Data is available.

Library of Congress Catalog Card Number: 91–35688
ISBN: 0–89930–693–4

First published in 1992

Quorum Books, 88 Post Road West, Westport, CT 06881
An imprint of Greenwood Publishing Group, Inc.

Printed in the United States of America

The paper used in this book complies with the
Permanent Paper Standard issued by the National
Information Standards Organization (Z39.48–1984).

10   9   8   7   6   5   4   3   2   1

To the Honorable
L. Welch Pogue

CONTENTS

PREFACE

Deregulation is a rather peculiar phenomenon. Its most fervent proponents continue to embrace it, not merely as an abstract economic theory, but with political, almost theological, devotion. No matter what evidence is adduced of widespread failure (and there is plenty), they tenaciously insist such evidence can be reinterpreted as proof of the success of deregulation. Some go so far as to assert that its failures can be attributed to not carrying deregulation far enough. Privatize the airports and let foreign airlines in, they insist, and we will at last achieve textbook levels of perfect competition.

Airlines were among the first of the major infrastructure industries to be deregulated. With the promulgation of the Airline Deregulation Act of 1978, Congress took the unprecedented step of sunsetting a major regulatory agency—the Civil Aeronautics Board—which had been established four decades earlier.

Beginning in the Carter administration, and reaching its zenith in the Reagan administration, federal oversight of industries as diverse as airlines, buses, railroads, trucking, telephones, cable television, radio and television broadcasting, banking, savings and loans, and oil and gas was significantly trashed. The transformation and radical shrinking of government proceeded along two, sometimes independent planes. Congress passed major legislation mandating various forms of deregulation between 1976 and 1985, while successive presidents appointed free market theologians to the regulatory agencies with the mission to exceed their legislative mandates and ignore their oaths of office.

The laissez-faire economists who convinced Congress to promulgate the Airline Deregulation Act of 1978 assured us that deregulation would result neither in increased concentration nor in destructive competition. This was true, they insisted, because the industry was structurally competitive, possessed few economies of scale, and was impeded by few barriers to entry. This book compares those predictions and assurances with the unfortunate results of deregulation:

- Under deregulation, the airline industry lost all of the money it made since the Wright Brothers' inaugural flight at Kitty Hawk in 1903, and $1.5 billion more.
- After more than 200 bankruptcies and 50 mergers, we now fly the oldest and most repainted fleet of aircraft in the developed world.
- Of the 176 airlines to which deregulation gave birth, only one remains and, as of 1992, it too was in bankruptcy.
- In 1991, fully 30 percent of the nation's fleet capacity was in bankruptcy or close to it.
- All the U.S. airlines together are now worth less than Japan Airlines individually.
- Despite predictions to the contrary, deregulation has produced the highest level of national and regional concentration in history.
- Although more people are flying than ever before, the percentage increase in domestic airline passenger boardings was lower during the first decade of deregulation than in every decade that preceded it.
- While most passengers now fly on a discounted ticket, the full fare has risen sharply under deregulation, far exceeding the rate of inflation, and the discounts are now encumbered with onerous prepurchase, nonrefundability and Saturday-night-stayover restrictions. Today's airline ticket is therefore an inferior product compared to its counterpart under regulation, which provided passengers with considerable flexibility.
- Despite allegations to the contrary, average real fuel-adjusted ticket prices are higher than they would have been had the pre-deregulation trend continued. Pricing has not only increased above pre-deregulation trend levels, it has grown monstrously discriminatory.
- Industry costs increased sharply under deregulation, while the long-term trend in productivity improvements fell flat.
- Hubbing-and-spoking, the dominant megatrend on the deregulation landscape, has caused some air travel to regress back to the DC-3 era, robbing aviation of its inherent advantage and people's most precious commodity—time.
- Business travelers lose billions of dollars in productivity as a result of circuitous and time-consuming hub-and-spoke operations.
- Service has declined under deregulation, while consumer fraud has increased.
- Although fatality statistics do not reflect it, the margin of safety has also declined.
- Labor-management relations have deteriorated.
- Americans now rate airlines as the industry in which they have the least confidence.

Neither economic nor equity goals have been advanced by deregulation. The assumptions on which it was based—that there were few economies of scale in aviation, that destructive competition was unlikely, that "contestability" of markets (the purported ease of potential entry) would discipline pricing—have proven false.

The time has come to reconsider the experiment of deregulation. Air transport is too critical to the productivity of the economy and the well-being of our citizens to abandon it to private concentrations of market power. This book examines the industry, its history, and the metamorphosis of deregulation. It also sets forth an agenda for legislative reform.

We are friends of this industry, not enemies of it. We recognize and appreciate the fundamental role commercial aviation plays in supporting the nation's commerce, communications, and national defense. We do not believe that government should apply command economy–type restrictions over price and supply. We do believe that somewhere between the regulatory regime established for airlines in 1938, and the contemporary environment of laissez-faire market Darwinism, lies the appropriate level of government oversight for this critical infrastructure industry.

We recognize the unpopularity of our position, and the power and strength of the opposition to any meaningful restoration of the public interest to airlines. But we would rather be right than loved.

Despite its profound economic losses, the industry itself opposes reregulation. The few survivors are on the verge of realizing the dream of every industrialist—to control an unregulated oligopoly (and in many markets, a monopoly) providing an essential infrastructure service that the public cannot do without.

The free market, laissez-faire movement is a passionate one. Indeed, not since the Bolshevik Revolution has the discipline of economics embraced an ideology with such zeal. Most of the Washington-based laissez-faire think tanks will fight to the death the restoration of any responsible government oversight of this industry. Notwithstanding their inability to predict the future with any respectable degree of accuracy, economists tend to view their discipline as a science ("dismal" though they concede it to be), and tend to see truth clearly. Only deeply religious people appear able to make the leap of faith from belief (economists call it theory) to reality that some economists do.

One particular Washington think tank has been at the forefront of the tenacious "deregulation-is-a-success" movement. Housed in a massive concrete edifice on Massachusetts Avenue in Washington, D.C., the Brookings Institution developed a reputation as a leading liberal think tank in an earlier part of this century. Paradoxically, its founder was a man who understood the importance of government as a modest participant in the economy. In the 1930s, Brookings' economists were at the forefront of New Deal efforts to regulate a number of important infrastructure industries, including transportation.

If Rip Van Winkle had fallen asleep in 1938 and awoken a century later, he would have been astounded at the transformation that has overcome Brookings. Beginning in the 1980s, Brookings published a series of studies proving conclusively that transportation deregulation saved consumers billions and billions of dollars. These findings were applauded on the editorial pages of major American newspapers, including the *New York Times* and *Wall Street Journal*. We devote a full chapter to Brookings.

In May 1990, Brookings put on a conference entitled "Deregulation: Success or Failure?" A balanced program it was not. It should have been entitled: "Deregulation: A Splendid, Magnificent, Unbelievable Success!"

All the nation's prominent deregulators were assembled to praise the remarkable achievements of deregulation. One after another, Alfred E. Kahn, Darius W. Gaskins, Jr., Steven Morrison, Michael E. Levine, and James C. Miller III patted each other on the back for their tremendous public policy contributions in dismantling government. All lavished accolades on their own and each other's brilliance in leading the country to the pinnacle of economic truth. They were the High Priests—the repositories of all truth to be known. It was deregulation's day in the sunshine. Not one representative of a contrary viewpoint had been invited to speak.

They spoke with contempt about those foolish enough to oppose them. Fred Kahn scornfully observed, "When they search for someone to appear on the MacNeil-Lehrer News Hour to oppose deregulation, they have to look under rocks."

So be it. We crawled out from under rocks to write this book because the story needs to be told: the Emperor has no clothes. Deregulation is a failure.

ACKNOWLEDGMENTS

The authors are deeply indebted to Kevin Quinn, staff economist at the Economic Policy Institute, who provided the airline pricing analysis and supporting appendix to chapter 21 of this study. The authors would also like to thank Theodore P. Harris and Harvey Wexler for their insights on airline economics and corporate policy. The authors also wish to thank the Economic Policy Institute for allowing them to publish material originally appearing in their study "Flying Blind: The Failure of Airline Deregulation," authored by Paul Stephen Dempsey.

Part I

AN INTRODUCTION TO THE DEREGULATED AIRLINE INDUSTRY

1

INTRODUCTION

Few industries inspire the passion that airlines do. The romantic allure of exotic destinations to which airlines provide access makes commercial aviation among the most glamorous of industries. Defying the law of gravity still gives many travelers sweaty palms on takeoff and landing. And few industries are as fundamentally important to the nation's commerce, communications, and national defense as is aviation.

All that has made this an industry with sex appeal, attracting suitors like Frank Lorenzo, Carl Icahn, Alfred Checchi, Donald Trump, Peter Ueberroth, Jay Pritzker, and Marvin Davis. Like the railroad robber barons of the nineteenth century, they all want membership in an exclusive and dwindling club of powerful entrepreneurs, dominating the travel patterns of 250 million Americans and their largest cities. And the profound change in government policy during the last decade—deregulation—let them loose while sending service, ticket prices, route patterns, the margin of safety, and the identity of the carriers painted on the fuselages of aircraft on an unprecedented roller-coaster ride. This book takes the reader on that ride.

Transportation is among the nation's most important industries, accounting for nearly 18 percent of the gross national product.[1] Of those infrastructure industries that have been traditionally either regulated or operated by industrialized nations (i.e., transportation, communications, and energy), in the United States, transportation has been deregulated more thoroughly than any other. And among the several modes of transport (i.e., air, rail, water, bus, and motor), airlines have been subjected to more comprehensive deregulation than any other.

It was not always so. The metamorphosis of governmental policy in this industry has been profound. The first airlines were encouraged by governmental subsidies, mainly to carry the mail. But much of the economic regime in which they developed was laissez-faire before 1938, when our federal government regulated the industry, and has been laissez-faire since 1978, when the industry was deregulated. Sandwiched between are four decades of economic regulation.

Paradoxically, transportation was the nation's first industry to be regulated by our federal and state governments and, a century later, the first to be deregulated. One can only observe with fascination that the transportation industry has come full circle, from its genesis in an unrestrained laissez-faire economic environment, through almost a century of comprehensive governmental regulation of entry, rates, and other corporate activities, and now back again to the unconstrained free market. The excesses of the market preceded deregulation, and those excesses have reappeared under deregulation.

Market failure gave birth to economic regulation. In the late nineteenth century, pricing discrimination and destructive competition in the railroad industry prompted Congress to establish our nation's first independent regulatory agency, the Interstate Commerce Commission, in 1887.[2] During the Great Depression, Congress concluded that the economic condition of the airline industry was unstable and that a continuation of its anemic condition could imperil its tremendous potential to satisfy national needs for growth and development. To avoid the deleterious impact of "cutthroat," "wasteful," "destructive," "excessive," and "unrestrained" competition and to avoid the economic "chaos" that had so plagued the rail and motor carrier industries, Congress sought to establish a regulatory structure similar to that devised for those industries that had also been perceived to be "public utility" types of enterprises.[3] Three years after motor carriers were brought under the regulatory umbrella, Congress added airlines to the regulatory scheme, promulgating the Civil Aeronautics Act of 1938. In so doing, Congress created a new agency to regulate this industry, the Civil Aeronautics Board (CAB).[4]

Beginning in the late 1970s, regulatory failure became the catalyst for deregulation. Various forms of de jure and de facto interstate deregulation resulted both from legislation passed by Congress in the mid-1970s and early 1980s and from the appointment by Presidents Carter and Reagan to the federal regulatory commissions of individuals fervently dedicated to deregulation.

The movement in favor of a reduced governmental presence found support on both ends of the political spectrum. America became infected by a mass psychology of antagonism toward government, stimulated on the Right by the Great Society and the growth of government and taxation and on the Left by Watergate and the war in Vietnam. For once, both sides viewed government as an enemy, rather than a friend.

In the 1960s and early 1970s, economists published a generous volume of literature critical of economic regulation.[5] Principal among their criticisms was that pricing and entry restrictions gave consumers excessive service and insufficient pricing competition, inflated airline costs, and denied the industry adequate profits. Senator Edward Kennedy chaired subcommittee hearings that served as the political genesis of congressional reform. The Kennedy Report concluded that deregulation would allow pricing flexibility, which would stimulate new and innovative offerings, allow passengers the range of price and service options dictated by consumer demand, enhance carrier productivity and efficiency, increase industry health, and result in a superior allocation of society's resources.[6]

With the inauguration of Jimmy Carter as president in 1976, the deregulation movement had a disciple in the White House. Carter appointed Alfred Kahn, a Cornell economics professor, to be chairman of the Civil Aeronautics Board.[7] Kahn criticized traditional CAB regulation as having "(a) caused air fares to be considerably higher than they otherwise would be; (b) resulted in a serious misallocation of resources; (c) encouraged carrier inefficiency; (d) denied consumers the range of price/service options they would prefer, and; (e) created a chronic tendency toward excess capacity in the industry."[8] As CAB chairman, Kahn implemented a number of initiatives that liberalized entry and pricing. In the late 1970s, the immediate results of the relatively modest efforts at regulatory reform were quite positive, creating in Washington and in the media a general euphoria that we were on the right course. Carriers in the late 1970s filled capacity, stimulated new demand by offering low fares, and enjoyed robust profits.[9]

Working with the White House, Kahn put his charismatic personality solidly behind the legislative effort for reform.[10] By the late 1970s, Congress had embraced deregulation as a major policy objective. The most sweeping such legislation was the Airline Deregulation Act of 1978. That statute abolished the Civil Aeronautics Board (as of January 1, 1985), which had regulated the airline industry for four decades. The legislation received overwhelming bipartisan legislative support. But the economic health of the industry was soon to spiral downward.[11]

The Airline Deregulation Act of 1978 was intended to provide a gradual transition to deregulated entry and rates, although the CAB quickly dropped any notion of "gradual" deregulation under Chairman Marvin Cohen. What had begun as a program of modest liberalization became an avalanche of abdication of responsible government oversight. Implementation of the new policy was immediate and comprehensive, and as the 1970s came to a close, the industry entered the darkest financial period in its history.[12] These problems were then exacerbated by the worst national economic recession since the Great Depression and an escalation in fuel prices. The Deregulation Act also called for the "sunset" of the CAB in 1985, when its remaining responsibilities were transferred to the U.S. Department of Transportation (DOT), an executive branch agency, which during the Reagan administration was wedded to the ideology of laissez-faire.[13] But even the

Bush administration's DOT tenaciously insists that, despite growing evidence to the contrary, "transportation deregulation has been a notable success."[14] It is this conclusion with which this book takes issue.

It was assumed that deregulation would create a healthy competitive environment, with lots of airlines offering a wide array of price and service options and a high level of safety. We now have more than a decade of empirical evidence to compare with those sanguine predictions.

Deregulation has become one of America's most important contemporary legal and political phenomena, dominating the domestic policy of recent presidents. Because deregulation has been implemented far more aggressively than anyone would have dared dream at its inception, it has had profound social and economic consequences.[15]

This book examines where the great American airline deregulation experiment has been, where it is, and where it appears to be going. We begin with an introduction to the contemporary airline industry—the identity and corporate cultures of the megacarriers and the men who dominate them. We then proceed chronologically, beginning with a historical review of the political, legal, and economic dimensions of airline regulation and reviewing the events that led our nation to establish a regime of economic regulation on the transportation industry and, a century later, to dismantle it. We will examine the metamorphosis of deregulation, focusing on several of the areas in which there has been a significant adverse impact, including an unprecedented level of national concentration, discriminatory pricing, fares that today are higher than the pre-deregulation trend, a deterioration in service, and a narrower margin of safety. Throughout this book, we will compare the empirical results of deregulation with the theories and assumptions of its major proponents, particularly Alfred E. Kahn, its principal architect and, on balance, a staunch defender.[16] We will also examine the issue of whether the fundamental theories of deregulation rested on false assumptions. The reader will begin to see strong parallels between the conditions preceding regulation and those following deregulation.

We will also address the issue of whether a bit more regulation might be in the public interest and, if so, what form it should take. We will conclude with an analysis of the public interest in transportation and the need for a new national transportation policy.[17] After a decade of deregulation, it seems appropriate to evaluate the empirical evidence and determine whether the policy has achieved desirable social and economic ends and, if not, how we might correct its course.

Since airline deregulation was the prototype for a decade of aggressive deregulation throughout the American economy (in industries such as telecommunications, broadcasting, railroads, trucking, buses, and banking), the results of our examination of this industry may have wider implications. It would be a mistake, for instance, to take the experience of the early years of airline deregulation—when low, simply structured fares and dramatic competition from new entrants seemed to justify the wildest claims of its proponents—as a model of the benefits that deregulation can bring generally. These short-term gains were followed by medium-term and, arguably, long-term pain.

So let us examine the theories, the myths, and the realities of the airline industry in the 1990s. We begin by introducing you to the megacarriers and the men who rule them.

NOTES

1. *See Gridlock!,* TIME, Sept. 12, 1988, at 52, 55.

2. P. DEMPSEY & W. THOMS, LAW & ECONOMIC REGULATION IN TRANSPORTATION 7–17 (1986) [hereinafter cited as P. DEMPSEY & W. THOMS].

3. Dempsey, *The Rise and Fall of the Civil Aeronautics Board—Opening Wide the Floodgates of Entry,* 11 TRANSP. L.J. 91, 95 (1979) [hereinafter cited as *The Rise & Fall of the CAB*].

4. The agency was initially named the Civil Aeronautics Authority.

5. *See* Hardaway, *Transportation Deregulation (1976–1984): Turning the Tide,* 14 TRANSP. L.J. 101, 106 n.17 (1985) [hereinafter cited as Hardaway] and articles cited therein. *See also* L. KEYES, FEDERAL ENTRY CONTROL OF ENTRY AND EXIT INTO AIR TRANSPORTATION (1951); R. CAVES, AIR TRANSPORT AND ITS STUDY: AN INDUSTRY STUDY (1967); and W. JORDAN, AIRLINE DEREGULATION IN AMERICA: EFFECTS AND IMPERFECTIONS (1970).

6. *Civil Aeronautics Board Practices and Procedures, Senate Subcomm. on Administrative Practice of the Judiciary Comm.,* 96th Cong., 1st Sess. (1976). *The Rise & Fall of the CAB, supra* note 3 at 114–18.

7. Alfred Kahn is perhaps more responsible for transportation deregulation than any other individual. It was he, as Jimmy Carter's chairman of the Civil Aeronautics Board, who forcefully lobbied in support of the Airline Deregulation Act of 1978, which, after a transition period, abolished airline entry and price regulation and terminated the Civil Aeronautics Board. It was Kahn, as Jimmy Carter's chairman of the Council on Wage and Price Stability, who lobbied strongly on behalf of trucking regulation, ultimately leading to the promulgation of the Motor Carrier Act of 1980. *Trucking Deregulation: Is It Happening? Hearing before the Joint Economic Committee,* 97th Cong., 1st Sess. 3 (1981). Along the way, Kahn made various predictions as to the benefits likely to flow from deregulation. Laced throughout this book is a comparison of Kahn's early assumptions and predictions with his more recent admissions and the empirical results of deregulation.

The predictions of what deregulation would bring were quite optimistic, despite strong misgivings by most of the industry. Kahn assured a skeptical public that the benefits of deregulation would be universally shared: "I am confident that . . . consumers will benefit; that the communities throughout the nation—large and small—which depend upon air transportation for their economic well·being will benefit, and that the people most closely connected with the airlines—their employees, their stockholders, their creditors—will benefit as well." *Statement of Alfred E. Kahn before the Aviation Subcommittee of the House Public Works and Trans-*

portation Committee on H.R. 11145, 95th Cong. 2d Sess. 8 (Mar. 6, 1978). *Aviation Regulatory Reform, Hearings before the Subcomm. on Aviation of the House Comm. on Public Works and Transportation*, 95th Cong., 2d Sess. 124 (1978). It is clear, however, that many of these constituencies, including stockholders and labor (and, arguably, much of the traveling public), have not benefited. Take labor. In his book, Kahn recently wrote, "In several of the industries, especially in the airlines and trucking, competition has exerted powerful downward pressure on egregiously inflated wages—*painful for the workers* affected but healthy for the economy at large." A. KAHN, THE ECONOMICS OF REGULATION xx (1988) [emphasis supplied] [hereinafter A. KAHN]. So, employees are not better off as he predicted, but he justifies that on grounds that they must suffer while the rest of us benefit. In his book, he also acknowledges that several of the deregulated industries have "sharply reduced their work forces." *Id.* Elsewhere, Kahn has conceded, "Labor unrest and the insecurity and downward pressure on the wages of the pre-existing labor force have been an undeniable cost of deregulation." Kahn, *Surprises from Deregulation*, 78 AEA PAPERS AND PROCEEDINGS 316, 317 (1988) [hereinafter *Surprises from Deregulation*].

8. Quoted in P. DEMPSEY, LAW & FOREIGN POLICY IN INTERNATIONAL AVIATION 24 (1987) [hereinafter P. DEMPSEY]. *See also* Kahn, *The Theory and Application of Regulation*, 55 ANTITRUST L.J. 177, 178 (1986) [hereinafter *Theory and Application*], and Kahn, *Transportation Deregulation . . . and All That*, ECON. DEVELOPMENT Q. 91, 92 (1987) [hereinafter *All That*].

9. As a young CAB attorney, Paul Stephen Dempsey was also swept up in the movement. In 1978, he praised the benefits of partial deregulation:

The objective of [deregulation] has been to provide the consumer . . . with improved service at reduced fares. In general, the theory has been that increased competition among air carriers will lead to improved quality and an increased variety of services available to the public at competitive prices reasonably related thereto, and that the price elasticity of the passenger market will ensure more efficient utilization of capacity for the carriers and, consequently, increased revenue. Enhanced reliance upon competitive market forces has tended to lower air fares and stimulate innovative price/service options. It has also tended to fill empty seats and thereby increase carrier revenue. The policies appear to have had an affirmative impact upon both consumers and the regulated industry that serves them.

Dempsey, *The International Rate and Route Revolution in North Atlantic Passenger Transportation*, 17 COLUM. J. TRANSNAT'L L. 393, 441 (1978).

10. Although most of the airline industry opposed deregulation, it was supported by Federal Express and United Airlines, the latter the largest airline in the free world.

11. By 1981, Paul Stephen Dempsey had reevaluated the deregulation experiment, looked into his crystal ball, and concluded that the metamorphosis would proceed through three major phases:

In the first, price and service competition are increased, carriers become innovative and imaginative in the types of price and service combinations they offer, and consumers thereby enjoy lower priced transportation. Carriers are free to maximize their profits by leaving unprofitable markets and investing their equipment in more lucrative ones. In the airline industry, lower prices initially generated increased passenger traffic, thereby enabling air carriers to fill seats which might have otherwise flown empty. . . . [A]ir carriers left many of the small, remote, isolated communities of our nation and transferred their aircraft to the more heavily traveled markets. Passengers in these dense markets enjoyed intense pricing and service competition. Airlines generally enjoyed higher profits, at least during stage one. . . . The first stage is the one to which deregulators point to demonstrate the attributes of deregulation.

The second stage is an embarrassment to deregulators. . . . Because of excess capacity and unrestrained price and service competition, air carrier profits have plummeted; indeed, the industry is experiencing the worst losses in the history of aviation. . . . [Economist Michael Evans] succinctly summarized the market effects of deregulation upon the airline industry:

"In the short run, deregulation does indeed seem to be the promised land. Prices rise more slowly, productivity increases, service expands, and everyone is happy. However, after the initial euphoria, it turns out that profits are not really increasing after all.

"As a result, rationalization of the route structure begins, which turns out to mean price-cutting on primary routes, coupled with higher prices and less service on secondary routes.

"When this happens, the gain in productivity slows or even reverses, thereby negating much of the benefits of deregulation. We end up with no improvement, or even higher prices and lower productivity in that industry. . . ."

The continued inability of many carriers to balance their books due to the intensive competition they are forced to endure under deregulation will force many carriers to float "belly up" in bankruptcy. . . . During the second stage, prices will continue to be set at reasonable levels in highly competitive markets, and will continue to grow at unreasonable rates in monopolistic or oligopolistic markets. Service will begin to deteriorate in both.

Stage three of deregulation will constitute the ultimate transportation system with which the nation is left. The carriers which have suffered most during stages one and two will, by this point, have gone bankrupt, leaving many markets with very little competition. A monopolistic or oligopolistic market structure will result in high prices, poor service, and little innovation or efficiency. Potential entrants, having witnessed the economic calamity of destructive competition, may be unwilling to enter so cutthroat an industry. . . . Small communities will receive poorer service and/or higher rates than they enjoyed under regulation. . . . In the end, the industry structure created by the free market may be much less desirable than that which was established by federal economic regulation.

Dempsey, *The Experience of Deregulation: Erosion of the Common Carrier System,* 13 Transp. L. Inst. 121, 172–75 (1981) [citations omitted]. The author continued, "Let us pray that this pessimistic portrait the author has painted of deregulation turns out to be inaccurate, for if the effects of deregulation are less than desirable, can all the king's horses and all the king's men ever put it back together again?" *Id.* at 176.

12. *See The Rise & Fall of the CAB, supra* note 3.

13. *See* P. Dempsey & W. Thoms, *supra* note 2, at 28–29.

14. U.S. Dep't of Transportation, Moving America 6 (1990).

15. *See* P. Dempsey, The Social and Economic Consequences of Deregulation (1989).

16. In theory, deregulation should have brought us lots of new entry and a healthy competitive environment—things that, in general, have not resulted. In fact, deregulation was premised on several false assumptions, including the contestability of airline markets. By the late 1980s, Kahn had become somewhat conciliatory about the problems that had emerged under deregulation and the inability of pro-deregulation economists to have predicted them. He insisted that the Department of Transportation was largely to blame for these ills by, for example, approving every merger submitted to it and not sufficiently expanding airport capacity. Nonetheless, Kahn noted, "There have of course been severe problems and reasons for concern even from the public's standpoint: most prominently sharply increased

congestion and delays, increased concentration at hubs, monopolistic exploitation of a minority of consumers, and possibly a narrowing of the margin of safety." Kahn, *Airline Deregulation—A Mixed Bag, But a Clear Success Nevertheless,* 16 TRANSP. L.J. 229, 251 (1988) [citation omitted] [hereinafter *A Mixed Bag*]. To his credit, Kahn has also become quite candid about his and his compatriots' failure to foresee the "explosion of entry, massive restructuring of routes, price wars, labor-management conflict, bankruptcies and consolidations and the generally dismal profit record of the last ten years." Kahn, *Surprises of Airline Deregulation,* 79 AEA PAPERS AND PROCEEDINGS 316 (1988).

17. Because the economic rationales and regulatory structure created for economic regulation of the other infrastructure industries (i.e., communications and energy) have so many parallels with those of transportation, the empirical experience with deregulation in this industry should provide insights into the impact of additional deregulation in telecommunications, broadcasting, oil and gas, and electric power.

2

CORPORATE PIRATES AND ROBBER BARONS IN THE COCKPIT

During the 1980s, several of the nation's largest airlines became targets for leveraged buyouts (LBOs): Continental, Eastern, Frontier, People Express, TWA, Ozark, Northwest, United and American. Only the last two acquisition efforts failed. The failure of the United LBO sent the Dow Jones Industrials skidding 190 points on Friday, October 13, 1989—the twelfth most serious collapse in Wall Street history.

Two reasons accounted for the interest in airline acquisitions. First, after more than 150 bankruptcies and 50 mergers, the industry became an oligopoly. Eight megacarriers dominate 94 percent of the domestic passenger market. With fortress hubs and shared monopolies, ticket prices are ascending.

Second, the glamor of the industry has always attracted men with huge egos. In the old days, it was buccaneers like Howard Hughes, Eddie Rickenbacker, and Juan Trippe. These days it is Marvin Davis, Donald Trump, and Peter Ueberroth. Owning an airline is more prestigious than owning an NFL franchise, for there are fewer airlines. Owning an airline also means becoming emperor of several fiefdoms, for the fortress hubs wield a stranglehold over the cities they serve.

For example, in buying Northwest for $3.7 billion, Alfred Checchi became king of Minneapolis, Detroit, and Memphis—Northwest's hubs. If Marvin Davis's $6.2 billion bid for United had been successful, he would have been lord of Chicago (O'Hare is the world's busiest airport), Denver, San Francisco, and Washington—United's hubs.

Prior LBOs reveal that corporate raiders leverage airlines to the teeth to pay for their acquisitions. In the mid-1980s, Frank Lorenzo gobbled up Continental and Eastern while Carl Icahn grabbed TWA and Ozark. Both added millions in indebtedness to these once proud airlines while stripping them of assets. Before Eastern fell into bankruptcy, it carried $2.5 billion in long-term debt; its debt service was a crushing $575 million. TWA carries $2.4 billion in debt and lease obligations and has a negative net worth of $30 million. Checchi may load Northwest with more than $3 billion in debt. United would have carried more than $6 billion in debt had its LBO been successful. This chapter will introduce the reader to several of the major actors in the Monopoly game, their enormous egos, and their ruthless game plan.

Foreign airlines are also gobbling up significant shares of U.S. airlines. Already Northwest, Delta, Texas Air, America West, and Hawaiian Airlines have significant foreign equity. Whereas debt poses significant problems for the long-term viability of airlines, foreign ownership raises national security concerns.

Criticism of LBOs centers on the impact of massive amounts of debt on the ability of airlines to make new aircraft purchases or maintain existing aircraft properly, expand operations, maintain competition, and withstand the vicissitudes of the market cycle. This debt, coupled with the recession of the early 1990s, produced a new round of bankruptcies and consolidations among debt-ridden airlines, which left the industry even more concentrated. Finally, foreign ownership of U.S. airlines raises competition and national security concerns. We begin with a look at deregulation.

DEREGULATION

The Airline Deregulation Act of 1978 was designed to create a more competitive environment in commercial aviation.[1] But as deregulation has matured, the industry has become more highly concentrated than at any other point in its history, and the horizon is devoid of new competitors. Deregulation has proceeded through several stages.

Price Wars

In the beginning, deregulation sent fares tumbling as new entrepreneurs, such as People Express and Air Florida, emerged to rival the megacarriers. Although the new entrants never accounted for more than 5 percent of the domestic passenger market, with lower costs they drove prices down. But industry profitability plummeted to the worst losses in the history of domestic aviation. These losses were exacerbated in the early 1980s by the worst recession since the Great Depression. During the first decade of deregulation, the industry as a whole made enough money to buy two Boeing 747s.[2]

Two economic characteristics of airlines lead to destructive competition when carriers compete head to head. First, airlines sell a product that is instantly perishable. Once a scheduled flight closes its door and pulls away from the jetway, any empty seats are lost forever. They cannot be warehoused and sold another day, as can manufactured goods. It is as if a grocer was selling groceries with the spoilage properties of open jars of unrefrigerated mayonnaise. The grocer would be forced to have a fire sale every afternoon, for any unsold inventory would have to be discarded.[3]

Second, the short-term marginal costs of production are nil. Adding another passenger to an empty seat costs the airline another bag of peanuts, a cup of Coke, and a few drops of fuel. Thus, adding nearly any bottom is profitable in the short term. Head-to-head competition between carriers usually results in destructive competition, for carriers price at the margin and fail to cover long-term and fixed costs.[4]

The hemorrhaging of dollars led management to slash wages, trim maintenance, reduce service, and defer new aircraft purchases. It also led to a massive shakeout of smaller firms. During the first decade of deregulation, more than 150 carriers collapsed into bankruptcy.[5]

Consolidations

To stave off bankruptcy, carriers began to reconfigure their operations. The entry and exit freedom produced by deregulation enabled them to establish hub and spoke systems. Four hubs (Atlanta Hartsfield, Chicago O'Hare, Dallas/Ft. Worth International, and Denver Stapleton) became duopolies, whereas all the rest became effective monopolies, with a single airline controlling more than 60 percent of the takeoffs and landings, gates, and passengers.[6]

A rash of mergers also produced greater concentration. During the first decade of deregulation, there were more than 50 mergers, acquisitions, and consolidations, the major ones concluded in 1986 and 1987 when the Reagan administration's Transportation Department embraced an exceptionally permissive antitrust policy.[7] Indeed, the Department of Transportation approved each of the 21 mergers submitted to it.[8]

As the 1990s dawned, the eight largest airlines dominated 94 percent of the domestic passenger industry and almost all hubs. Not only are passenger airlines highly concentrated: mergers in the cargo industry have reduced it to a duopoly. Federal Express acquired Flying Tigers, which itself had consumed Seaboard when deregulation was young. Consolidated Freightways, one of the nation's largest trucking companies, acquired Emery Air Freight, which itself had consumed Purolator.

Profitability

With such tremendous concentration, carriers were able to raise ticket prices significantly. In 1989, the General Accounting Office reported that prices were 27 percent higher at monopoly or duopoly hubs than at competitive airports.[9]

The oligopoly that emerged from deregulation grew increasingly profitable. The two years ending June 30, 1989, represented the most profitable period for airlines in history.[10] One source noted, "After a decade of turbulence, [the industry] is entering a new period of prosperity: a period where tight airport space and increasing demand for air travel will produce the steady cash flow necessary for a smooth buyout."[11] But it has been a roller coaster ride, for the two-year period ending December 1991 was the most dismal in the industry's history, with losses totaling $5.7 billion.[12]

Leveraged Buyouts

With unprecedented profitability, and the innate glamor of the industry, three of the nation's four largest airlines became targets for LBOs in 1989. The Denver oil king Marvin Davis launched a $2.7-billion bid for Northwest Airlines. Northwest ultimately fell to a $3.7-billion bid by Alfred Checchi.[13] Davis enjoyed a $30-billion profit on the Northwest raid, then turned around and put a siege on United. That raid was preempted by a management/pilot bid for United led by CEO Stephen Wolf for $300 a share, or nearly $7 billion. In October 1989, Donald Trump, former suitor of United[14] and purchaser of the Eastern Air Lines New York–Washington–Boston shuttle,[15] launched a $7.54-billion bid for American Airlines.[16]

One source summarized the principal reasons motivating airline LBOs:

1) The belief that the significant earnings and earnings potential demonstrated during the [late 1980s], and the concurrent strong level of cash flow generation is sustainable. Inherent in this tenet is the expectation that the degree of cyclicality and even seasonality airline earnings and cash flow have historically demonstrated will be absent or lessened in the future.

2) The realization of premium values for used aircraft, facilities as well as new aircraft delivery positions, which has increased the liquidity (and enhanced the equity capital) of many carriers. Included in the strong market for airline assets is premium values being accorded gates, slots, real estate and other tangible and intangible assets.

3) The availability of capital, both equity and debt, due in part to the renewed interest in airline lending by commercial banks and the current favorable interest rate environment. Included in this tenet is the tremendous increase in leasing capital, which has provided, and is expected to continue to provide more than half the capital expenditures in the 1990s.[17]

Financing for the $7-billion management/labor bid for United collapsed, and Donald Trump withdrew his $7.5-billion bid for American.

Some LBOs can be justified on grounds that they rid companies of ineffective management and improve productivity, profitability, and performance by paring unrelated assets and squeezing labor. But American and

United are generally viewed as among the best-managed and most-efficient companies in the business. Let us examine America's two largest megacarriers, the assaults by corporate raiders, and the entrepreneurs who battle for control of the nation's aviation system.

American Airlines and CEO Robert Crandall

American Airlines has been the most vocal opponent of LBOs, describing Trump's bid as "ill considered and reckless"[18] and insisting that "excessive levels of debt in the airline industry are not in the public interest."[19] Said CEO Robert Crandall, "The disadvantages of excessive leverage, and its effects are heightened by the continuing volatility of airline earnings."[20] American called for congressional protection against LBOs.

Although initially a critic of deregulation, Crandall moved quickly to capitalize on its opportunities for growth. His aggressive policies of reinvesting earnings, growing from within, establishing new hubs from scratch (Nashville, Raleigh-Durham, and San Jose) and thereby outflanking the dominant southeast hub of Atlanta, aggressively managing yield, inventing frequent-flyer programs, and getting out early with a computer reservations system have made American Airlines the largest airline in the United States in terms of revenue passenger miles.

The man is a chain-smoker and an avid jogger—two packs and four miles a day, respectively.[21] One commentator noted, "His tough stance on union wages, his bare-knuckled price-cutting and his proclivity for salty phrases have all contributed to Robert Crandall's public image as a hard-nosed street fighter."[22] Above all, Crandall is a fierce competitor. As one acquaintance noted: "He doesn't want anybody to beat him. . . . He's in business to put his competition out of business."[23] Crandall views the deregulated environment as one in which he can wage "legalized warfare in the industry."[24]

After a series of price wars that left both American and Braniff bleeding in their Dallas hub,[25] Crandall sought to discuss prices with Braniff's president, Howard Putnam. Crandall and Putnam had the following conversation on February 1, 1982:

CRANDALL: I think it's dumb as hell for Christ's sake, all right, to sit here and pound the shit out of each other and neither one of us making a fucking dime.

PUTNAM: Well—

CRANDALL: I mean, you know, goddamn, what the fuck is the point of it?

PUTNAM: Nobody asked American to serve Harlingen. Nobody asked American to serve Kansas City, and there were low fares in there, you, know, before. So—

CRANDALL: You better believe it, Howard. But you, you, you know, the complex is here—ain't gonna change a goddamn thing, all right. We can, we can both live here and there ain't no room for Delta. But there's, ah, no reason that I can see, all right, to put both companies out of business.

PUTNAM: But if you're going to overlay every route of American's on top of over, on top of every route that Braniff has—I can't just sit here and allow you to bury us without giving our best effort.

CRANDALL: Oh sure, but Eastern and Delta do the same thing in Atlanta and have for years.

PUTNAM: Do you have a suggestion for me?

CRANDALL: Yes. I have a suggestion for you. Raise your goddamned fares 20 percent. I'll raise mine the next morning.

PUTNAM: Robert, we . . .

CRANDALL: You'll make more money and I will too.

PUTNAM: We can't talk about pricing.

CRANDALL: Oh, bullshit, Howard. We can talk about any goddamned thing we want to talk about.[26]

Putnam taped the conversation and turned the tape over to the Justice Department for antitrust prosecution. Price-fixing is, after all, a per se violation of the Sherman Act, one that could have landed Crandall in prison. Most convicted wealthy white-collar criminals actually end up in Club Fed, as did Ivan Boesky, working on their muscles and tans in minimum-security institutions. It is, nonetheless, an embarrassing way to spend your time. The Justice Department was less ambitious. It initially sought a court order prohibiting Crandall from working in any responsible airline position for two years and prohibiting American Airlines from discussing pricing for a decade.[27] Ultimately, the Reagan administration settled for less still—a consent decree in 1986 in which Crandall neither admitted nor denied guilt.[28]

But nothing was to save Putnam from demise. Braniff entered bankruptcy in 1982.[29] After scaling down significantly, selling its Latin American routes to Eastern and selling many of its aircraft, the new Braniff emerged from reorganization under the control of the Pritzker family of Chicago (who control the Hyatt Hotel chain) and reassumed its Dallas/Ft. Worth operations. But head-to-head competition with the two megacarriers that dominated Dallas—American and Delta—proved infeasible. With a generous loan from American to buy new aircraft, Braniff abandoned Dallas and moved its hub to Kansas City, where it died.

Southwest Airlines dominates tiny Dallas Love Field while American dominates Dallas/Ft. Worth International Airport. Southwest's Chairman Herb Kelleher once joked to Crandall that their relationship was analogous to that of tiny Finland and mighty Russia. "There's only one difference," Crandall retorted with a Siberian stare, "I ain't reducing troops."[30]

Crandall has adopted an extremely aggressive approach to capitalizing on the opportunities afforded by airline deregulation. American had adopted the philosophy of, in its words, "competitive anger." As Crandall put it:

"We like to be successful. When we're not, we're angry with ourselves, our colleagues and the world at large."[31] He has repeatedly insisted: "My friends call me Mr. Crandall. My enemies call me Fang."[32]

Destroying competitors means more to Crandall than running them out of town. It includes assailing their character. In 1987, Crandall bought 15,000 copies of a scathing *Texas Monthly* article about Texas Air's Frank Lorenzo to distribute at employee meetings.[33] For his part, Lorenzo describes Crandall as "hypocritical" and "afraid of competition"—the pot calling the kettle black, so to speak.[34]

But Crandall didn't like it when the shoe was on the other foot. In response to John Nance's book about Braniff, in which Crandall was portrayed unfavorably, Crandall bought 25,000 copies, to take them out of circulation, then paid the publisher $150,000 to discard existing inventory and print a reworded edition.[35]

Crandall's aggressive character also strongly manifests itself in his internal domination of American. He has a fiery temper. Richard Murray, a former American Airlines executive, recalls being fired at several meetings, only to be rehired before adjournment. Once, Crandall became so angry at a competitor that he flew into a rage and accidentally pulled some blinds off a window and onto his head. When aides rushed to help, he responded: "To hell with my head. What are we going to do about this problem?"[36]

Crandall loves detail. He likes to immerse himself in the numbers. Crandall was once spotted humped over paperwork three inches high on an American flight on Christmas morning.[37] He brags that he cut $40,000 in operating expenses by removing olives from American's dinner salads.[38] When Crandall took over as chief operating officer in 1980, he reduced the number of guards at an American facility from three to one. The lone guard was then replaced with a part-time guard and later with a guard dog. Finally, Crandall inquired whether it might be possible to replace the dog with a loudspeaker system broadcasting a tape recording of barking dogs.[39]

Crandall's tight-fisted managerial style, entrepreneurial bravado, marketing acumen, and streetwise shrewdness made American the largest airline in the free world, second in number of aircraft only to the former Soviet Union's Aeroflot. Under Crandall, American's revenue passenger miles grew steadily since 1981; its market share increased steadily since 1980; it turned a profit every year between 1983 and 1991; and its debt-to-equity ratio was superior to that of the Dow Jones airlines since 1985.[40] That such a lean, mean flying machine as American would be assaulted in a leveraged buyout left most analysts stunned in disbelief in October 1989 when Donald Trump made a bid of $120 a share, or $7.54 billion.[41] It was like a minnow swallowing a whale. Trump purchased the Eastern shuttle, which flies 21 aircraft between three cities; American has 480 aircraft.[42] In 1988 American earned $476.8 million on revenue of $8.8 billion.[43] Trump's acquisition would have added $6.5 billion in debt to American.[44] Perhaps Trump's ego got the best of him. As one source noted:

Mr. Trump, a billionaire with a towering ego who made his fortune with glitzy skyscrapers and casinos, entered the airline business last Spring by buying Texas Air Corp.'s Eastern shuttle for $365 million and renaming it the Trump shuttle. He owns New York City's famed Plaza Hotel, plus buildings named Trump Tower, Trump Parc, and Trump Palace.[45]

Trump promised not to rename American Airlines, Trump Airlines, however. But after the stock market collapse of Friday, October 13, 1989, Donald Trump withdrew his bid for American. His financial empire began to crumble around him in the 1990s.

United Airlines and CEOs Richard Ferris and Stephen Wolf

Stephen Wolf is presently chief executive officer of United. But much of its corporate culture was shaped by his predecessor, Richard Ferris. Ferris was one of the major actors in the quest for deregulation. As United's chief from 1976 until 1987, Ferris led the carrier to break ranks with the rest of the industry and promote deregulation.

As the nation's largest carrier, United believed that the deregulated skies would be friendly to it. United worked long and hard behind the scenes to persuade Congress and the Carter administration to pass the Airline Deregulation Act of 1978. "If the truth be known," said a former United executive, "Monte Lazarus [a lieutenant of Ferris's] wrote the Airline Deregulation Act." Ironically, Lazarus, a former assistant to CAB Chairman Secore Brown, was known as the consummate Washington bureaucrat even after joining United.

Under regulation, United had been hindered from growing. In 1938, United enjoyed about 22 percent of the domestic passenger market. By the mid-1970s, its share had declined slightly, to 20 percent. Under regulation, the CAB had long favored smaller and weaker airlines in route awards, depriving United of new markets. Ferris believed that United would have few opportunities for expansion under a benevolent CAB.[46] Deregulation would be the means for United to grow.

Once deregulated, United pulled out of many of its thin markets, abandoning a large number of small and medium-size cities, and concentrated on dense, long-haul routes. But United soon found that it needed regional feed into its hubs to fill the long-haul capacity, and it reversed course, repurchasing many of the small 737s that is had just sold. Today, United serves at least one airport in each state so that it can boast, "We serve all 50 states."

Working from a stand-up desk, Ferris was known to be a tough, hot-tempered competitor.[47] Take his role in the demise of Frontier. In the mid-

1980s, Denver's Stapleton Airport was the only airport in the country to be used as a hub by three airlines. As a consequence, Denver consumers enjoyed some of the lowest airfares in the country. But for the three airlines—United, Continental, and Frontier—the results were disastrous. Profitability in the market plummeted.

So in 1985, United bought 30 of Frontier's jets for $360 million. Later that year, Donald Burr's People Express bought the rest of Frontier for $307 million. People's "no frills" fares were matched by United and Continental, and an economic bloodbath ensued.

In July 1986, United agreed to take Frontier off Burr's hands for $146 million, with the contract condition that United negotiate with Frontier's unions to reach an agreement satisfactory to United. United met with the pilots, but not the other four Frontier unions. After several weeks, during which additional Frontier assets were transferred to United, United announced that the labor negotiations were at an impasse.[48] Burr had little choice but to put Frontier into bankruptcy in late August 1986. And then there were two in Denver. Prices and profitability began to climb.

Ferris began his reign at United with good rapport with labor, frequently visiting the cockpits and taking the time to earn a pilot's license.[49] But a 29-day strike by United's pilots in 1985 began a seething relationship that caused Ferris to begin flying private jets, avoiding his own company's planes. At a dinner in 1986, Ferris was overheard boasting to American's CEO Robert Crandall that United would one day have some of the lowest labor costs in the industry.[50]

Ferris came to head United through the ranks of its Westin Hotel chain, which may explain his obsession with creating a vertically integrated travel conglomerate. Already owning Westin, United went on a binge under Ferris in which airline profits were spent on developing a computer reservations system (Apollo) and on buying a rental-car company (Hertz) and yet another hotel chain (Hilton International, formerly owned by TWA). In 1986, the combined company flew 50 million passengers, controlled about one-third of the car-rental business, and owned 150 hotels. To reflect its scattered emphasis, United dropped the UAL label and renamed the holding company Allegis, a bastardization of the words "allegiance" and "aegis".

Not only was the name wormy, but the combination made United ripe for a hostile takeover, for its dismembered parts were worth more than its barely unified whole. Although the idea of a unified full-service travel empire was not a bad one (selling a customer an airline ticket, hotel room, and rental car as a package intuitively seemed an attractive market concept), it never really got off the ground before the vultures began to circle.

In 1987, Donald Trump, who owned 5 percent of the company, urged Ferris to break up the conglomerate and sell all its parts separately.[51] The pilots, angry with Ferris for different reasons, began to put together their own $2.3-billion bid for the company.[52] And other suitors were waiting in the wings, including the Coniston Partners. As one analyst noted, "If the pilots wanted to stir up a hornet's nest, it looks like they have."[53]

Ferris was a fiery-tempered executive who attacked problems by promptly moving on the offensive.[54] In addition to the usual poison pills and golden parachutes, he concluded a unique financial arrangement with Boeing that gave it some unusual powers over the business operations.[55] When that wasn't enough, he proposed to saddle the company with a $3-billion re-capitalization to thwart the takeover attempts, distributing the proceeds as a $60-a-share dividend.[56]

Shareholder resistance and difficulty in financing led the board of direc-tors to balk. Ferris resigned, red-faced, in June 1987. He was succeeded for a short term by Frank Olson, chairman of the Hertz unit.[57]

Although the company had spent $7.3 million on the name change (to which Wall Street gave a thumbs down), United abandoned the Allegis title in 1987.[58] United also sold off the hotel and car-rental businesses, took on $3 billion in debt, and paid shareholders a hefty dividend. Olson was subsequently replaced by Stephen Wolf, a former chairman of Flying Tigers.

In early August 1989, the Denver oil king Marvin Davis offered $240 a share, or $5.4 billion, for United, later raising his bid to $275.[59] Manage-ment responded with a $300 a share, or $6.75-billion, buyout of its own, involving the pilots. British Airways was also a partner, putting up $750 million, or about 78 percent of the equity.[60] Management was to have owned 10 percent, British Airways 15 percent, and the pilots 75 percent.[61] To pay for their share, the pilots would take pay cuts of up to 10 percent, less overtime pay, and fewer vacation days.[62] The debt would have created interest payments of $600 million to $700 million annually.[63] The machin-ists union criticized the deal as unrealistic, saying, "Placing billions of dol-lars of additional debt on the carrier . . . would seriously jeopardize the carrier's operation, safety and future existence."[64]

The financing fell through on Friday, October 13, 1989, sending the Dow Jones Industrial averages tumbling 190 points.[65] Oddly, the stock market panic was motivated, at least in part, by anxiety over junk bonds. But the United financing had none, to which the Japanese banks ob-jected.[66]

Shortly thereafter, Marvin Davis withdrew his bid, and British Airways backed out of the management/pilot buyout.[67] Under the deal that col-lapsed, United CEO Stephen Wolf was to have earned $76.7 million and new UAL stock options.[68] Management would have then spent $15 mil-lion for a 1 percent stake and been given 9 percent more in stock op-tions.[69] Everyone's eyes had become filled with dollar signs. The board of directors voted lifetime first-class passes for themselves and their spouses and $20,000 a year for life.[70] The investment bankers would get $59 million and lawyers $45 million.[71] United's 25,000 machinists and 25,000 noncontract employees criticized Wolf's greed in pursuing an LBO that would enrich him while forcing pay cuts and benefit reductions on labor, and they called for his resignation.[72]

The failed management/pilot bid saddled UAL with $58.7 million in expenses, nearly enough to buy two B-737s.[73] The action of the board of directors to pay this indebtedness was challenged by some as a waste of corporate assets, and their decision to pay themselves $20,000 for life was challenged as a breach of fiduciary responsibility to stockholders.[74] Wolf earned some $17 million in 1990, despite the fact that United's profits fell more than 60 percent that year.

DEBT—ON BALANCE SHEET AND OFF

As the 1990s dawned, four of the nation's largest airlines—Continental, Eastern, Pan Am and TWA—had a negative net worth, a debt-to-equity ratio in excess of 100 percent.[75] Three of these companies were owned by two corporate raiders—Frank Lorenzo's Texas Air controlled Continental and Eastern while Carl Icahn owned TWA. By 1992, Eastern and Pan Am had both ceased operations, while Continental was struggling to emerge from bankruptcy.

Northwest was saddled with $3.3 billion to pay for the Checchi acquisition (quadrupling its long-term debt).[76] The industry as a whole is burdened with excessive debt, which makes it difficult for it to weather recessions, expand operations, modernize fleets, and maintain older equipment.[77] Such economic difficulties enhance public concerns over airline safety. Table 2.1 depicts the huge amounts of debt with which the nation's airlines have been burdened by virtue of gluttonous acquisitions, mergers, and buyouts in recent years. By reducing competition, the acquisition or merger of one airline by another enhances the survivor's profitability. But the acquisition by corporate raiders produces no such benefits.

Not only LBOs but also aircraft acquisitions are burying airlines in debt. Media attention has focused on the geriatric jets—the peeling skin and the exploding doors (known in the industry as Kahndoors, after the father of deregulation, Alfred Kahn). The fear of flying, as well as the cost of operating the aging aircraft, has prompted many airlines to order huge new fleets of aircraft. The conventional wisdom also identifies mass as a key ingredient of survival. So fleets grow.

As the 1990s dawned, the airline industry had $150 billion in orders or options for more than 2,500 new aircraft.[78] In contrast, the foreign debt of Brazil, which has the highest debt of all Latin American nations, is a paltry $114 billion.[79] The industry as a whole had operating cash of less than $5 billion in 1988, which was a very good year.[80] The industry's

Table 2.1
Major Mergers and Acquisitions since 1986

Date Completed	Acquirer (Acquired)	Value (in Millions)
Aug. 86	NWA (Republic)	$884
Sept. 86	TWA (Ozark)	250
Sept. 86	Texas Air (Eastern)	676
Dec. 86	Texas Air (People Express)	112
Mar. 87	AMR (Air Cal)	225
Apr. 87	Delta (Western)	860
May 87	USAir (Pacific Southwest)	400
Oct. 87	USAir (Piedmont)	1,590
Nov. 88	Carl Icahn (TWA)privatization	N.A.
May 89	Trump (Eastern Shuttle)	365
July 89	Checchi Group (NWA) buy out	3,650
Withdrawn	Management/Labor(UAL)buy out	6,790
Withdrawn	Trump (AMR)	7,540

N.A. = not applicable
Source: "Airlines Restructure," WALL STREET JOURNAL, (Oct. 6, 1989), at A3.

capital expenditures between 1991 and 1994 were estimated to be $15 billion per year.[81]

In 1989, United placed a record $15.7-billion order for 370 Boeing 737s and 757s (180 firm orders and 190 on option). American had 259 aircraft on order and 302 on option, totaling $14.5 billion.[82] In late 1988, Delta placed $13 billion in options or orders for 215 jets, including 40 giant MD-11s, and expanded that with a $10-billion order in November 1989 for up to 260 aircraft (firm orders for 50 new MD-90s and 50 B-737-300s and options for 110 MD-90s and 50 B-737s).[83] Texas Air placed an order for 100 jets in early 1989—50 firm and 50 on option—and then a second order, on behalf of Continental in November 1989, for 40 Airbus medium- and long-range jets—20 firm and 20 on order.[84] Even debt-saddled Northwest signed a $5.2-billion contract with Boeing for 80 757s (half of which are options) and 10 747-400s (4 of which are options).[85] Northwest placed a $3.2-billion order for 50 Airbus A320s in 1986.[86]

In part, airlines may be trading in aircraft options. Their huge orders enable them to enjoy volume discounts from the manufacturers. Before delivery, should they need the cash more than they need the planes, they can sell their delivery positions, as financially strapped Pan Am did in 1988 when it sold deliveries of 50 Airbus A320s to Braniff for $115 million. (Braniff overreached and consequently found itself in bankruptcy for the second time.) But aircraft futures bring a profit only during a bull market for planes, an environment that exists only when growth in passenger demand exceeds existing capacity.

Adding new jets will mercifully reduce the age of the nation's fleet. That will be a welcome blessing for the margin of safety. But it saddles the industry with even more debt. What's worse, unlike the days before deregulation when airlines actually owned most of their aircraft, today they lease. For example, American Airlines owns only about a third of its 476 aircraft outright.[87] Even solid carriers like Delta have sold large numbers of aircraft only to lease them back. That increases debt but decreases value. Potential and successful LBOs will accelerate this trend.

Lease obligations usually don't show up on balance sheets as debt, but like accumulated frequent-flyer mileage, they should. Including these obligations reveals that the industry's debt-to-equity ratio is significantly worse than it was in the mid-1980s, although the industry's performance has dramatically improved since then. For example, Delta's balance sheet debt as a percentage of total capital is only 31 percent, but adding the debt equivalent of aircraft leases (about $3 billion to balance sheet debt of $1.2 billion) increases the debt-to-equity ratio to 61 percent.[88]

Leasing has become an increasingly popular means of retiring debt assumed in LBOs or, for LBO targets, as a means of reducing the availability of assets that could be liquidated, thereby making them less attractive targets. The increased operating costs of leasing and the loss of residual aircraft values on retirement from the U.S. system (many aging Boeing 747s today sell for more than their purchase price when new) are partially offset by flexibility and the sharing of risk that leases offer. Leasing companies are stimulated by the underlying margins in the interest rate environment and the tax advantages of a leasing portfolio.[89]

Whether purchased outright or leased, new aircraft not only impose tremendous debt but also flood the market with capacity. For example, American Airlines may have a fleet of more than 800 aircraft by the late 1990s. If we learned nothing else from deregulation, we should have learned that excess capacity causes prices to spiral downward and leaves the airlines hemorrhaging red ink. A soft economy may dissuade the airlines from retiring the geriatric jets.

So now the wild cards—fuel prices, aerial terrorism, or recession. The former raises industry costs, particularly with fuel-guzzling hub-and-spoke operations (a 10-cent-per-gallon increase shaves $1.3 billion from the industry's operating earnings, which were $2.3 billion in 1988).[90] The latter two curtail demand.

When recession rears its ugly head, watch out. Few industries are as susceptible to downward turns in the economy as are airlines. Recessions prompt travelers to cancel their vacations and businesspeople to tighten their belts. Passenger demand plummets.

As noted above, the seats airlines sell resemble an instantly perishable commodity, and short-term marginal costs (another meal and a few more drops of fuel) are nil. So during slack demand periods, ticket prices spiral downward. Undoubtedly, falling prices will cause Alfred Kahn to babble on about how thankful we should be that he deregulated the airlines. But carrier profitability crumbles.

Couple a prolonged recession with excess capacity and high debt service and we see another round of bankruptcies and mergers like the one we endured in the early 1980s. When the dust settles, the industry will be even more concentrated. Recession and a modest spike in fuel costs caused the industry to suffer its worst losses ever ($3.9 billion) in 1990, and to lose nearly half as much again ($1.8 billion) in 1991. These combined losses consumed all profits made by the industry since the flight of the Wright Brothers at Kitty Hawk, N.C., in 1903.

CORPORATE PIRATES AND ROBBER BARONS

The airline industry has always attracted people with huge egos. Millionaires like Howard Hughes and flying aces like Eddie Rickenbacker found the allure of the heavens irresistible. These were men who built and pioneered the industry and nurtured its technological development. They came from a class of pilots and engineers who appreciated the beauty and necessity of flight and were awed by its technology. They were buccaneers, explorers, and brash entrepreneurs. But unlike their contemporary counterparts, they saw aviation as strongly grounded in the public interest.

What attracts the likes of Marvin Davis, Carl Icahn, Frank Lorenzo, Jay Pritzker, Donald Trump, and Peter Ueberroth to an industry like airlines? Is it the glamor of flight, the defiance of gravity, the sweaty palms many passengers still get on takeoff and landing, the allure of exotic destinations, or the raw sex appeal of the industry? Yes, partly that.

Owning an airline is a terribly prestigious endeavor, more prestigious today than owning an NFL franchise, for there are far fewer airline clubs playing in the league for domination of the heavens and America's largest cities. Only the very elite can afford to belong to the exclusive and dwindling club of airline entrepreneurs. And it is much less regulated.

So it attracts men with extraordinary egos, as it always has. From the earliest days of deregulation, the prevailing wisdom has been that after the dust settles, only a small handful of gargantuan carriers will dominate the industry. Each chief executive officer recognized that the pile of airline corpses would be high, but each believed he would rise to the top of the heap. Much chest beating and bravado was exhibited by CEOs under deregulation, even as their firms went bankrupt or as they were gobbled up by larger airlines.

A century after the railroad robber barons appeared,[91] the same thirst for wealth and power has motivated a new generation of robber barons to dominate airlines and use this industry's tremendous market power to pillage the nation. The primordial desire to dominate the nation's transportation industry, it seems, is nearly as old as the invention of the wheel.

But the original airline entrepreneurs were honest, devoted to aviation and its role in serving the needs of a great nation. These men built the great service-oriented airline companies and ran them from the 1930s until the 1960s: William (Bill) Patterson of United; Cyrus (C. R.) Smith of American; Edward V. (Cap'n Eddie) Rickenbacker of Eastern; Juan Trippe of Pan American; Howard Hughes of TWA; and C. E. Woolman of Delta. These men were "giants among a band of intuitive executives who counted few pygmies in their numbers."[92]

The new generation of airline entrepreneurs are giants too. But under deregulation, their devotion to the public interest is an anathema to their lust for wealth. A senior executive of Boeing predicted, "The only guys who'll survive [under deregulation] are those who eat raw meat."[93]

Under the stewardship of Frank Lorenzo and Carl Icahn, the once proud Continental, Eastern, and TWA have been stripped of assets and have little cash, aging fleets, a sliding reputation, and declining market shares. Let us introduce you.

Francisco Anthony Lorenzo of Texas Air

Frank Lorenzo, the Darth Vader of the airline industry, feared by his competitors and despised by labor, is among the greatest robber barons of all time. In a decade of bold acquisitions, adept financial maneuverings, mergers, bankruptcies, union busting, asset stripping, and old-fashioned wheeling and dealing, his Texas Air empire amassed nine different airlines, becoming, for a short while, the largest airline company in the nation. Only the former Soviet Union's Aeroflot flew more aircraft. As the *Wall Street Journal* observed: "Mr. Lorenzo is widely viewed as a master at acquiring airlines and a genius at high finance. No one questions his vision in creating the nation's largest and lowest-cost airline-holding company from a rag-tag assemblage of operations."[94]

Like Crandall, Lorenzo is an avid jogger.[95] The son of Spanish-born immigrants who ran a beauty parlor in Queens, New York, young Frank grew up in the flight path of LaGuardia Airport. Lorenzo was given the nickname "Frankie Smooth Talk" while a student at Columbia University. At Columbia, Lorenzo resigned a dorm council position after he and several other students allegedly attempted to rig a student election.[96] While he has a reputation of being pleasant and charming in personal encounters, an Eastern pilot noted, "He shakes your hand and smiles, and then as you start to walk away, he slaps you."[97]

Lorenzo worked and borrowed his way through Harvard Business School, ironically as a card-carrying teamster driving a Coca-Cola truck.[98] After graduating, Lorenzo became a financial analyst for TWA and then Eastern.

In 1969, Lorenzo and a classmate, Robert Carney, created Jet Capital Corporation. Jet Capital became an advisor to nearly bankrupt Texas International Airlines (called Trans-Texas before 1968).[99] Lorenzo and Carney acquired Texas International in 1972 by helping to refinance it.[100] Lorenzo became president and chief executive officer at the age of 32. Lorenzo's headquarters have been in Houston ever since, although curiously neither Texas Air nor its many subsidiaries were listed on the directory of the skyscraper he occupied.[101]

Lorenzo initially opposed deregulation, arguing that small firms like his would be gobbled up or driven under by the big boys. But once deregulation became a *fait accompli,* Lorenzo jumped aboard with some enthusiasm, offering discount "Peanuts fares" to fill his planes and passing out peanuts to customers. Texas International billboards showed flying peanuts grinning from ear to ear. Somehow it all seemed appropriate. Jimmy Carter, the former peanut farmer from Plains, Georgia, was President, and it was he who had blindly championed deregulation.

The more savvy analysts and industry executives predicted that when the dust of deregulation finally settled, the industry would be dominated by a handful of megacarriers, perhaps no more than four or five firms. Neither Jimmy Carter nor his CAB chairman, the economist Alfred Kahn, could afford to agree with so dire a prediction, for that would mean that deregulation would be an imprudent experiment. But most of the industry's elite knew better. No one understood more clearly than Frank Lorenzo that only giant gorillas would rule the jungle.

And no one enjoyed the Monopoly game better than Lorenzo. As a former associate said: "Frank's into making money and doing deals. He's the classic entrepreneur. Every morning when he wakes up he's got a better idea than the one he had the day before."[102] He has a reputation of successfully executing complex transactions that put him on top of the heap. As one commentator noted, "In his 16-year campaign to build his vision of an airline for the future, he has taken no prisoners, using adroit maneuvers, leveraged buyouts and tough negotiating to conquer one airline after another."[103] But after the Eastern bankruptcy, another observer pointed out that although his strength lies in making deals, his inability to manage people may be his undoing: "I see Lorenzo as a deal-maker, a guy who has never been noted for having a very clear strategy for how to build the human organization and is now reaping the [results of] that lack of vision."[104]

In 1979, Lorenzo began a hostile takeover attempt of National Airlines, a company three times the size of Texas International.[105] National was a carrier with a route structure radiating north and west from Florida and east to London. At $26 a share, National offered a stable of used aircraft at a premium price. After Lorenzo began his raid, a number of other airlines jumped in, including Pan American, Eastern, and Air Florida. Pan

Am, which wanted National for the domestic feed it could supply for Pan Am's international routes, ultimately concluded a nonhostile "white knight" acquisition for $55 a share, or a total of $400 million, and swallowed National. National would give Pan Am an almost fatal bout of indigestion, but "Frankie Smooth Talk" walked away with a cool $46 million in arbitrage.[106]

The money was not to sit in his icy hands for long. He invited Edwin Smart, TWA's chairman, to breakfast at the Hotel Carlyle in New York and offered to buy TWA, then ten times the size of tiny Texas International. An insulted Smart left abruptly without eating.[107]

Rebuffed by TWA, Lorenzo soon began a hostile acquisition of Continental Airlines, whose stock was selling at less than the book value of the aircraft it owned. Continental had tried mergers with Western Airlines but had not been able to conclude them. In a desperate move to avoid Lorenzo's assault, Alvin Feldman, Continental's CEO, desperately tried to arrange an employee buyout. But it was too little, too late. Lorenzo had 51 percent of Continental for $100 million.[108] Feldman put a revolver to his head and pulled the trigger.[109]

Lorenzo also believed that just being big was not enough. He felt that the key to long-term success in the deregulated airline industry was to be a large low-cost carrier, one with a computer reservations system. He began his assault on labor by letting contracts with Texas International pilots drag on for a year and a half before settling them, refusing to negotiate, and appealing over the heads of the union chiefs to labor.[110]

After acquiring Continental, Lorenzo established a nonunion subsidiary, New York Air, to fly in the northeastern United States. The threat of transferring aircraft out of unionized Texas International and Continental into nonunion New York Air gave him additional leverage in reducing wages and revising work rules with the unions.

Although deregulation meant that Washington's role would be reduced, it still was important, particularly in approving mergers and in acquiring international routes. So Lorenzo began recruiting the Washington airline establishment. He lured Alfred Kahn, who had been the misguided chairman of the CAB at the time the Airline Deregulation Act was enacted, and Kahn's two principal deputies, Michael Levine (CAB director of pricing and domestic aviation) and Phil Bakes (CAB general counsel), to the Texas Air empire. Kahn would repeatedly testify before Congress in favor of deregulation. Levine would head New York Air while Kahn would sit on its board of directors and Bakes would lead both Continental and Eastern into bankruptcy. Bakes had served on Teddy Kennedy's Senate Judiciary Committee staff when deregulation was on the table, and after joining Lorenzo, he recruited many Kennedy deputies and prominent Democratic staffers as Texas Air lawyers and lobbyists, as impressive an array as had ever been seen on Capitol Hill.[111] Lorenzo also picked up the head of the transportation section of the Antitrust Division of the U.S. Department of Justice, Elliott Seiden. Seiden, as father confessor of the industry's antitrust sins, perhaps more than any government official was privy to the darkest secrets of Lorenzo's competition and, indeed, Lorenzo himself. With friends in high places, Lorenzo could proceed without the government breathing down his neck.

Texas Air spent more money on political action committees than any other airline.[112] It has been estimated that Texas Air spent at least $2 million a year on lobbying and public relations alone.[113]

In September 1983, Lorenzo made his most infamous move. After two years of wrangling over wages with the machinists union and six weeks after the union's strike, Lorenzo led Continental into Chapter 11 reorganization bankruptcy proceedings. Three days thereafter, he tore up all his labor agreements, including those of the nonstriking pilots, fired all of Continental's 12,000 employees, and unilaterally cut wages between 40 and 60 percent.[114]

Labor felt betrayed. At no time during negotiations with pilots had management ever suggested cutting wages below the average for large established trunk-line carriers. Continental was hardly near liquidation, with several hundred million dollars in ready cash. The pilots and flight attendants began their strike in October.

It was a bitter strike. At one point a scab pilot, sleeping in his home in Evergreen, Colorado, was wakened abruptly at about three o'clock in the morning by the sound of crashing glass. Someone had thrown an elk head through the plate-glass window of his living room. At about the same time, Lorenzo flew into Denver's Stapleton Airport aboard a Continental jet whose pilot missed the runway, landing instead on the parallel taxiway. The union ended their strike in 1985. But by then, their backs had been broken. Lorenzo had earned the reputation of being a union buster.[115]

Lorenzo's reputation as a union buster was to cost him other acquisitions, including runs at Frontier and TWA in 1985. At both airlines, the pilots surrendered millions of dollars in wage and work-rule concessions to avoid the dreaded Lorenzo.

In late 1986, Lorenzo swept in with an offer to buy People Express and its Frontier, Britt, and PBA subsidiaries for $298 million, less than the $307 million that People had paid for Frontier the year before.[116] After the offer had been accepted, as the People Express position became increasingly untenable, Lorenzo tendered an even lower counteroffer to People's CEO Donald Burr on a take it or leave it basis. Burr had no choice but to accept. He rejoined the Texas Air empire but soon left, his tail between his legs. Lorenzo folded all the airlines—New York Air, People Express, and Frontier—into Continental in a messy overnight transition on February 1, 1988.[117]

Also in 1986, Lorenzo made his boldest purchase of all—Eastern Air

Lines, for $615 million.[118] Eastern had cash of $463 million, more than Lorenzo's outlay.[119] Lorenzo had Eastern borrow about $300 million to finance his purchase of it.[120]

Eastern had been managed, badly, by the former astronaut Frank Borman. Eastern lost about a billion dollars during the first decade of deregulation. Borman had tried to trim costs by rolling back wages, concessions he exchanged with labor for 25 percent of Eastern's stock and labor presence on the board of directors. While the other unions had taken salary cuts of about 28 percent, the machinists union, headed by Charlie Bryan, a small but feisty and contentious Irishman, would stand for none. One Eastern executive described Bryan as "an 800-pound gorilla."[121] The presence of Bryan on Eastern's board made life for Borman a living hell.[122]

Borman criticized labor for failing to see the "big picture." To that, one labor leader responded: "I know why we can't see the big picture. We can't see the big picture because it's written on the far side of the moon. And [former astronaut] Borman is the only one who's seen the far side of the moon."

Eastern's serious financial problems led Borman to three options: "Fix it, sell it, or tank it."[123] He concluded that he couldn't fix it and didn't want to tank it, so he sold it . . . to the monster Lorenzo. But, it seems, what he really did was to tank the unions, with a vengeance.

Borman was gone, but the unions were no happier. The battle between Lorenzo and Eastern's unions began almost from day one. There is one episode the unions love to tell:

It is March 1986, and Lorenzo is locked in ferocious battle with the unions over the future of Eastern Air Lines. The negotiations have been punctuated by loud noises and nasty words. Insults have been exchanged. Then Charlie Bryan, head of the Eastern machinists union and Lorenzo's chief antagonist, undergoes an epiphany. He extends an olive branch, sending Lorenzo a telegram suggesting that they meet and calmly discuss their differences with an eye toward working together. Lorenzo's reply is swift and clear: "I do not talk to union leaders."[124]

After acquiring Eastern, Lorenzo prepared for the siege. He had barbed wire stretched along the top of fences around the Miami headquarters. He had closed-circuit cameras mounted in hangars to monitor mechanics. He also had manhole covers in the base welded shut.[125] Lorenzo wanted major wage concessions from the machinists, and the machinists weren't about to surrender them without a fight.

As Lorenzo began to turn up the heat, the unions began their own assault on the man they loved to hate, Lorenzo, "the devil incarnate, a hardheaded, hard-hearted wheeler-dealer intent on destroying their unions, their airline and their lives."[126] They would paint him as the Great Satan—the Antichrist.

Lorenzo's Texas Air corporate structure was complicated, and intentionally so. Lorenzo owned 52 percent of Jet Capital Corporation. With 1 percent of Texas Air's equity, Jet Capitol enjoyed 34 percent voting control of Texas Air and the right to elect seven of Texas Air's directors through a special class of stock.[127] Texas Air, in turn, had more than 20 subsidiaries.[128] In the late 1920s, financial pyramiding of a similar nature gave birth to federal public utility regulation.

Shortly after acquiring Eastern, Lorenzo looted it of some of its more valuable assets. He began by stripping Eastern of its computer reservations system (System One) for a paltry $100 million at a time when Eastern's bankers estimated that its worth was between $200 million and $320 million and it was generating $255 million a year in cash.[129] To finance the acquisition, Texas Air gave Eastern a 25-year note at 6.5 percent interest.[130] Eastern, of course, was left without a computer reservations system, and was required to buy services from System One, for which it paid $130 million *a year* to Texas Air.[131]

Lorenzo controlled a fuel-brokerage firm, from which Continental and Eastern had to buy all their fuel, at a 1 percent commission, or about $30 million a year.[132] Eastern was forced to buy a $25-million unsecured note from People Express, bringing Texas Air a $4-million profit.[133] Thus, Texas Air upstreamed cash to the parent in the form of management and service fees charged the subsidiaries, Continental and Eastern.

Lorenzo transferred 11 of Eastern's gates at Newark to Continental for an $11-million promissory note paying 10 percent interest. In contrast, Piedmont paid $25 million to Eastern for eight gates and related facilities at Charlotte.[134] Lorenzo also transferred the lucrative Miami-to-London route and 20 aircraft to Continental.[135] Eastern also paid Continental $30 million to train 400 pilots to keep Eastern flying in the event of a strike.[136]

Lorenzo closed Eastern's Kansas City hub and laid off about 25 percent of the work force. He also proposed to transfer the lucrative Boston–New York–Washington shuttle to a Texas Air subsidiary for $225 million, a transaction for which Jet Capital arranged a juicy $1.25-million fee for itself for advising Texas Air.[137] The shuttle was responsible for one-third of Eastern's profits. Its transfer was abated only when blocked by court order.[138]

Lorenzo leveraged Eastern heavily with debt, mortgaging its unencumbered assets. In 1988, its annual debt-service burden was a staggering $575 million.[139] Before Eastern's bankruptcy, its long-term debt was estimated to be $2.5 billion.[140] Although secured by equipment, the debt had interest rates as high as 17.25 percent—radically higher than the 10 percent note accepted by Eastern from Continental for 11 gates and the 6.5 percent note accepted by Eastern from Texas Air for the System One computer reservations system.[141] But Eastern's creditors could reach Texas Air for only about 10 percent of the debt, for Lorenzo has carefully shielded the parent.[142] As one source noted:

Mr. Lorenzo has built one of the most leveraged major corporations in the nation while insulating Texas Air—and himself—from most of the cost and much of the risk. . . . Mr. Lorenzo presides over some of the nation's sickest airlines. . . . All are losing money at some of the fastest rates in aviation history and rank as the industry's biggest debtors. As a group, the Texas Air companies have piled up $5.4 billion in debt. Last year they had to pay $623 million simply to service the long-term part of that debt—an interest bill higher than the annual revenue of each of nearly 100 companies at the bottom of the Fortune 500.[143]

The unions, which owned 25 percent of Eastern's stock, complained that Lorenzo was draining off its assets for his own benefit.[144] In one lawsuit, the pilots union alleged that Lorenzo intended to "loot Eastern for the benefit of Texas Air."[145] An Eastern pilot noted, "I think it's clear to even the most casual observer that they're engaged in union-busting by spinning off the airline's most valuable assets."[146] When asked whether he intended to bust Eastern's unions so that he could enjoy a cost structure comparable to that of Continental, Lorenzo insisted, "That's utter bullshit."[147]

Before Eastern's bankruptcy, a Texas Air spokesman promised, "Frank [Lorenzo] and [Eastern President] Phil Bakes have absolutely no plans for a Chapter 11 filing at Eastern."[148] In response to inquiries by reporters as to whether Eastern would be placed in bankruptcy, Bakes himself said: "We've ruled that out. Bankruptcy never has been an option."[149] No doubt, these false assurances were designed to calm nervous passengers booking flights and buying tickets.

Hatred for Lorenzo galvanized the unions. As an Eastern pilot said, "As long as money is flowing up into a tornado called 'Jet Capital,' I see no reason why I or any other employee should feed this whirlwind with money out of our pockets."[150] When the machinists struck, the pilots honored their picket lines, and Eastern was shut down. Despite the earlier assurances, Lorenzo quickly flew Eastern into Chapter 11 bankruptcy, further dismembering its assets. But Lorenzo could not tear up the union contracts at Eastern, as he had in the Continental bankruptcy. Partly in response to Lorenzo's use of bankruptcy in 1983 to shed Continental of its union contracts, Congress had amended the Bankruptcy Code in 1984 to make such an action impossible without permission of the bankruptcy judge.

Offers were made for Eastern by the TWA raider Carl Icahn and Major League Baseball Commissioner Peter Ueberroth. Both were rejected by Lorenzo. Eastern was dismembered by selling the shuttle to Donald Trump, its Latin American routes to American Airlines, and its Philadelphia gates and Canadian routes to Midway Airlines, along with scores of aircraft.

Eastern employees burned Lorenzo in effigy. As one commentator noted: "Among many of them, a sense of betrayal runs deep. And the lightning rod for their anger is Frank Lorenzo, the steel-willed chairman of Texas Air Corp."[151]

An editorial summed up the mark Lorenzo has made on the airline industry:

The trouble with Lorenzo is that his only genuine successes have been in creating an empire of misfits which has accumulated debts of over $5 billion, in attracting undiluted hatred from his workforce, in bringing on an unprecedented investigation by the DOT into his fitness to manage an airline, and in his blatant efforts in asset-stripping.[152]

Lorenzo's self-image is more positive. Said he, "I'm not a guy associated with a lot of ego."[153] If not that, he is associated with lots of other things. Lorenzo left the airline industry in 1990, taking $30 million from Scandinavian Airline System (SAS) for his shares in Texas Air (renamed Continental Airline Holdings) and agreeing to leave the industry for at least seven years. Shortly thereafter, Continental entered bankruptcy, and in 1991, Eastern ceased operations.

Carl Icahn of TWA

Unlike many of the other robber barons, Carl Icahn is not a builder of great airlines. He is a corporate raider, a financial pirate, pure and simple, whose interest in companies focuses on what they can produce at the bottom line, in nice crisp dollars. A TWA union leader summarized the difference between Icahn and Lorenzo: "Mr. Lorenzo wants to own the largest airline in the world. Mr. Icahn wants to be the richest man in the world."[154]

As noted above, an attempted takeover of Trans World Airlines by Frank Lorenzo led its unions to give major wage and work-rule concessions to Carl Icahn, who had acquired TWA in 1986, paying $440 million for 22 million shares.[155] He soon took TWA private and moved its headquarters out of Rockefeller Center in Manhattan to Mt. Kisco, New York, near his home.[156] The Mt. Kisco facilities are adorned with gilded chandeliers hanging from high ceilings and with oil paintings of dueling cavalrymen, Napoleon with his marshals, and ferocious sea battles.[157] The thrill of battle consumes Icahn. So too does the glamor of the airline industry.

In 1986, TWA concluded a $224-million agreement to acquire Ozark Airlines, which shared TWA's St. Louis hub. The merger gave the consolidated firm 76 percent of the gates at Lambert International Airport and 86 percent of passenger enplanements.[158] This enabled TWA to raise ticket prices, which it promptly did.[159]

In 1987, Icahn made a $1.6-billion bid for USAir at a time when USAir was attempting to acquire Piedmont. USAir rejected the bid, but there was speculation on Wall Street that what Icahn really wanted was to force USAir to buy TWA.[160] As one analyst noted: "He's gone everywhere trying to sell TWA. There aren't any takers."[161] Others speculated that Icahn wanted to "green mail" USAir into buying back his 14.8 percent stock interest at a premium.[162]

In 1987, the Securities and Exchange Commission (SEC) began an investigation of Icahn's activities as part of a wider insider-trading probe created by the Ivan Boesky scandal. In particular, the SEC was looking at Icahn's proposed bid for Phillips Petroleum in 1985 and his stake in Gulf & Western.[163]

Icahn leveraged TWA to the teeth, doubling its long-term debt to raise cash for other acquisitions, including USX and Texaco. TWA's $2.5-billion debt and lease obligations crushes the airline's earnings with annual interest charges of $375 million.[164] TWA has a negative net worth of $30 million.[165] The company has a 15-1 debt-to-equity ratio.[166] In 1988, the Consumer Federation of America became so concerned about these manipulations that it alleged, "After running up huge amounts of debt, the Icahn-led group now proposes to take all its money (and then some) out of TWA, leave the company with absolutely no equity, and leave the airline on the brink of bankruptcy."[167] Ironically, Carl Icahn recently noted, "There is no question that leverage during the past year has gotten out of hand—it was almost a feeding frenzy."[168]

TWA flies the oldest fleet of aircraft in the industry. Icahn's failure to reinvest TWA's capital in the airline led the pilots union to charge that Icahn had betrayed them at TWA's acquisition when he had assured them that he would not dismember the airline. But until the spring of 1989, TWA had placed no orders for new aircraft. TWA President Joseph Corr resigned when he became convinced that Icahn would not buy the planes the airline needed.[169] (The hulking, blunt-talking Corr subsequently became Continental's CEO for a short while.)[170] To stem the criticism, TWA ordered a few Airbus A-330 widebodies. But there was some speculation that they might be sold off before their scheduled delivery in 1994. Former TWA executives claim that the company also needs another 50 to 100 narrowbodied aircraft to replace aging aircraft.[171] Any order for the Boeing or McDonnell-Douglas aircraft TWA desperately needs would also not see delivery until 1994.

Nonetheless, to fatten his war chest for future raids, Icahn proceeded to leverage TWA still further. With long-term debt of $2.5 billion, TWA had already pledged its aircraft and engines to existing lenders. In June 1989, Icahn announced a $300-million high-interest junk bond offering secured on TWA's spare parts such as light bulbs, gaskets, and landing slots.[172]

Ironically, Icahn acquired TWA with generous concessions from the pilots, who were intent on avoiding the union-busting Lorenzo's hostile acquisition. But soon after climbing into TWA's cockpit, Icahn was to crush a union himself—the flight attendants, who struck in 1986. He had trained an army of scabs to pass out the dinner trays and pour drinks. (Actually, flight attendants are on board because the FAA requires their presence to protect passenger safety.) Soon he had a union on its knees, anxious to return to work at sharply reduced wages and benefits and under stiffer work rules. Icahn, the union buster.

Icahn is not without his dirty linen. He so slashed costs that TWA reduced the frequency with which it washed its blankets, with malodorous results.[173] Something is rotten at TWA.

FOREIGN OWNERSHIP

Not only is the debt caused by LBOs of serious public concern, so too is the rapidly growing phenomenon of foreign ownership. Foreign alliances with U.S. airlines began around the frequent-flyer programs created by the U.S. carriers.[174] The second wave of alliances occurred when foreign airlines affiliated with U.S. carriers' computer reservations systems. The most recent round of foreign interest in U.S. airlines has involved direct ownership.[175]

Several public-policy concerns arise over foreign ownership of U.S. airlines. The first surrounds national security. America depends on its Civil Reserve Aviation Fleet (CRAF) for airlift capacity in time of war. Foreign ownership may jeopardize access to the fleet. The second concern surrounds the integrity of air transport negotiations between the United States and foreign governments. International routes are traded by nations on a bilateral basis, usually with candid input from their carriers.[176] Multiple allegiances may well jeopardize the integrity of that process. Third, these alliances may significantly reduce competition in international aviation. How strongly will United and British Airways compete in the U.S.-U.K. market, for example, if the two carriers have common ownership? These issues will be developed in chapter 25.

NOTES

1. See P. DEMPSEY & W. THOMS, LAW & ECONOMIC REGULATION IN TRANSPORTATION 28–29 (1986).

2. *See, generally,* Dempsey, *Transportation Deregulation—On a Collision Course?*, 13 TRANSP. L.J. 329 (1984); Dempsey, *The Rise and Fall of the Civil Aeronautics Board—Opening Wide the Floodgates of Entry*, 11 TRANSP. L.J. 91 (1979).

3. See Dempsey, *The Empirical Results of Deregulation: A Decade Later and the Band Played On*, 17 TRANSP. L.J. 31 (1988).

4. See P. DEMPSEY, THE SOCIAL AND ECONOMIC CONSEQUENCES OF DEREGULATION 95–104 (1989).

5. *Id.* at 86–92.

6. *Id.* at 86–92.

7. Goetz & Dempsey, *Airline Deregulation Ten Years After: Something Foul in the Air,* 54 J. AIR L. & COM. 927 (1989).

8. Dempsey, *Antitrust Law & Policy in Transportation: Monopoly I$ the Name of the Game,* 21 GA. L. REV. 505 (1987).

9. GENERAL ACCOUNTING OFFICE, AIR FARES AND SERVICE AT CONCENTRATED AIRPORTS (1989).

10. Statement of Timothy Pettee before the U.S. Senate Committee on Commerce, Science and Transportation, Oct. 4, 1989, at 14.

11. Ellis, *United's Buyers May Be Wearing Rose-Colored Goggles,* BUS. WEEK, Oct. 16, 1989, at 36.

12. Dempsey, *supra* note 3, at 31. *U.S. Airlines Will Lose Another $1.8 Billion in 1991,* AVIATION DAILY, Dec. 10, 1991, at 429.

13. Hughes & Smith, *Failed Bid for NWA Leaves Marvin Davis Richer and Still Ready,* WALL ST. J., June 21, 1989, at 1.

14. In March 1987, Trump purchased 4.9 percent of UAL, selling it in April for $73 a share, making him a $55-million profit. *Donald Trump's Investment Track Record,* WALL ST. J., Oct. 6, 1989, at A3.

15. The Trump Shuttle operates 21 aircraft between three cities. In contrast, American Airlines has 480 planes. *An Ego as Big as American,* NEWSWEEK, Oct. 16, 1989, at 56. Another source reports that American has 683 aircraft. *Here Comes Donald, Duck!,* TIME, Oct. 16, 1989, at 52.

16. O'Brian & Valente, *Crandall's American Is Unlikely Recipient of $8 Billion Trump Bid,* WALL ST. J., Oct. 6, 1989, at 1.

17. Statement of Timothy Pettee before the Senate Comm. on Commerce, Science and Technology (Oct. 4, 1989), at 14.

18. Phillips, *AMR Earnings Decrease 8.8% in Third Period,* WALL ST. J., Oct. 19, 1989, at A3.

19. *Here Comes Donald, supra* note 15, at 52.

20. *With Trump Gone, AMR Targets Growth,* USA TODAY, Oct. 18, 1989.

21. Loeffelholz, *Competitive Anger,* FINANCIAL WORLD, Jan. 10, 1989, at 28.

22. *"I'm Going to Run This Joint,"* N.Y. TIMES, Dec. 8, 1985, at 8F.

23. Brown III, *American Airlines Boss Blossoms as Champion of the Poor Passenger,* WALL ST. J., Mar. 4, 1988, at 1.

24. S. DAVIS, DELTA AIR LINES: DEBUNKING THE MYTH 166 (1988).

25. Salpukas, *The Braniff-American Air Duel,* N.Y. TIMES, Mar. 29, 1982, at 25.

26. Complaint of U.S. Dep't of Justice in United States v. American Airlines, Inc., and Robert Crandall (N.D. Tex.).

27. P. DEMPSEY & W. THOMS, LAW & ECONOMIC REGULATION IN TRANSPORTATION 214 (1986).

28. Brown III, *supra* note 23.

29. Karr, *Airline Deregulation after Braniff's Fall,* WALL ST. J., June 14, 1982, at 16.

30. *American Aims for the Sky,* BUS. WEEK, Feb. 20, 1989, at 54.

31. Loeffelholz, *supra* note 21, at 28.

32. *Id.* at 33.

33. B. NASH & A. ZULLO, THE MISFORTUNE 500 140 (1988).

34. *See* Easterbrook, *Lorenzo Braves the Air Wars,* N.Y. TIMES, Nov. 29, 1987, at 17.

35. *Id.*

36. Brown III, *supra* note 23.

37. *American Aims for the Sky, supra* note 30.

38. *Id.* at 55.

39. *Loeffelholz, supra* note 21, at 30.

40. *Sizing Up AMR Corp.,* WALL ST. J., Oct. 6, 1989, at A3.

41. O'Brian & Valente, *supra* note 16, at 1.

42. *An Ego as Big as American,* NEWSWEEK, Oct. 16, 1989, at 56.

43. *Smith, Trump Bid $7.54 Billion to Acquire American Air,* WALL ST. J., Oct. 6, 1989, at A3.

44. *Id.* at A3.

45. *Id.* at A3.

46. S. DAVIS, *supra* note 27, at 12.

47. *Rising UAL Turmoil Threatens Ferris's Job as the Chief Executive,* WALL ST. J., Apr. 17, 1987, at 1.

48. Amended Complaint of Frontier Airlines in In re Frontier Airlines Reorganization, Case No. 860B-802IE (Jan. 2, 1987).

49. *Rising UAL Turmoil, supra* note 47.

50. *Id.*

51. *Id.*

52. Cohen & Kilman, *Talk of a Possible Takeover of UAL Inc. Is in the Air,* WALL ST. J., Apr. 9, 1987, at 6.

53. *Id.*

54. *Rising UAL Turmoil, supra* note 47.

55. *Allegis Shakeup Came as Shareholder Ire Put Board Tenure in Doubt,* WALL ST. J., June 11, 1987, at 1.

56. *Id.*

57. *Id.*

58. B. NASH & A. ZULLO, *supra* note 33, at x.

59. Storch & Jouzaitis, *Little Hope Seen in UAL Bid's 2d Wind,* CHICAGO TRIBUNE, Oct. 22, 1989, at C1, C12.

60. *Id.* at C12.

61. *British Air May Balk at Any Haste in Reformulating a UAL Buy-Out,* WALL ST. J., Oct. 18, 1989, at A3.

62. Valente & Smith, *United Air Pilots Face Cuts in Wages, Overtime Pay, Vacation to Finance Bid,* WALL ST. J., Sept. 11, 1989, at A4.

63. *In UAL Bid, Trust May Be Elusive between Management and Pilots,* WALL ST. J., Sept. 18, 1989, at A10.

64. Smith & Valente, *UAL Machinists Attack Proposal Backed by Pilots,* WALL ST. J., Sept. 6, 1989, at A3.

65. *Storch, Bankers Bring Down United Buyout,* CHICAGO TRIBUNE, Oct. 14, 1989, at 9.

66. *Banks Rejecting UAL Saw Unique Defects in This Buy-Out Deal,* WALL ST. J., Oct. 16, 1989, at 1.

67. Carroll, *UAL Drops; Bid Is Still Up in the Air,* USA TODAY, Oct. 18, 1989,

at 3B; Valente, *British Airways Won't Revive UAL Buy-Out*, WALL ST. J., Oct. 20, 1989, at A3.

68. *After Buyout Diet, Wolf May Be Dessert*, CHICAGO TRIBUNE, Oct. 22, 1989, at 12.

69. Storch & Jouzaitis, *United Execs Could Score Buyout Bonanza*, CHICAGO TRIBUNE, Sept. 23, 1989, at 1, 6.

70. Bailey, *Two UAL Officers Get No Parachutes in Stock's Free Fall*, WALL ST. J., Oct. 18, 1989, at A3.

71. *United*, CHICAGO TRIBUNE, Sept. 23, 1989, at 6.

72. *After Buyout Diet*, supra note 68.

73. Smith, *In Failed Bid for UAL, Lawyers and Bankers Didn't Fail to Get Fees*, WALL ST. J., Nov. 30, 1989, at A1.

74. *UAL Stockholder Calls for Directors to Resign*, CHICAGO TRIBUNE, Nov. 28, 1989, at 3–7; *Stockholder Proposes Resolution Calling for UAL Board to Quit*, ATLANTA JOURNAL & CONSTITUTION, Nov. 28, 1989, at B-6.

75. Valente, *Transportation Agency May Rein in Airline Buy-Outs, Foreign Investments*, WALL ST. J., Aug. 31, 1989, at A3.

76. Dallos, *U.S. Conducts Fitness Exam of Northwest Airlines; Debt Cited*, L.A. TIMES, Sept. 5, 1989, at 1.

77. McGinley, *Skinner Warns about Airlines Piling Up Debt*, WALL ST. J., Sept. 20, 1989, at A4.

78. Salpukas, *Airlines' Big Gamble on Expansion*, N.Y. TIMES, Feb. 20, 1990, at C1, C5.

79. *Brady Strategy: Rest in Peace*, WALL ST. J., Jan. 22, 1990, at 1, col. 1, 5.

80. *Statement of Philip Baggaley to the House Subcomm. on Aviation of the House Comm. on Public Works and Transp.*, 101st Cong., 1st Sess. (Oct. 4, 1989), at 3.

81. *Statement of Timothy Pettee before the Aviation Subcomm. of the House Comm. on Public Works and Transportation*, 101st Cong., 1st Sess. (Oct. 4, 1989), at 4.

82. O'Brian & Valente, supra note 16, at 1.

83. Waldman & Wartzman, *Delta Air Sets Orders, Options for $10 Billion*, WALL ST. J., Nov. 15, 1989, at A3.

84. Manges, *Texas Air's Continental Unit Set to Buy Up to 40 Airbus Jetliners for $4.5 Billion*, WALL ST. J., Nov. 17, 1989, at A3.

85. Nomani, *NWA to Unveil Major Order with Boeing*, WALL ST. J., Oct. 11, 1989, at A4.

86. *NWA Orders 90 New Jets*, MSP AIRPORT NEWS, Oct 19, 1989, at 1.

87. In contrast, United owns 80 percent of its $3-billion fleet outright. Morris, *Soaring Airline Stocks: A Leveraged Way to Fly*, Sept. 3, 1989, at V-3.

88. *Statement of Philip Baggaley*, supra note 80, at 3.

89. *Statement of Timothy Pettee*, supra note 81, at 19.

90. *Statement of John F. Peterpaul before the Aviation Subcomm. of the House Comm. on Public Works and Transportation*, 101st Cong., 1st Sess. (Oct. 4, 1989), at 11; *Statement of Timothy Pettee*, supra note 81, at 15.

91. See P. DEMPSEY, supra note 4, 6–12.

92. R.E.G. DAVIES, AIRLINES OF THE UNITED STATES SINCE 1914 532–33 (1972).

93. S. DAVIS, supra note 24, at 10.

94. Thomas, *Frank Lorenzo, Builder of Airlines, Now Faces Task of Running One*, WALL ST. J., July 24, 1987, at 1.

95. *See* Easterbrook, *Lorenzo Braves the Air Wars*, N.Y. TIMES MAGAZINE, Nov. 29, 1987, at 17.

96. B. NASH & A. ZULLO, *supra* note 33, at 140. *See The New Master of the Skies*, FORTUNE, Jan. 5, 1987, at 72.

97. Christensen, *Unions, Eastern Spent 2 Years Gearing for Strike*, ATLANTA JOURNAL & CONSTITUTION, Mar. 5, 1989, at 11A.

98. *The New Master of the Skies, supra* note 96.

99. *See* R.E.G. DAVIES, *supra* note 92, at 416.

100. Thurow, *Frank Lorenzo Tries to Navigate 3 Airlines through Stormy Skies*, WALL ST. J., Feb. 18, 1982, at 1.

101. Christensen, *supra* note 97.

102. Thurow, *supra* note 100.

103. Christensen, *supra* note 97.

104. Bennett, *Personalizing the Conflict at Eastern Air*, WALL ST. J., Mar. 9, 1989, at B1.

105. Hamilton, *A Tale of Two Airlines: Texas Air, USAir Survive at Different Speeds*, WASHINGTON POST, May 22, 1988, at H1, H7.

106. Easterbrook, *supra* note 95, at 18.

107. Ennis, *Sky King*, BUSINESS MONTH, Sept. 1988, at 27, 32.

108. Witkin, *Texas Air's Continental Takeover*, N.Y. TIMES, Mar. 25, 1982, at 31.

109. *See* Easterbrook, *supra* note 95, at 62.

110. Thurow, *supra* note 100. *See Easterbrook, supra* note 95.

111. Abramson & Sarasohn, *On Board with Frank Lorenzo*, LEGAL TIMES, May 16, 1988, at 1, 10, 11.

112. *See* AVIATION DAILY, Mar. 7, 1988, at 349.

113. Abramson & Sarasohn, *supra* note 111, at 12.

114. *How the Continental Strategy Worked*, WALL ST. J., Mar. 10, 1989, at A6. Easterbrook, *supra* note 95, at 64. Ennis, *supra* note 107, at 32.

115. O'Brian, *A Look at Troubled Eastern's Options*, WALL ST. J., Mar. 8, 1989, at B1.

116. Dempsey, *Antitrust Law & Policy in Transportation: Monopoly I$ the Name of the Game*, 21 GA. L. REV. 505, 540–41 (1987).

117. Hamilton, *supra* note 105, at H7.

118. O'Brian & Dahl, *Ueberroth Group Plans to Acquire Eastern from Texas Air Under $464 Million Pact*, WALL ST. J., Apr. 7, 1989, at A3.

119. Ennis, *supra* note 107, at 32.

120. Nelson, *Lorenzo Has His Cake and Sells It Too*, DENVER POST, Mar. 29, 1988, at 6B.

121. Stockton, *Tearing Apart Eastern Airlines*, N.Y. TIMES MAGAZINE, Nov. 6, 1988, at 36, 39.

122. *See* Borman, *Showdown in Miami*, BUS. MONTH, Sept. 1988, 39.

123. Castro, *Eastern Goes Bust*, TIME, Mar. 20, 1989, at 52, 53.

124. *The New Master of the Skies, supra* note 96.

125. Christensen, *supra* note 97.

126. *Id.*

127. *House of Mirrors: Lorenzo's Texas Air Keeps Collecting Fees from Airline Units That Have Continuing Losses,* WALL ST. J., Apr. 7, 1988, at 14.

128. Ennis, *supra* note 107, at 33.

129. *House of Mirrors, supra* note 127.

130. *Id.*

131. *Lorenzo Has His Cake, supra* note 120. *House of Mirrors, supra* note 127.

132. Nelson, *Lorenzo's Empire under Scrutiny,* DENVER POST, Apr. 18, 1988, at 6B.

133. *The Other Side of the Coin,* AIRLINE BUS., Dec. 1988, at 24.

134. *House of Mirrors, supra* note 127.

135. Castro, *supra* note 123, at 53.

136. Hamilton, *Eastern Head Tells of Paying Continental,* WASHINGTON POST, Aug. 18, 1988, at B1.

137. Thomas, *Texas Air Drops Plan to Transfer Eastern Shuttle,* WALL ST. J., July 5, 1988, at 4. *House of Mirrors, supra* note 127.

138. Ennis, *supra* note 107, at 33; *EAL Recalls Shuttle Plan to Clear Contempt Ruling,* TRAVEL WEEKLY, Apr. 14, 1988, at 5.

139. *Going for Broke: Lorenzo's Effort to Remake Eastern Could Fail,* WALL ST. J., Mar. 10, 1989, at A6.

140. O'Brian, *supra* note 115.

141. *See* Luke, *Next Two Weeks Could Decide Eastern's Fate,* ATLANTA JOURNAL & CONSTITUTION, Mar. 5, 1989, at 11A.

142. *Going for Broke, supra* note 139.

143. *Statement of John F. Peterpaul, supra* note 90, at 9–10.

144. Nelson, *supra* note 132.

145. *House of Mirrors, supra* note 127.

146. Knox, *Lorenzo, Eastern's Unions Square Off,* ROCKY MOUNTAIN NEWS, Feb. 21, 1988, at 72.

147. *Lorenzo, Being Frank,* AIRLINE BUS., Jan. 1989, at 18, 21.

148. Knox, *supra* note 146.

149. Hamilton & Swoboda, *Lorenzo Faces Turbulent Skies at Eastern,* WASHINGTON POST, Mar. 20, 1988, at H1, H8.

150. Stockton, *Tearing Apart Eastern Airlines,* N.Y. TIMES MAGAZINE, Nov. 6, 1988, at 36, 39.

151. Solis & de Cordoba, *Eastern Strike Bares Long-Festering Anger over Sense of Betrayal,* WALL ST. J., Mar. 17, 1989, at 1.

152. *Gratitude or Vitriol?,* AIRLINE BUS., Aug. 1988, at 5.

153. Stockton, *Tearing Apart the Airlines,* N.Y. TIMES MAGAZINE, Apr. 11, 1988, at 36, 86.

154. Hamilton, *TWA's Unions Try to Ground Icahn,* WASHINGTON POST, Sept. 30, 1988, at F1, F2.

155. Carley & Agins, *TWA to Go Private in $1.2 Billion Plan That Would Boost Icahn's Stake to 90%,* WALL ST. J., July 23, 1987, at 3.

156. *Id.*

157. *Carl Icahn Has Lots of Cash: Will He Spend It on TWA?,* BUS. WEEK, July 17, 1988, at 86.

158. Dempsey, *supra* note 116, at 511–12.

159. GENERAL ACCOUNTING OFFICE, AIRLINE COMPETITION: FARE AND SERVICE CHANGES AT ST. LOUIS SINCE THE TWO-OZARK MERGER (1988).

160. *US Air Rejects TWA's Takeover Proposal of $52 a Share, or More Than $1.6 Billion,* WALL ST. J., Mar. 6, 1987, at 4.

161. *Id.*

162. Cohen & Agins, *TWA Ends Bid for USAir, Confirms Icahn Is Under Investigation by SEC,* WALL ST. J., Mar. 17, 1987, at 3.

163. *Id.*

164. *Carl Icahn Has Lots of Cash, supra* note 157, at 87.

165. Power, *Raiders May Not Make the Best Airline Pilots,* BUS. WEEK, May 15, 1989, at 35.

166. *Statement of Bill Hoffman before the Aviation Subcomm. of the House Comm. on Public Works and Transportation,* 101st Cong., 1st Sess. (Oct. 4, 1989), at 2.

167. Consumer Federation of America, Press Release of September 20, 1988.

168. *Statement of Bill Hoffman, supra* note 166, at 2.

169. *Carl Icahn Has Lots of Cash, supra* note 157, at 87.

170. *A Mr. Fix-It Goes to Work on Lorenzo's Continental,* BUS. WEEK, May 22, 1989, at 134.

171. *Carl Icahn Has Lots of Cash, supra* note 157, at 87.

172. Sandler, *TWA to Sell $300 Million Notes Secured in Part by Light Bulbs,* WALL ST. J., June 2, 1989, at C1.

173. *Carl Icahn Has Lots of Cash, supra* note 157, at 87.

174. *Statement of Timothy Pettee before the Senate Comm. on Commerce, Science and Technology,* 101st Cong., 1st Sess. (Oct. 4, 1989), at 4.

175. See, generally, Winter, *Congress Questions DOT's Role in Airline Leveraged Buy-Outs,* TRAFFIC WORLD, July 24, 1989, at 35.

176. See, generally, P. DEMPSEY, LAW & FOREIGN POLICY IN INTERNATIONAL AVIATION (1987).

3

THE MEGACARRIERS

The initial development of commercial aviation in the United States, as well as the identity of America's largest airlines and their geographic emphasis, was largely determined by the original airmail contracts let in the first third of this century. Indeed, without the airmail subsidies, commercial aviation would not have become ubiquitous so early. Mail went in the belly of the aircraft while passengers rode on top.

On New Year's Day 1914, commercial air transportation was born with the world's first regularly scheduled air passenger flight of a Benoist flying boat on a 22-mile route from Tampa to St. Petersburg, Florida. The operation was a financial failure, but it afforded the world a glimpse into the future of commercial air transportation.[1]

By 1926, 14 airmail routes were in operation by the forerunners of today's airlines. However, passenger air travel lagged in development until Charles Lindbergh's solo flight across the Atlantic on May 20–21, 1927, which inspired the world to fly. Soon thereafter, it was realized that only coherent route systems would develop solid passenger revenue. The McNary–Watres Act of 1930 gave Postmaster General Walter F. Brown wide-ranging control over the 43 existing airlines, through airmail contract awards. These contracts developed a pattern of airline routes that formed the basis for today's system.[2]

Government sponsorship and regulation are inextricably bound with the birth and growth of U.S. airlines. Through the airmail contract awards, Brown fostered the amalgamation of small, financially weak airmail contractors and passenger carriers into three transcontinental airlines—the predecessors of United, American, and TWA.

United won the airmail contract between New York and San Francisco, via Chicago, and between Seattle and Los Angeles, via San Francisco. American flew between Boston and New York to Los Angeles via Chicago,

Table 3.1
Domestic Airlines Ranked by Revenue Passenger Miles under Deregulation,
1978–1989

	1978	79	80	81	82	83	84	85	86	87	88	89	90
AMERICAN	2	2	3	3	2	2	2	1	2	2	2	1	1
UNITED	1	1	1	1	1	1	1	2	1	1	1	2	2
DELTA	5	5	4	6	6	6	6	5	4	3	3	3	3
NORTHWEST	11	8	7	7	7	7	7	7	5	4	5	4	4
CONTINENTAL	9	10	12	10	8	10	8	8	8	5	4	5	5
USAIR	12	11	10	12	11	11	11	12	9	9	9	7	6
TWA	3	4	6	5	5	5	4	4	6	7	6	6	7
PAN AM	6	6	5	2	3	3	5	6	7	8	7	9	8
EASTERN	4	3	2	4	4	4	3	3	3	6	8	8	9

Source: AVIATION DAILY, From Department of Transportation data.

Figure 3.1
U.S. Industry Traffic Market Share in Percentage of Revenue Passenger Miles

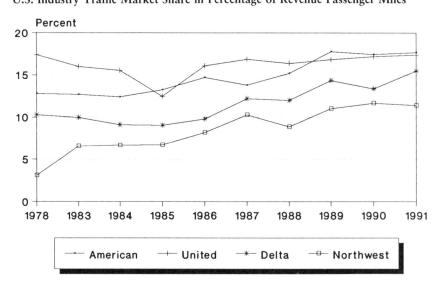

Table 3.2
Market Shares of the Largest Airlines
(in Revenue Passenger Miles, January 1990)

Airline	RPMs (in millions)	Share (%)
1. American	5,676,675	16.720
2. United	5,428,222	15.988
3. Delta	4,418,700	13.015
4. Northwest	3,935,990	11.593
5. Continental	2,988,344	8.802
6. USAir	2,522,339	7.429
7. TWA	2,423,400	7.138
8. Pan Am	2,342,000	6.898
9. Eastern	1,520,000	4.477

Source: AVIATION DAILY, Feb. 21, 1990, at 365.

Figure 3.2
Comparison of Eight Airlines, 1967–1979 (in Revenue Passenger Miles)

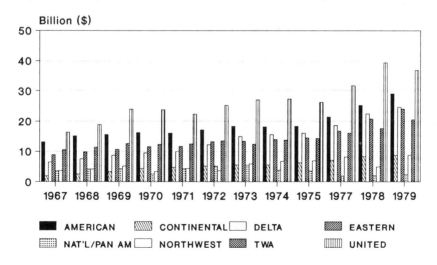

No 1976 stats; Pan Am begins 1977-79

Figure 3.3
Comparison of Eight Airlines, 1980–1988 (in Revenue Passenger Miles)

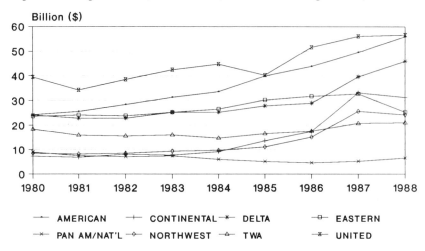

Pan Am Includes Nat'l Results (1980-88)

Nashville, and Dallas/Ft. Worth. TWA carried the mail between New York and Los Angeles via St. Louis and Kansas City. Eastern was the only one of the "big four" without a transcontinental route, flying between New York and New Orleans and between Chicago and Miami via Atlanta.[3]

Figure 3.4
United versus American Total Seats

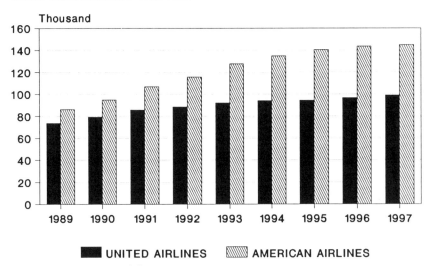

1990–1997 are estimated figures.

Figure 3.5
Airline Comparison—Top 8 No. of Aircraft/Total Aircraft

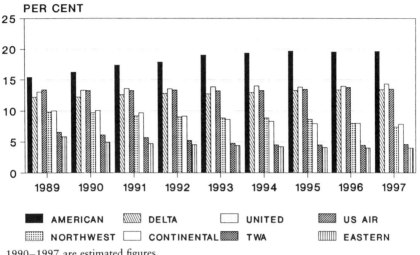

1990–1997 are estimated figures.

Table 3.3
The World's Largest Airlines, 1988
(in Revenue Passenger Kilometers)

Rank	Airline	RPKs (000,000)
1.	Aeroflot	213,171
2.	United	111,184
3.	American	104,216
4.	Delta	83,266
5.	Qantas	66,223
6.	Continental	66,077
7.	Northwest	65,728
8.	British Airways	58,200
9.	TWA	55,998
10.	Pan Am	47,273

Source: AIR TRANSPORT WORLD, June 1989, at 91.

These "big four" airlines were grandfathered in—issued certificates of "public convenience and necessity"—when the Civil Aeronautics Board was created in 1938. These airlines retained their original geographic emphasis during the ensuing decades.

By the time of deregulation in 1978, the big four had become the "big

Table 3.4
U.S. Carrier Market Share at Concentrated Major U.S. Airports
(January–June 1989)

Airport	Dominant Carrier(s)	% of total passengers
Atlanta Hartsfield	Delta	78.6
	Eastern	12.3
Charlotte Douglas	USAir	94.1
Chicago O'Hare	United	49.0
	American	33.3
Dallas/Fort Worth	American	63.8
	Delta	28.5
Denver Stapleton	United	48.8
	Continental	36.2
Detroit Metro Wayne	Northwest	65.3
Houston Intercontinental	Continental	76.1
Memphis	Northwest	82.2
Minneapolis/St. Paul	Northwest	79.3
New York Kennedy	Pan Am	31.2
	TWA	31.1
	American	22.6
Newark	Continental	50.2
	USAir	13.9
	United	10.7
	American	10.1

Table 3.4
Continued

Orlando	Delta	34.4
	USAir	18.5
Philadelphia	USAir	47.3
	American	11.3
	Delta	10.5
	United	10.0
Phoenix Sky Harbor	America West	40.8
	Southwest	22.3
Pittsburgh	USAir	88.3
Salt Lake City	Delta	82.6
San Francisco	United	37.2
	USAir	15.9
St. Louis Lambert	TWA	81.8

Note: Airports listed are those at which one or two airlines are responsible for more than 50 percent of passengers enplaned. Piedmont's market share data were folded into that of USAir.

five," with Delta entering their ranks, having acquired Central & Southern in 1953 and Northeast Airlines in 1972.[4] Delta flew the mail between Charleston and Dallas/Ft. Worth via Atlanta. The predecessor of Chicago & Southern Airlines (Pacific Seaboard) carried the mail between Chicago and New Orleans while Northeast's predecessor (Boston & Maine) flew the mail in New England.[5]

The smaller airlines too owe their geographic emphasis to the original airmail contracts. Northwest flew the mail between Chicago and Seattle via Minneapolis. The predecessors of Western Air Lines (General Air Lines and Wyoming Air Service, merged in 1944) carried the mail between Los Angeles and Salt Lake City and north and south from Denver. The predecessor of Continental (Varney) flew in Colorado and New Mexico. Braniff had the contract between Dallas/Ft. Worth and Chicago via Kansas City.[6] Note that these airlines, by and large, continued to dominate the geographic regions and cities (today, hubs) for which they originally earned their airmail contracts.

In 1938, when Congress passed the Civil Aeronautics Act, the four largest domestic carriers were American, United, TWA, and Eastern. Another

Table 3.5
Fleets of the Major Airlines (as of January 1, 1990)

Aircraft	AMR	CON	DEL	EAL	NWA	PAN	TWA	UAL	USA
B767	45	–	30	–	–	–	11	19	16
B757	8	–	52	24	33	–	–	5	–
B747	2	8	–	–	44	35	19	34	–
B737	11	94	72	–	–	–	–	147	217
B727	164	94	129	50	71	89	69	140	44
DC-10	59	15	–	2	20	–	–	55	–
DC-9	–	40	36	65	105	–	41	–	74
DC-8	–	–	–	–	–	–	–	27	–
MD-80	180	65	–	–	8	–	–	–	31
MD-82	–	–	–	–	–	–	29	–	–
MD-83	–	–	–	–	–	–	4	–	–
MD-88	–	–	48	–	–	–	–	–	–
L-1011	–	–	40	14	–	–	35	–	–
A320	–	–	–	–	6	–	–	–	–
A310	–	–	–	–	–	7	–	–	–
A300	25	12	–	18	–	24	–	–	–
BAE146	6	–	–	–	–	–	–	–	21
F28	–	–	–	–	–	–	–	–	45
FOKKER100	–	–	–	–	–	–	–	–	8
SHORTS360	–	–	–	–	–	–	–	–	5
TOTAL	500	328	407	173	287	155	208	427	461

Source: AVIATION DAILY, Jan. 26, 1990, at 188.

U.S. carrier, Pan American World Airways, had no domestic routes but was developing a monopoly in rapidly expanding international markets.[7] By 1972, the 10 largest airlines were United, American, TWA, Delta, Eastern, Western, National, Continental, Braniff, and Northwest.

Table 3.6

Major Air Carrier Mergers, Acquisitions, Purchases, and Consolidations since Promulgation of the Airline Deregulation Act of 1978

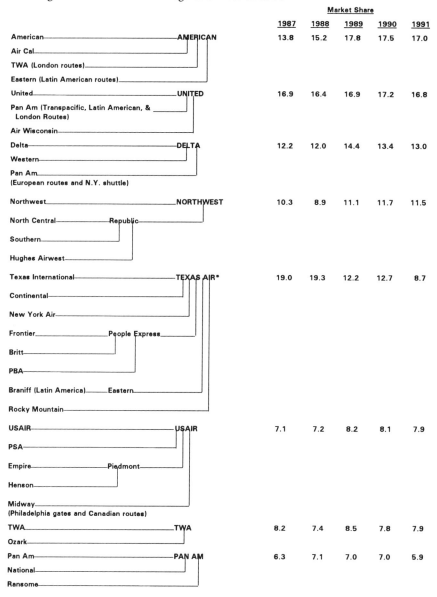

	Market Share				
	1987	1988	1989	1990	1991
American ——— AMERICAN	13.8	15.2	17.8	17.5	17.0
Air Cal					
TWA (London routes)					
Eastern (Latin American routes)					
United ——— UNITED	16.9	16.4	16.9	17.2	16.8
Pan Am (Transpacific, Latin American, & London Routes)					
Air Wisconsin					
Delta ——— DELTA	12.2	12.0	14.4	13.4	13.0
Western					
Pan Am (European routes and N.Y. shuttle)					
Northwest ——— NORTHWEST	10.3	8.9	11.1	11.7	11.5
North Central ——— Republic					
Southern					
Hughes Airwest					
Texas International ——— TEXAS AIR*	19.0	19.3	12.2	12.7	8.7
Continental					
New York Air					
Frontier ——— People Express					
Britt					
PBA					
Braniff (Latin America) ——— Eastern					
Rocky Mountain					
USAIR ——— USAIR	7.1	7.2	8.2	8.1	7.9
PSA					
Empire ——— Piedmont					
Henson					
Midway (Philadelphia gates and Canadian routes)					
TWA ——— TWA	8.2	7.4	8.5	7.8	7.9
Ozark					
Pan Am ——— PAN AM	6.3	7.1	7.0	7.0	5.9
National					
Ransome					

Sources: 1987 statistics, *Business Week*, Oct. 5, 1987, at 40.
1988-89 statistics, *Wall Street Journal*, Mar. 10, 1989, at A8.
1990 statistics, *Aviation Daily*, Jan. 29, 1991, at 189.
1991 statistics (First Ten Months only), *Aviation Daily*, Nov. 26, 1991, at 859.

* Renamed Continental Airline Holdings

In the post-deregulation environment, several characteristics have been identified as critical to survival: (1) hub/spoke route structures; (2) yield (pricing) management; (3) capacity (aircraft) management; (4) low labor costs; (5) computerized reservation system; and (6) ability to take advantage of size.[8] Table 3.1 displays how well the various carriers have fared under deregulation. Note that whereas American, United, and Delta remain strong, TWA and Eastern dropped sharply from the leading pack. And today's "big three"—American, United and Delta—account for more than 50 percent of the domestic market.[9] The metamorphosis of America's largest airlines under deregulation is shown in table 3.1.

One interesting phenomenon emerging after a decade of deregulation is that the "big four" (American, United, Delta, and Northwest) are pulling away from the pack, dominating more than 60 percent of the market (as measured in revenue passenger miles) (figure 3.1). American, Delta, and United are pulling away from the pack in terms of profitability (accounting for nearly 82 percent of the industry's third-quarter 1989 profits, for example).[10] This occurred at a time when three of the original airlines (TWA, Pan Am, and Eastern), with dwindling market shares, were on the endangered species list—candidates for bankruptcy and/or liquidation (Eastern and Pan Am, in fact, expired in 1991). Table 3.2 reveals the market shares of the megacarriers. (See also figures 3.2 to 3.5.)

The combined market share of the eight largest airlines in 1990 was 92 percent and in late 1991, 95 percent, compared with between 80 and 82 percent before deregulation.

Many proponents of deregulation pointed gleefully to the entry of People Express, Air Florida, Midway, and America West. The former four no longer exist, and the combined market share of the latter two was not quite a puny 3.5 percent.[11]

The collapse of the time-space continuum is, of course, the premier contribution of aviation. In a global economic environment, the international picture is necessary for perspective. Compare the above figures with the global data of international airlines, as measured in passenger kilometers traveled (table 3.3). America's airlines still account for seven of the top ten carriers in the world, but their relative share of the international market is slipping.[12]

Another measure of market strength is dominance of the nation's largest airports. Table 3.4 reveals the dominant airlines at the leading airports.

Still another measure of size is the number and type of aircraft in each of the megacarriers' fleets. Table 3.5 depicts the fleet composition and size of the largest airlines.

The next several chapters will take a closer look at the megacarriers that have emerged from deregulation, in alphabetical order. Table 3.6 graphically depicts the mergers that have transpired since deregulation, and the carriers' relative market shares.

NOTES

1. *Air Transport,* COLLIER'S ENCYCLOPEDIA I, 387 (1985).

2. In February 1934, all airmail contracts were canceled by the government on the grounds that there had been collusion in route awards. The army started flying the mail, but a string of fatal crashes occurred and the airlines regained their mail business, providing that there be complete reorganization of the airlines. The Civil Aeronautics Act of 1938 was eventually passed and later insured such reorganization. See Air Transport, I ENCYCLOPEDIA AMERICANA 402 (1988).

3. R. DAVIES, AIRLINES OF THE UNITED STATES SINCE 1914 168 (1972).

4. *Id.* at 566.

5. *Id.* at 168–69, 604–5.

6. *Id.* at 168–69, 604–5.

7. P. DEMPSEY, LAW & FOREIGN POLICY IN INTERNATIONAL AVIATION 8 (1987).

8. AIRLINE ECONOMICS, INC., AIRLINE CONSOLIDATION 2 (1987).

9. *See An Industry Poised for Expansion,* N.Y. TIMES, February 20, 1990, at C5.

10. *American, Delta, United Account for Most of Industry Profits,* AVIATION DAILY, Jan. 9, 1990, at 57.

11. AVIATION DAILY, Feb. 21, 1990, at 365.

12. *See* DEMPSEY, *supra* note 7, 86–91.

4

AMERICAN AIRLINES

Hubs: Dallas/Ft. Worth, Chicago, Nashville, Raleigh/Durham
Mini Hubs: Miami, San Juan, San Jose
Post-deregulation Merger: Air Cal (1987)
Computer Reservations System: SABRE
Rank and Market Share: 1978—second, 12.8%; 1990—first, 17.5%

American Airlines has been described as the most "widely regarded of late as the industry's best managed and most innovative carrier."[1] Headed by tenacious Robert Crandall, American pioneered two-tier wage rates for labor in 1984. This made expansion a lower-cost endeavor, an opportunity on which American capitalized by creating new hubs in Nashville and Raleigh/Durham, San Juan and San Jose, and by expanding geographically into Europe, the Caribbean, and Mexico, as well as buying Eastern's Latin American and TWA's London operations. In just six years, American doubled its payroll, to 67,000 workers (from 42,000 in 1984).[2] Today, about half of its workers earn lower, B-scale wages.[3]

In 1989, American purchased Eastern's Latin American routes for $349 million at the bankruptcy sale (which, ironically, Eastern had bought from bankrupt Braniff for $30 million in 1983).[4] To reach these international destinations, American has 446 aircraft on order or option.[5]

American has not paid dividends since 1982, reinvesting all its profits and throwing labor a bone or two in profit sharing. In 1988, American paid each worker an average of $2,000 in profit sharing.[6] Thus, it has enjoyed meteoric growth, in 1990 surpassing United as the nation's largest airline.

American was also an innovator in computer reservations systems. Its

SABRE system is dangerously successful, hooked up to more than 50,000 locations. American earns a 20 percent return on SABRE operations, far more than it earns on its airline services.[7] American also inaugurated frequent-flyer programs.[8] Let us explore American's long and proud history, beginning in the 1920s.

On March 1, 1929, "The Aviation Corporation (Avco)" was established.[9] Avco acquired five airlines, three of which were holding companies themselves, controlling eleven individual carriers. To these eleven, Avco added two independent airlines: Embry-Riddle[10] and the fast-growing Colonial Airways Corporation, which then added another holding company and three more airlines to the roster. The biggest acquisition and the third independent airline was Universal Air Lines System, a combination of six airlines.[11]

Next, Avco acquired a third holding company in the South: Southern Air Transport, which was a combination of two airlines, Gulf and Texas Air Transport. The Avco network was extended even further when the holding company bought out Interstate Airlines, which possessed another prime mail route.[12] By the end of 1929, Avco had acquired thirteen airlines, eleven mail contracts, a fleet consisting of virtually every type of civil aircraft available, and an impressive route system, which lacked a cohesive structure.[13]

Avco's first president was Graham Grosvenor. Since starting its acquisition spree, Avco had lost $1.4 million, which was massive red ink in those depression-plagued days. Avco's board of directors then decided to consolidate all of Avco's domestic airline holdings into a new subsidiary corporation.[14] The Avco board elected a new president, Frederic G. Coburn, who was expected to straighten out the mess and bring order to the chaotic 9,100-mile system. Coburn became president of both Avco and the airline on January 25, 1930, and on that date, American Airways, Inc., came into existence.[15]

In the first three years of American's existence, Coburn brought some order out of the chaos by trimming Avco's 80 subsidiaries down to 20 and cutting operating losses from $3.4 million in 1930 to $1 million in 1931. Still, Coburn failed to centralize control of American's operations.[16]

Postmaster Brown felt that the existing 44 airlines were too many. He therefore persuaded Congress to promulgate the McNary–Watres Act in 1930, which eliminated the smaller airlines from the market.[17] Brown also summoned the heads of the large airlines to Washington for the "Spoils Conference." Brown's main thrust at the conference was for three transcontinental routes. American Airways won the southern route, which collided with the routes of Southwest Air Fast Express (SAFE).[18] After a lot of haggling, SAFE sold out to American.[19]

In 1932, unable to stem the flow of red ink or bring unity to the airline's four divisions, Coburn resigned. Lamotte T. Cohu succeeded Coburn as American Airway's president.[20]

Cohu lasted for only nine months, but while president, he was able to consolidate American Airways' four unwieldy divisions into two: Southern and Eastern. The Southern headquarters was in Fort Worth, and the Eastern headquarters was in New York. Under this arrangement, C. R. Smith headed the Southern Division but also commanded both, as vice president of operations based in St. Louis.[21]

After a proxy fight, Errett Lobban Cord became the new chairman of the board, and Avco became a subsidiary of the Cord Corporation.[22] Cord proceeded to chop heads in the company. Cohu was the first to go, and others followed. Eddie Rickenbacker left voluntarily.[23] Lester D. Seymour became the new president of American Airways.[24]

American still lacked a direct New York–Chicago route. American's purchase of Transamerican, with it strong Chicago-Detroit-Cleveland-Buffalo service, was the first step in closing the New York–Chicago gap. The next was the acquisition of Martz Airlines, the possessor of a passenger route from Newark and Buffalo. In 1932, the Chicago–New York route was complete. American no longer had to fly north to Albany and then west to serve Chicago, competing hopelessly against United's and TWA's more direct routing.[25]

In 1934, the hearings involving Postmaster Brown's award of American's southern transcontinental route came under close scrutiny. Brown's defense was that the mail contract awards were merely a means of developing the passenger traffic, which only the larger, more experienced airlines could accomplish.[26]

The domestic airmail contracts were cancelled on February 19, 1934.[27] Theoretically, American had the most to lose from the cancellation, since it operated more airmail mileage than anyone else. Cord kept American from protesting, since he foresaw the blunders the army would make when it took over the airmail routes.[28] Stripped of their mail revenues, the airlines were forced to furlough hundreds of employees, mostly pilots, and American was no exception. American alone was running about $300,000 per month in the red.[29]

The Air Mail Act of 1934 set new standards for the newly opened bids for the airmail routes.[30] During this time, American Airways was incorporated as American Airlines, and American bid for the same routes it had held before.[31] American was awarded its old routes with a few modifications.[32] Cord was now looking ahead to a bright future for passenger traffic.[33] C. R. Smith was formally elected president of American Airlines in 1934.[34]

Smith's master plan was to develop passenger traffic through safety, marketing, customer service, and equipment innovations. Safety and reliability of aircraft would be the thrust of his marketing plan, since the fear of flying was predominant among the public in the 1930s.[35]

By the end of 1936, American had 20 DC-3s in operation.[36] United had only 10 DC-3s, and TWA was still waiting for the first of 8 to be delivered.[37] The impact of the DC-3 can best be measured by the number of people who began flying them. In 1936 alone, the U.S. airlines, for the first time in history, carried more than a million passengers, double the 1934 total, and the traffic curve was to point steadily upward for decades. In 1939, the airlines flew 42.2 percent more people than in the previous year, a staggering rate of growth that had to be credited almost solely to the DC-3, which by then was carrying 90 percent of the nation's air traffic.[38] The DC-3 freed the airlines from complete dependency on government mail pay. It was the first airplane that could make money by just handling passengers.[39]

In 1939, C. R. Smith decided to move American's headquarters to New York, where the nation's most modern airport was being built, LaGuardia Field.[40] On July 5, 1945, the CAB simultaneously approved American's control of American Export Airlines (AEA) and granted the merged carrier the routes across the North Atlantic to the United Kingdom, Scandinavia, the Netherlands, and Germany.[41] In 1950, after observing the handicaps of AEA and its detriment to American, American sold AEA to Pan American.[42] With the jet age looming over the horizon, American bought 58 piston-driven DC-7s.[43] The DC-7s would serve American for only 10 years.[44]

The Air Line Pilots Association's (ALPA) 1958–59 strike was symptomatic of airline unions' growing strength, accompanied by increasing militancy. There were 10 strikes in the industry in 1958 alone. Labor was trying to solidify its position in anticipation of the expected jet-age travel boom. Over the next five years there would be 25 airline shutdowns by various unions, with some carriers hit more than once.[45]

In anticipation of further labor problems and also knowing that the industry could not afford crippling strikes or equally crippling settlements, C. R. Smith developed a Mutual Aid Pact, which was signed by virtually every scheduled carrier in the United States. The Mutual Aid Pact was an agreement that would come into effect during labor strikes; the airlines benefiting from a competitor's strike would turn those strike-generated extra revenues over to the shutdown carrier. The pact did not stop strikes, but it eased the financial pains of an affected airline and increased management's bargaining power. The pact was repealed with the promulgation of the Airline Deregulation Act of 1978.[46]

American began transcontinental jet service between New York and Los Angeles on January 25, 1959.[47] But in 1961, American lost its rank as the nation's largest airline as United absorbed Capital in what was then the biggest merger in U.S. airline history.[48]

In 1959, American took the lead in the airline industry in another area when it and IBM announced the development of a Semi-Automated Business Reservations Environment (SABRE). Until the 1970s, SABRE's unduplicated efficiency was a major weapon against United's size.[49]

In the mid-1960s, the airline industry suffered from an overcapacity problem due to the enormous influx of larger aircraft.[50] American initiated reduced-fare plans to fill its aircrafts' empty seats to solve the capacity problem.[51]

Between June 3 and October 30, 1970, there were 10 separate contacts between officials of American and Western Airlines over a possible merger between the two companies.[52] Before the American/Western merger agreement was signed, American had acquired Trans-Caribbean Airlines (TCA). The TCA merger was submitted to the CAB in January 1970 and was approved the following December.[53] The unions, along with Continental, United, the Justice Department's antitrust division, and the CAB's Bureau of Operating Rights, strongly opposed the American/Western merger.[54] The CAB rejected the merger on July 28, 1972, and President Richard Nixon upheld the decision.[55]

Several of American's officers were charged with taking illegal kickbacks in the early 1970s, which eventually cost them their jobs and seriously hurt the airline's image.[56] As a result, employee morale sank even further. The average employee was getting the impression that many officials had been stealing large amounts of company funds. They had seen fellow workers furloughed, labor relations deteriorate, a succession of mistakes made by management, a highly touted merger go down the drain, millions spent on new planes with consistently empty seats, and international routes with no traffic potential expensively promoted. The employees reacted with indifference, sloppy performance, and a poor attitude. Management had lost their respect, and they had lost their pride. By January 1973, service had seriously deteriorated and was further damaged by labor troubles. Hundreds of flights in late December had been cancelled by a combination of weather and pilots' deliberate slowdowns over dragging contract negotiations.[57] The pilots' subtle sabotage tactics were a reaction to President George Spater's indecision and his inability to communicate.[58]

American's image as a professionally run carrier had been badly damaged. And the causes went far beyond the excuses of recession and overcapacity. In trying to cut costs, American had indiscriminately cut into its greatest strength, customer service.[59]

In 1973, Spater lured Bob Crandall away from his position as senior vice president and treasurer of Bloomingdale's to the post of American's senior vice president of finance. Crandall was joining an airline in serious trouble. Spater remained optimistic, until he assumed full responsibility for a possible illegal contribution to President Nixon's reelection campaign in violation of campaign-financing laws. His admission was still another blow to the airline's morale, and it spelled the end of his career at American.[60]

Figure 4.1
Top Three Airline Comparison 1967–1978 (in Revenue Passenger Miles)

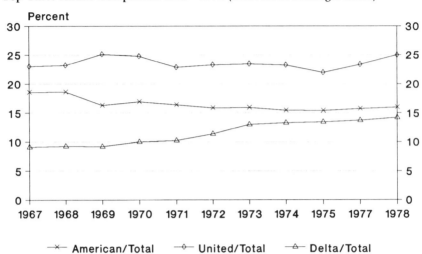

C. R. Smith was Spater's temporary successor. In the late 1960s, Smith had left American to be secretary of commerce under President Lyndon Johnson, and he had turned over American to George Spater.[61] Now Smith was back to help American Airlines clean up its mess. Smith served American for only six months, as he had promised, but he accomplished what the directors wanted. Smith helped improve the morale among American's employees, which glued the airline back together and made the company a more viable enterprise for the man who would succeed him.[62]

Albert Vincent Casey became Smith's successor on February 20, 1974. Casey did not find the airline in as sorry a state as he had expected. Smith already had started the turnaround with cost-cutting moves, including the furloughing of some 1,300 employees and the dumping of the South Pacific albatross. Furthermore, the airline had a healthy positive cash flow from equipment depreciation while capital spending had been light since 1972.[63]

In 1976, American began to establish SABRE's data-processing capabilities among travel agents. SABRE would eventually serve more than 10,000 travel agents and corporate travel departments, or 41 percent of computerized travel agencies, compared with United's 39 percent.[64]

The financial pendulum began to swing back in 1976. Profits that year hit $76.3 million and jumped to nearly $81 million the following year, reaching $134.4 million in 1978, a record figure despite a $101-million hike in fuel costs. Part of the comeback in the mid-1970s was due to Crandall's innovative marketing schemes, such as the famous Super Saver fares

Figure 4.2
United vs. American, 1979–1989 (Revenue Passenger Miles)

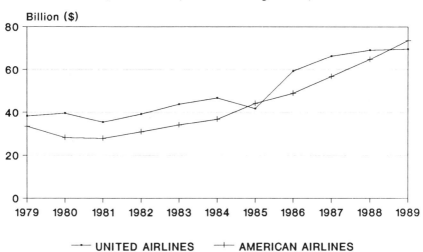

program. Super Saver was a device used to discount fares without diverting traffic from regular fares by giving discounts to passengers who purchased tickets well in advance and stayed at their destinations at least 14 days.[65] But throughout the decade, American remained second to United. (See figure 4.1.)

In 1978, Casey announced that American's general offices were moving

Figure 4.3
United vs. American, 1979–1989 (Available Seat Miles)

Figure 4.4
United vs. American, 1979–1988 (Total Aircraft Fleet)

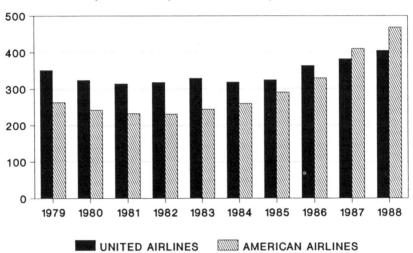

from New York to Dallas/Ft. Worth, Texas. The move took place in the summer of 1979.[66]

Congress passed the Airline Deregulation Act in 1978. Virtually overnight, the established carriers were playing a different ball game under different rules, with an invasion of key markets by new low-cost airlines, an epidemic of cutthroat fare wars, and the end of the industry's traditional way of life. To meet this challenge, Chairman Al Casey picked a new leader, Robert Lloyd Crandall.[67] Crandall became president of American on July 16, 1980.[68]

At the time of deregulation, American faced soaring fuel prices, a fleet almost 10 years old on average, and expensive contractual commitments to its unions. The only bright spot on the horizon was the 1978 order for 30 fuel-efficient Boeing 767s.[69] In 1980, American put its entire 707 fleet up for sale, withdrew from a number of unprofitable routes in the northeast, focused its operations on the Dallas/Ft. Worth hub, and reconfigured all aircraft to provide more seats.[70]

Crandall directed American back to the industry leadership through innovative programs such as A-Advantage, Super Saver, Ultimate Super Saver, and a two-tier wage structure.[71] Crandall also had his mind on buying new planes. So American worked out a lease deal with McDonnell-Douglas for the new MD Super 80s, introducing the Super 80 into service in May 1983.[72] American took its B-747s out of passenger service, and Crandall sold the B-747 freighters and traded the remaining passenger B-747s to Pan Am in exchange for an equal number of DC-10s.[73]

Casey retired in February 1985, with Crandall becoming chairman, chief executive officer, and president of American Airlines.[74] Crandall turned American into the nation's most formidable airline.

As the 1990s began, American appeared poised to expand rapidly, particularly into international markets. It inaugurated service to Australia. It purchased Eastern's Latin American routes and Miami-London and Miami-Madrid authority and TWA's Chicago-London route. It was building its Miami hub to 100 flights a day. It was seeking authority to Japan and from Chicago to Milan and Rome.[75] And it bought TWA's primary routes to London's Heathrow Airport. Under Robert Crandall, American began to pull ahead of United, as shown in figures 4.2 through 4.4.

NOTES

1. S. Davis, Delta Air Lines: Debunking the Myth 26 (1988).

2. Woodbury, *How the New No. 1 Got There*, Time, May 15, 1989, at 57.

3. Salpukas, *Steve Wolf's Big Test*, N.Y Times, Jan. 8, 1989, at 24.

4. P. Dempsey, Law & Foreign Policy in International Aviation 88 (1987).

5. Salpukas, *Airlines' Big Gamble on Expansion*, N.Y. Times, Feb. 20, 1990, at C1, C5.

6. Woodbury, *supra* note 2.

7. S. Davis, *supra* note 1, at 46, 49.

8. Hayes, *American Joins the Low-Cost Ranks*, N.Y. Times, Dec. 8, 1985, at 8F.

9. R. Serling, Eagle: The Story of American Airlines 11 (1985).

10. The prospect of commercial aviation as a potentially viable enterprise attracted investors such as the Embry-Riddle Company of Cincinnati, which had just won a contract to fly the mail between Cincinnati and Chicago. Embry-Riddle was founded by John Paul Riddle and T. Higbee Embry in 1925. Embry, a businessman, provided most of the capital, and Riddle was a pilot. *Id.* at 10.

11. *Id.* at 12–13.

12. *Id.* at 15.

13. *Id.* at 15. The largest and most critical gap in Avco's route system was the lack of a Chicago–New York link.

14. *Id.* at 16.

15. *Id.* at 16.

16. *Id.* at 17.

17. *Id.* at 32.

18. *Id.* at 33.

19. *Id.* at 34.

20. *Id.* at 37.

21. *Id.* at 39, 45.

22. *Id.* at 44.

23. *Id.* at 44.

24. *Id.* at 53.

25. *Id.* at 53.

26. *Id.* at 65.

27. *Id.* at 67.

28. *Id.* at 68.

29. *Id.* at 69.

30. *Id.* at 69.

31. *Id.* at 70–71.

32. *Id.* at 71.

33. *Id.* at 72.

34. *Id.* at 73. Avco today is one of America's best-run diversified conglomerates; its roots to American withered away until only dusty files of the past remain. *Id.* at 79.

35. *Id.* at 81–82.

36. C. R. Smith placed an order with Douglas for a greatly modified DC-2, which, when built, was the Douglas Sleeper Transport (DST) for the night traveler and the DC-3 for the daytime model, the most famous single airplane in the history of commercial aviation. The DC-3s were called "Flagships," which was to become an American Airlines trademark for years to come. *Id.* at 89–91. The DC-3 era began June 25, 1936. *Id.* at 101.

37. *Id.* at 110.

38. *Id.* at 109.

39. It cost American more per mile to operate the DC-3 than the DC-2, but the DC-3's additional seven seats made up the difference. Eventually, American would be operating an all-DC-3 fleet of 94 aircraft, but it was the first 20 that made aviation history. The DC-3s also made possible American's first employee bonus, which American gave to its 2,000 employees in late December 1938. *Id.* at 110.

40. *Id.* at 131, 133. By 1940, the airline was carrying 32 percent of the nation's air traffic in the industry's largest fleet, with 64 DC-3s and 15 DSTs. C. R. Smith went off to the war as an army colonel and left Ralph Damon to run the airline while he was gone. *Id.* at 164. By 1944, American was the second-largest international air carrier in the world; its ATC operations were topped only by Pan American. *Id.* at 179. Those who returned after the war found American Airlines far bigger. American was operating 93 DC-3s, and personnel had nearly tripled since 1940, from slightly over 4,000 to 11,450. The increase in American's fleet size stemmed from aircraft the military had turned back to the airlines as Douglas assembly lines spewed out some 10,000 C-47s specifically designed for military cargo work. American started getting planes back in the summer of 1944 and discovered that they had to almost completely rebuild. *Id.* at 183.

41. DEMPSEY, *supra* note 4, at 19. The price tag for the deal was $3 million, with American obtaining 51.4 percent control via stock purchases. With AEA's C-54s converted to DC-4 Flagships, American became a full-fledged international carrier operating AEA as a separate subsidiary under a new name: American Overseas Airlines. R. SERLING, *supra* note 9, at 188–89.

42. *Id.* at 225–26.

43. The first Boeing 707s rolled off the assembly line in 1954. *Id.* at 254.

44. *Id.* at 261.

45. *Id.* at 305.
46. *Id.* at 305.
47. *Id.* at 306.
48. *Id.* at 333.
49. *Id.* at 347–48.
50. With the arrival of the Boeing 747 in 1970, the airlines suffered their worst overcapacity problem in the industry's history. American's gimmick of a piano bar on its 747s helped alleviate the overcapacity problem until passenger traffic started an upswing. *Id.* at 402. The DC-10 went into service the first week in August 1971, between Chicago and Los Angeles. The DC-10, whose capacity fell somewhere between that of the 707 and the 747, was vindicated in just this one key market where American was competing with TWA, United, and Continental. The 707 had been too small and the 747 too large. The DC-10 was just right. *Id.* at 404.
51. *Id.* at 368–69.
52. *Id.* at 399.
53. *Id.* at 399.
54. *Id.* at 400–401.
55. *Id.* at 401.
56. *Id.* at 405.
57. *Id.* at 412–13.
58. *Id.* at 413.
59. *Id.* at 413–14.
60. Spater resigned on September 18, 1973. *Id.* at 415.
61. *Id.* at 416–18.
62. *Id.* at 418.
63. *Id.* at 439–41.
64. *Id.* at 443–44. Also in 1976, Casey thought of the twice-tried merger route with Pan Am, after a $20-million loss in 1975. *Id.* at 445.
65. *Id.* at 446.
66. *Id.* at 453.
67. *Id.* at 454.
68. *Id.* at 455.
69. *Id.* at 455.
70. *Id.* at 457.
71. *Id.* at 459–60.
72. *Id.* at 463.
73. *Id.* at 465.
74. *Id.* at 466.
75. *American Positions Itself for the Nineties*, AVIATION DAILY, Feb. 7, 1990, at 263.

5

CONTINENTAL AIRLINES

Hubs: Houston, Denver, Newark, Cleveland

Post-deregulation Mergers: Texas International (1988), People Express (1988), Frontier (1988), and New York Air (1988)

Computer Reservations System: System One, owned by Texas Air (renamed Continental Airline Holdings)

Rank and Market Share: 1978—seventh, 3.8%; 1990—fifth, 8.8%

CONTINENTAL AIRLINES CORPORATE CULTURE

Continental Airlines, a Texas Air subsidiary, is a blend of companies and a blend of cultures. Before Frank Lorenzo consumed it, Continental was the "Proud Bird with the Golden Tail"—the premium-service long-haul carrier formed by its flamboyant chairman, Robert Six, with a route structure focused on the western United States. It had an excellent reputation among business travelers. Chivas Regal was served in economy class.[1]

With the mergers of the 1980s, Continental is today a low-cost, low-service blend of one national airline (Continental), two former local-service carriers (Texas International and Frontier), and two post-deregulation upstart airlines (People Express and New York Air). Texas International was Frank Lorenzo's first airline, acquired in 1972. Its operations focused on Dallas, then Houston. New York Air was Lorenzo's creation in the early 1980s, a nonunion airline flying in the northeastern United States.[2]

All of these disparate cultures were rammed together on February 1, 1987, under the Continental banner. It came as no surprise that Continental became difficult to manage, with hundreds of flights cancelled, thousands of passengers stranded, and a mountain of luggage warehoused. The reasons for factionalization and deteriorating service in the merged airline were obvious:

The mergers have thrown together young, enthusiastic employees and jaded veterans of a bitter strike that lasted through Continental's three year reorganization under the bankruptcy laws. The labor force includes former Frontier Airlines employees, who have filed suit after being forced to the bottom of company seniority lists, and former People Express employees, whose almost fanatical devotion to their chairman, the imaginative Donald Burr, was lost when he resigned from Texas Air.[3]

But let us go back in time and observe an earlier, prouder Continental.

CONTINENTAL'S PROUD TRADITION

Robert Six was the man who built Continental Airlines into the prestigious airline of the West. A high school dropout, as a young man Six was fired by Pacific Gas & Electric for learning to fly on company time and was expelled from the United Airlines school for pilots for using company mechanics to prepare his plane for weekend races.[4]

In 1936, Six borrowed $90,000 from his father-in-law to purchase 40% of Varney Air Transport, a small western airline headquartered in El Paso and serving six cities (Denver, Colorado Springs, Pueblo, Santa Fe, Albuquerque, and El Paso).[5] By 1938, Six was president of Varney, had moved the corporate headquarters to Denver, and had changed its name to Continental Airlines.[6] Continental then had 29 employees and six aircraft, served 624 route miles, and had carried 2,316 passengers more than 1.1 million miles without accident or fatality.[7]

Six was Continental Airlines. He built it, he ran it, he dominated it. He had a fiery temper. But he ran it hands on—in the early years more like a family business than a large corporation.[8] Throughout, he remained loyal to Continental, rebuffing offers to leave and head TWA in the 1940s and in the 1960s.[9]

Continental grew robustly until World War II, when the War Department appropriated 50 percent of the airline's fleet for military service.[10] Six himself entered the war as a U.S. Army air corps captain in 1942.[11]

After the war, Six modernized Continental's fleet, expanded the route system, and hired more employees. Through interchange agreements with other airlines, Continental expanded into Houston, Los Angeles, and St. Louis.[12] In 1954, Continental merged with Pioneer Air Lines, giving the combined company access to 46 cities in six states.[13]

Although Continental lost money in 1958 (the first losses in 17 years),[14] by 1960 it had introduced Boeing 707s into its fleet and enjoyed the highest daily jet utilization in the world, the lowest operating costs, and the lowest break-even load factor.[15] By 1962, Continental had flown 10.5 million passengers more than 6 billion passenger miles without a fatality. But in that year, a 707 came apart over Missouri during a heavy storm (an accident later determined to be the work of sabotage).[16] Also that year, Six moved the corporation's headquarters from Denver to Los Angeles.[17]

In 1964, Continental expanded into the transpacific market, gaining access to Hawaii, Guam, the Philippines, Japan, Taiwan, Korea, Okinawa, South Vietnam, and Thailand.[18] In the 1960s, Continental enjoyed record profitability.

The relationship between management and labor remained healthy until 1976, when the pilots struck.[19] Six shut Continental down for 26 days before submitting to most of the union's demands.[20] Despite these difficulties, Continental employees generally enjoyed a sense of community with the company, with employee tenure rivaling that of a Japanese corporation.[21]

With the advent of deregulation in 1978, Continental had the fourth-strongest balance sheet of all major carriers.[22] But it needed to restructure its routes around hubs, since feed traffic from other airlines would dry up. Moreover, it was crippled in 1979 when the FAA ordered all DC-10s grounded for 38 days after an American Airlines crash in Chicago, when an engine fell off the plane. DC-10s accounted for 42 percent of Continental's capacity.[23] And fuel prices were escalating.

FRANK LORENZO ACQUIRES CONTINENTAL

As Continental entered the Brave New World of deregulation, Six passed the reigns of power to Alvin L. Feldman, an engineer who formerly, and successfully, headed Frontier.[24] Feldman began a dual-hub system, radiating primarily from Denver and secondarily from Houston (see figure 5.1).[25] But Continental's widebodied aircraft were ill suited for short hauls, and the company needed to merge with another carrier. It attempted to merge with Western Air Lines twice, the first try denied by the CAB and the second scuttled by an 18-month dispute with the flight attendants.[26]

While the second merger attempt with Western was pending, Frank Lorenzo began buying Continental's stock. After acquiring 48.5 percent, Lorenzo made a tender offer to Continental. Feldman tried several maneuvers to keep Continental free of Lorenzo, including an employee stock-ownership plan (ESOP) allowing Continental's employees to buy controlling interest in the airline.[27] Feldman insisted that the combination would simply not work. He told Lorenzo:

[Although you won 48.5% of Continental's stock, we] cannot accept a proposal that is not fair to our remaining public shareholders. Any merger must be fair as well to our employees. Continental's employees have borne the largest burden in building our great airline and must not be treated as pawns in a financial transaction. Finally, we cannot accept a proposal that will result in a company so weak that it cannot survive and prosper in today's environment. . . . As I understand

Figure 5.1
Airline Market Share at Houston

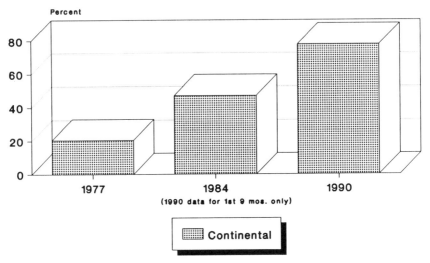

Sources: AVIATION DAILY, Apr. 19, 1985, at 28; Feb. 1, 1990, at 230; Apr. 29, 1990, at 628; Mar. 29, 1991, at 590; and CONSUMER REPORTS.

your proposal, you intend to effect the consolidation of the two airlines through a two-step transaction. In the first step, you would pass the 48.5% of Continental's stock now owned by Texas International up to Texas International's parent, Texas Air Corporation, while leaving the cost of that stock an obligation on the books of Texas International. The result of that first step is to reduce your equity in Texas International by about $93 million. Since Texas International has a present net worth of only $53 million, that would leave Texas International with a negative shareholders equity of $40 million.

In the second step of the transaction, Texas International and Continental would be merged into a new company in which you expect to receive 56% of the common stock—in exchange for Texas International's negative net worth and your minority position in Continental—while Continental's public shareholders would be diluted from their present majority position to 44% of the common stock, plus shares of a non-voting preferred. In effect, the new company would end up paying off the debt Texas International incurred to buy its Continental stock.

Your proposal hardly seems fair. But, more important, the resulting company would be very weak. I believe its chances for survival would be poor. Even before it purchased Continental's stock, Texas International was a highly leveraged company with $183 million in debt and only $53 million in equity. When you add to Texas International the burden of the additional borrowings you made to purchase the Continental stock, the situation becomes almost intolerable. The combined company would have long term obligations of $642 million, and equity of only $142 million. This results in an 82:18 debt to equity ratio, which is worse than

Braniff Airways at the end of 1979. More importantly, the debt service coverage requirement, including the dividend on the preferred stock you propose, approaches $150 million annually. The operating profit required to service this debt is more than our two companies together have ever earned.[28]

But Lorenzo was not to be stopped by Feldman's pleas or his eleventh-hour ESOP maneuver, and Continental was soon his. Feldman died from a bullet to the head, a suicide. But Feldman's prophecy was soon true—within two years, Lorenzo had led ailing Continental into bankruptcy. A labor impasse led the machinists to strike on August 13, 1983. On September 24, Lorenzo put Continental into bankruptcy.

Continental's soul was crushed when the unions' backs were broken as Lorenzo waltzed the company through Chapter 11 bankruptcy and shed himself of union contracts. Frontier had been known as a polished little airline serving the Rocky Mountain region with sharp service and a splendid safety record. Its soul too collapsed when People Express put the company in bankruptcy in 1986, after which both Frontier and People Express were consumed by Texas Air and folded into Continental.

FRONTIER AIRLINES

Frontier began service in 1946 as Monarch Airlines. Merging with Arizona Airways and Challenger Airlines in 1950, the newly named Frontier Airlines provided regional service with DC-3 aircraft in several western states. In 1962, Maytag sold its 67 percent interest in Frontier to Goldfield Corporation of San Francisco, a mining firm. In 1965, controlling interest was sold to RKO General, Inc., a subsidiary of General Tire and Rubber Company.[29] After its merger with Fort Worth's Central Airlines in 1967, Frontier covered 7,465 miles of routes in 14 states (the fourth-largest route system in the nation, after United, Eastern, and Delta).[30] By the mid-1970s, Frontier served more cities than any other airline.[31]

As deregulation dawned with the Airline Deregulation Act of 1978, Frontier was one of the few airlines that had a hub-and-spoke system (radiating from Denver's Stapleton International Airport), serving 89 cities in 20 states as well as Mexico and Canada. Between 1978 and 1982, Frontier restructured its route system, discontinuing service to 39 cities, adding 29 others, increasing frequencies, and expanding operations at Denver.[32] In 1979, under the leadership of Alvin Feldman, Frontier enjoyed record profitability and the second-lowest complaint rate of any airline (behind Delta).[33] In 1980, Feldman left to head Continental, and Glen Ryland became Frontier's CEO.[34]

Frontier was a technical leader, one of the earliest users of Boeing 727 and 737 aircraft and one of the first airlines to use a computer reservations system to handle its passengers.[35] For the 10-year period preceding 1982, Frontier enjoyed consistent profitability each and every quarter, with profits totaling $163 million.[36]

In the summer of 1982, Frontier offered 38 nonstop flights from Denver, compared with United's 24 and Continental's 16.[37] But that year United launched a campaign to dominate Denver's Stapleton Airport by increasing its flights by a third and invading many of Frontier's markets. Moreover, seven new carriers had come to Denver since deregulation, including American, Eastern, Northwest, Piedmont, and Southwest.[38] The recession of 1982 also hit the airline industry, and Frontier, hard. That year, the industry suffered the worst losses in its history.

Denver's Stapleton International Airport was the only airport in the nation with three carriers competing for hub dominance. By 1983, Denver was the most overserved market in the nation.[39] That year, United (which had dumped 100,000 seats per week into Denver since deregulation) had 40.8 percent of the market at Stapleton; Frontier had overtaken Continental for second place at 24.3 percent; and Continental was third at 18.5 percent.[40] United had 27 gates and 174 departures at Stapleton; Frontier had 51 gates and 138 daily departures.[41]

United, the nation's largest airline, was determined to dominate Denver. With a deep pocket that could cross-subsidize losses in competitive markets, a powerful computer reservations system that could discriminate against competitors, and an attractive frequent-flyer program that could lure business travelers (the most lucrative segment of the passenger market), United, the nation's largest airline, began to turn up the heat on Frontier. In September 1983, Continental entered the domain of Chapter 11 reorganization bankruptcy, allowing it to shed itself of its union contracts and emerge as a low-cost cut-rate airline. It too joined the Darwinian struggle for dominance.

Thus on the one side Frontier was squeezed by an 800-pound gorilla, United, and on the other by the new low-cost Continental. In 1983, Frontier suffered its first annual loss since 1971.[42] Frontier's employees agreed to substantial pay cuts in 1983 and 1984 to help the company restore profitability.[43]

Average fares in 1984 of $127.24 at Denver were among the lowest in the country (significantly lower than at the rival hubs of Chicago, Dallas, Detroit, Houston, New York, San Francisco, St. Louis, or Washington, for example). The averages then fell 8.3 percent in 1985 and another 4.6 percent in 1986.[44] Denver's consumers enjoyed the lowest air fares of all major markets, but the squeeze was on Frontier.

Although only 69 percent of Frontier's nonstop seat miles faced a competitor before deregulation, by 1984 more than half its seat miles had two or more competitors.[45] Frontier tried several attempts to avoid erosion of its market, including establishing an alter-ego airline, Frontier Horizon. In early 1985, Frontier launched this small nonunion carrier flying seven 727s on long-hauls out of Denver.[46]

Frontier was doing a lot of things right. Its load factors surpassed industry averages.[47] It had a history of lawful and safe compliance. Based on the number of takeoffs and landings, Frontier had the best safety record in the history of domestic aviation. Frontier's compliance with regulatory obligations was ranked "outstanding" by the CAB (in contrast to the "unacceptable" ranking of People Express).[48] It had exceptional labor-management relations, never having lost a day of service due to a strike.

But with the strain placed on Frontier's balance sheet by United and Continental, its principal owner, RKO General (which owned 45 percent of Frontier), concluded that it had two options—liquidation or sale. Frontier's President, Hank Lund, attributed the company's problems to competing "head to head against larger airlines with greater resources and new, or reorganized carriers with low cost structures."[49] It was a squeeze play by United and Continental. In early 1985, Frontier sold Frontier Horizon Airlines to Skybus and sold five MD-80 aircraft to United for $95 million.[50]

But labor was eager to own the company if RKO General wanted out. Preliminary discussions between management and labor in late 1984 and early 1985 led to a series of transactions in which some assets would be sold to make Frontier debt-free so that an ESOP could be consummated. In May 1985, Frontier sold 25 of its 51 Boeing 737-200 aircraft to United for $265 million.[51] The money was used to pay down about $92 million in long-term bank debt.[52] United leased 15 of these planes back to Frontier.[53] Frontier's assets were being cannibalized, and United was enjoying the feast. As Frontier delivered planes to United, it dropped service to several cities.[54]

The Frontier board of directors formally approved an ESOP in July 1985 (allowing Frontier's 4,750 employees to buy the company for $220.4 million, or $17 a share, with significant labor concessions).[55] But the employees were subsequently outbid, first by Frank Lorenzo's Texas Air (at $250 million, or $20 a share) and then by Donald Burr's People Express (for $307 million, or $24 a share, with a proviso that the concessions surrendered by labor for the proposed ESOP remain in effect).[56]

Continental's interest was predicated on "forestall[ing] tougher competition from a lower-cost Frontier operating under labor concessions" and significantly expanding Continental's presence at Denver.[57] Continental and Frontier together had 245 flights out of Denver, compared with United's 165.[58] Continental and Frontier together had 46 gates at Denver's Stapleton Airport, compared with United's 21.[59] The two had overlapping service to 17 cities, and Frontier served 36 cities that Continental did not.[60] Nationally, the combined carriers would have flown 2.2 billion revenue miles, compared with Northwest's 2.5 billion and Delta's 2.7 billion.[61]

But labor was willing to concede the wage reductions demanded by People Express in order to avoid being clutched by the union-busting Lorenzo.[62] In exchange, People Express made the following agreement with Frontier's employees:

1. Except for circumstances outside the control of the company, Frontier employees would be protected from furlough until August 1, 1989;

2. Frontier would not be merged or consolidated with another airline until at least February 1, 1990;

3. Frontier would not dispose of assets in excess of $25 million total and asset sales must be at fair market value;

4. People Express would make available to Frontier sufficient working capital to operate the company; and

5. People Express intended Frontier to be a viable entity and strengthened as an airline.

As we shall see, not one of these contractual obligations was fulfilled.

The fact that there were three bids for the company suggests that Frontier was indeed a potentially valuable asset. Frontier had the following strengths:

- A strong hub at Denver, geographically one of the most desirable locations in the nation, with sufficient gates (21) and maintenance facilities (including two hangars) to compete (Frontier had about 23 percent of the airport's passengers, about the same as Continental)[63]

- A productive and dedicated labor force that had surrendered major pay and work-rule concessions (worth about $32 million annually)[64]

- A good history of high service and dependability (Frontier had 1.11 complaints per 100,000 passengers in 1985, compared with People's 4.59 in 1985)[65]

- An impeccable safety record (the best in the industry)

- A dedicated and loyal corps of travelers and ticket agents

- A standardized fleet of 38 Boeing 737-200s and four MD-80s aircraft, with advanced thrust engines, in top condition[66]

DONALD BURR AND PEOPLE EXPRESS

To understand the corporate culture of People Express, one needs to understand the psychology of the man who built and destroyed it. As a young boy in South Windsor, Connecticut, Donald Burr had admired the organizational structure of his local Congregational church—free and fiesty, yet disciplined. He told everyone he wanted to grow up to be a clergyman.[67]

Burr joined Frank Lorenzo's Texas International in 1973, rising to the position of chief operating officer in three years.[68] He left Texas Air in 1980 to start his own airline, People Express. He later recalled: "In 1980 people told me it was the worst time in the world to start an airline. Interest rates were 21%, and we were in a recession. 'You're starting an airline in Newark? ' they'd say. 'You're insane!' "[69]

Burr started People Express with three used Boeing 727s (then selling for about $4 million each), flying from Newark to Buffalo, Cleveland, and Norfolk.[70] Raising $24 million in capital, he quickly expanded, making down payments on 17 used 737 jets and establishing a hub-and-spoke operation radiating from Newark.[71] Flyers were thrilled to see $19 fares from Newark to Buffalo. Burr's perceived "niche" for People Express was across-the-board ultra-low no-frills airfares. The absence of frills meant the nonexistence of advance seat reservations, a $3 charge to check a bag or buy a sandwich, 50¢ for a cup of coffee, and paying for tickets on board.[72] Flight attendants would come down the isle with a cart, accepting cash, credit cards, money orders, personal checks, or traveler's checks in exchange for an airline ticket.[73]

As his financier said, "Donald Burr is really operating from a philosophical base, rather than a financial one."[74] Burr used army style interrogation techniques to select just the right people to staff People Express. Burr had a vision of a low-priced airline with a loose ("you're the boss") management style. Every employee would be an "employee-manager" who, as a stockholder, would thrive on dividends while agreeing to hard work, long hours, small salaries, and no overtime pay.[75] Each had to buy 100 shares of the company at a 70 percent discount, borrowing the money to buy the stock and repaying it with paycheck deductions.[76]

Burr expected his employees to buy additional People Express stock. In the early days, People Express would post its stock price on bulletin boards around the system to raise morale.[77] The *Wall Street Journal* described People Express as a "cult-like company in which all employees were 'managers,' and to whom Mr. Burr often gave inspirational speeches via video screens."[78] Some called the company an "aerial 'Love Boat.' "[79] In the early days, People Express had an unusual esprit de corps.[80]

People Express had only three levels of management, with everyone given the title of "manager."[81] There were customer-service managers (ticket agents and flight attendants), maintenance managers (who kept up the aircraft), and flight managers (pilots).[82] The organizational chart was an inverted pyramid, with Burr at the bottom.[83] Burr eschewed hierarchy and specialization, emphasizing instead self-management and voluntary cooperation.[84]

Burr refused to recruit experienced airline executives, training his from within. Top managers, all making less than six figures a year, were denied secretaries and expense accounts.[85] Every employee was given two jobs (a pilot would sometimes double as an inventory manager, for example), an approach dubbed "cross fertilization" by Burr.[86] Harvard Professor John Meyers observed: "Burr broke rank by not following the hierarchical, paramilitary model. People is run more like a commune."[87]

Burr's corporate structure was relatively devoid of middle managers. Burr met personally with all 1,000 of his team leaders on a regular basis—about 20 at a time in sessions that could last up to eight hours.[88] "They think my meetings are too long," grumbled Burr. "I like that. It means we go into detail."[89]

But not everyone at People Express shared Burr's passion for his unusual management style. His efforts to instill his philosophy led several top aides to quit and one to be fired.[90] And not all workers were in the aviation business for a cause. After spending thousands of dollars to train pilots, People Express saw more than 200 leave to fly for other, better-paying, airlines. "Low pay, no work rules, no seniority," were described as reasons for leaving.[91] The exodus grew after People Express stock began to plummet.[92] (The stock reached a high of $25.875 in 1983, then fell below $5 by mid-1986, and dropped to about $0.50 by the time Frontier discontinued operations.)[93] In November 1986, People Express cut salaries by 12 percent.[94]

A former People Express executive remarked: "Burr has brainwashed employees into working 60-to-80-hour weeks by calling them all managers. They're in Disneyland. But his spell can go only so far."[95] Some employees began to refer to Burr's indoctrination as "Kool-Aid," presumably referring to the Jim Jones mass suicide in Guyana.[96] *Time* magazine observed:

People [Express] has been a kind of continuing social experiment. Much of its past success has come from the willingness of full-time employees, all shareholders, to slash overhead by performing many jobs. Talk around the airline's Newark headquarters centers on how the company operates like a family. . . .

But critics contend that much vital work—aircraft scheduling, for example— would be performed far better by full-time professionals than by pilots or flight attendants who delve into it intermittently. . . .

Last winter's economic slump revealed another potential weakness in the People Express system—the flip side of employee ownership. When the value of their stock plunged, some flight attendants became markedly unenthusiastic about taking an extra turn at the reservations counters, and some pilots weren't quite so eager to tackle a stack of paperwork between flights.[97]

Burr wanted all employees to have direct eyeball-to-eyeball contact with customers. Thus, People Express had no inside legal or accounting staff and shed itself of its computer reservations system.[98] Reservations and baggage handling were contracted out, and ticketing was turned over to part-time college students.[99]

As one source noted:

Burr's alma mater, the Harvard Business School, was . . . agog over its alumnus for putting in place such a radical and apparently profitable management structure. It made a great case study to teach aspiring business managers and was cited extensively in books like "In Search of Excellence" and "Re-Inventing the Corporation."[100]

Burr appeared to be America's dream of a resurgence in productivity and pride at a time when the country suffered from an inferiority complex over foreign producers. The media lapped up the seeming success of this avant-garde management philosophy. Burr's face graced the cover of *Time* and *Inc.* But unfortunately, Burr became intoxicated by the attention and the power. He become autocratic, surrounding himself with yes men.[101]

One of People's board members observed that Burr was "absolutely fearless" and took "business risks that are unbelievable."[102] Burr was consumed by ambition. As late as January 1986, Burr was predicting, "In five years, People Express will be a worldwide transportation company, carrying people and freight, and packaging hotels and rent-a-cars, the works."[103]

The People Express fleet was growing at lightning speed. In 1983, Burr doubled the size of his fleet from 20 to 42 aircraft, many purchased from bankrupt Braniff.[104] People Express grew to 55 planes by the end of 1984 and 80 aircraft by the end of 1986.[105] It took delivery of a 727 on the average of one every 20 days.[106] People Express doubled its capacity from 2.8 billion seat miles in the first quarter of 1984 to 5.5 billion in 1985.[107] Donald Burr described the People Express "mystique" in these terms: "We couldn't do anything wrong. We just bought planes, hired people, and put them in the air. Grow, grow, grow. . . . The expectations at this place are colossal. And self-generated. We went around telling everybody that we're going to be great, do great, and conquer the world."[108] The messianic fervor of Donald Burr was driving People Express to go boldly where no airline had gone before.

People Express was buying aircraft before it had decided where to fly them.[109] Between July 1984 and January 1986, People Express added service to 29 U.S. cities, as well as Montreal and Brussels, and was seeking authority to serve Shannon, Amsterdam, Frankfurt, Luxembourg, and Zurich.[110] Some noted the parallel between Donald Burr and Freddie Laker, whose cut-rate transatlantic carrier (Laker Skytrain) expanded too fast and went belly up.[111] One commentator observed: "The preoccupation with growth has left the carrier with huge expansion costs, which have led to losses and enormous capacity. It has also chastened Mr. Burr and some of his top managers, who now concede that the enormous expansion led to major problems and tactical mistakes."[112] Another noted, "Burr's management philosophy was not adaptable to a really large company."[113]

People Express was being squeezed out of its East Coast market niche by the big boys—the megacarriers. The larger airlines, with their deeper pockets, were not about to surrender their lucrative routes to People Express.[114] By shrewdly managing yields through their sophisticated computer programs, gargantuan carriers like United and American (with their powerful computer reservations systems, Apollo and Sabre, respectively) were able to offer a limited number of seats at fares competitive with People Express—including, for example, the "ultimate super saver" fare—thereby flooding the low end of the market with excess capacity, without significantly diluting yield.[115] Burr knew he was in trouble when his mother told him that she was about to book a flight on American because she didn't mind their 30-day advance-purchase requirement.[116] (Burr would ultimately blame the demise of People Express on these sophisticated computer programs, which allowed the large airlines to manage yield to fill empty seats at People Express ticket prices.)[117] The megacarriers were matching People's fares on some seats but were also offering full service. As one source noted:

> In 1984, it was increasingly obvious that competing strictly on the basis of low fares and no frills was a failing proposition. Big airlines spent hundreds of millions of dollars on advance computer booking and found ways to match People's fares.
> By sticking rigidly to no-frills, the airline played into the hands of competitors. They knew a customer's rationale would always be: Why should I pay 50 cents for a cup of coffee when I can get the same air fare somewhere else and get my bags checked for free?[118]

Business travelers, the most lucrative segment of the air travel market, steered clear of People Express. People had a reputation for overbooking and cancelling flights, and its rat-infested 50-year-old Newark North Terminal was decrepit, requiring passengers to tromp through the snow and up wet steps in order to enter the aircraft.[119]

The competition was turning up the heat. The megacarriers forced People to abandon service to Minneapolis, Detroit, and cities in South Carolina.[120] At its Minneapolis hub, Northwest adopted a "scorched-earth" policy.[121] So long as People Express was flying to cities off the beaten path, the megacarriers left People alone. But when Burr took on the major carriers on the most profitable routes, they made him suffer unmercifully.[122]

People's load factors and profits were plummeting. For the first 10 months of 1985, its planes were 62 percent full, compared with 71 percent the preceding year. Between October 1984 and March 1985, People suffered operating losses of $21 million. Its stock dropped from over $25 a share at its peak in 1983 to less than $10 a share in 1986.[123]

People Express was desperate to improve its fortunes. Donald Burr thought westward expansion would be just the ticket.

DONALD BURR ACQUIRES FRONTIER AIRLINES

Following the advice of Horace Greeley ("go west, young man"), Burr sought to expand nationally by acquiring another airline, hoping that the synergism between the two companies would improve the fortunes of both. In an interview during the summer of 1985, Burr noted: "We need to be able to take on other entities, either develop them ourselves or acquire them. . . . Our systems are becoming more and more McDonaldish, and as they become more and more replicable, it's my view that we could, relatively simply, install them on other properties."[124] Burr was beginning to believe that People Express was becoming the MacDonald's of the airline industry.

Not long thereafter, Burr was playing tennis with a friend who suggested he go after Frontier. Burr thought it was a splendid idea and made the decision to acquire Frontier without consulting the other People Express executives. Said Burr, "I was convinced that it was the brilliant thing to do."[125]

Burr also decided to purchase two small airlines, Britt and PBA.[126] Britt was the nation's third-largest commuter airline, serving 29 midwestern cities.[127] PBA, in Chapter 11 bankruptcy, and operating 41 aircraft in 20 cities, cost People Express about $10 million (unsecured creditors received $300,000 for their debt of $6 million, and holders of its 4.8 million shares got nothing). Britt floated 10-year $35-million notes, loaning the proceeds to People Express.[128] Here again, the acquired carriers provided most of the capital for their acquisition. Burr was emulating the tactics of Lorenzo, who had served as his mentor at Texas Air. Lorenzo had made raids at Texas International, National, TWA, Continental, and Eastern and, where successful, had used the acquired company's money to finance much of the purchase price. Burr was an observing protégé. As one source noted, "The two men . . . have become archrivals, fighting for supremacy in the chaos and cutthroat competition of the deregulated airline world."[129]

Deregulation brought the destruction of more than 200 airlines through bankruptcy or merger. The prevailing wisdom in the industry, then as now, was that size, or mass, was essential to survival. Burr deemed expansion necessary to gain "critical mass."[130] According to People Express, "The Company believes that an airline must have a national transportation system in order for it to emerge from the current period of consolidation in the industry as a strong and viable competitor."[131] People's acquisition of Frontier would combine the nation's ninth-largest airline with the fifteenth-largest.[132] Together, they would become the nation's fifth-largest airline in terms of numbers of passengers flown, behind United, American, Eastern, and Delta.[133]

Frontier looked like a choice prize. It was relatively debt-free. Through the first nine months of 1985, Frontier earned a $59.2-million net profit on $453.3 million in revenues.[134] (In contrast, People Express suffered a $27.5-million net loss during 1985, contrasted with a razor-thin $1.6-million profit during 1984.)[135] Frontier's coffers would provide the lion's share of its purchase price, and its labor was willing to surrender pay and work-rule concessions worth tens of millions of dollars. One source noted, "Some industry experts think Burr was also attracted by something else he could use: a fresh supply of managers."[136] And Frontier offered access to the strategically important hub of Denver—the nation's fifth-busiest airport. One analyst described People's acquisition of Frontier as

the airline coup of the year, possibly the entire post-deregulation era. This upstart airline, headquartered in the backwater of New York, snatched Frontier from the grasp of Texas Air and used Frontier's own money to finance the deal. . . . People, with 368 flights a day to 32 U.S. cities—plus London, has grown to a size where it's just big enough to challenge the majors, but lacks the routes to carry much clout.

Frontier's operations will give People 120 more flights. In addition—with its low cost, no-frills fares—People attracts travelers away from cars and buses, but lacks the loyal passenger base Frontier can provide.

Denver as a second hub for People gives it a door to lucrative western markets. It's an important hub for any airline because traffic can be routed through Stapleton International Airport from all directions, say airline sources.[137]

Multiple hubs were widely viewed as essential to survival, and People Express had been looking for another hub for several years.[138] Frontier's Denver hub would give People a national base (within a 1,000-mile radius lie major markets like Minneapolis, Chicago, St. Louis, Kansas City, Dallas, Houston, Los Angeles, and San Francisco). Ruth Hennefeld, an investment manager at Merrill Lynch, observed: "The deal is a bargain for People. The company bought into the Denver market for far less than it would have cost it to start from scratch."[139] Frontier's computer system would also offer new opportunities.[140] The combined company would serve 104 cities in 43 states.[141] People Express purchased Frontier on November 22, 1985, for $307 million. Donald Burr boasted, "We're one of the big boys now."[142]

Burr took 93 percent of Frontier's working capital ($193 million), mostly raised through an earlier sale of assets, as a "dividend" to help finance the transaction.[143] This left Frontier with less than $14 million in working capital.[144] As we shall see, stripping Frontier of its working capital would be fatal. Frontier's unencumbered assets (eight airplanes and seven engines) were used as collateral to secure a $50-million note to pay a due tax bill of $44 million from the proceeds of aircraft sold to support the proposed ESOP.[145] Of the $307-million purchase price, Frontier supplied $212 million and People Express put up only $95 million, or about 31 percent.[146]

But People Express did not halt its buying binge. The company's philosophy was expressed as follows:

Management believes that the airline industry may be entering a period of consolidation. In such an environment, opportunities to acquire scarce strategic resources, such as access to certain airport facilities, landing slots and passenger traffic flows, may present themselves at any time. Management believes that the future expansion of operations may be dependent upon its ability to make such acquisitions and, consequently, the Company continues to evaluate other possible acquisitions in addition to Frontier Airlines.[147]

As noted above, People Express soon bought Britt and PBA.

Burr made it clear from the outset that he wanted to have Frontier adopt People's organizational style and no-frills, low-cost product.[148] He wanted a "top-to-bottom revolution in the corporate culture at Frontier."[149] Burr replaced Frontier's CEO, Joe O'Gorman, with 36-year-old Larry Martin, described by a Frontier officer as "really an arm of People Express."[150] In cutting costs, Martin "gutted" Frontier's management.[151] Said Martin, "We intend to be the dominant carrier in the Denver market."[152] Young Martin would revamp Frontier's fare structure and marketing strategy in Burr's image.[153]

Martin was a member of both the Frontier policy committee and the People Express policy committee—the directing arm of the company.[154] But every major decision of both companies was made by the People Express policy committee.[155] Frontier executives began to jump ship in protest to People's dominance.[156] A Frontier vice president noted an example of how the company was run:

The Frontier policy committee was adamantly opposed to charging for coffee and bags and on-board snacks, which was the People Express way of doing business. We strongly urged them not to do it. We felt that it wouldn't work in Denver. Nonetheless, the People Express policy committee dictated to us that we would do that. And we did it.[157]

Burr turned Frontier's traditional full-service operations into a low-fare no-frills discount operation—an approach that had worked well in Newark but went over like a lead balloon in Denver. Westerners had grown accustomed to the high-quality service Frontier had consistently provided.

Frontier's fares were slashed (up to 60 percent systemwide initially, then 70 percent)[158] while the passenger paid for soft drinks and $3 extra for checking a bag or getting a cold snack. A first-class cabin was added by jamming the coach seats closer together rather than by reducing the number of coach seats.[159] Hot meals were discontinued, and ovens were removed from planes.[160] While coach passengers could buy a sack lunch, usually consisting of some crackers, cheese, and sausage (referred to as "Kibbles and Bits" in the industry), first-class passengers could spend $6 for a "multi-course, gourmet meal."[161] Both Frontier's coach and first-class passengers were unaccustomed to paying for food or soft drinks or for the privilege of checking a bag, and never had the quality of the product fallen so low. Continental scored a marketing coup by passing out "Sympathy Lunches" on Frontier's Denver concourse D, a free sack lunch with a sandwich and fruit for any poor soul who displayed a Frontier ticket. All this destroyed Frontier's traditional customer base.

Continental and United met the new low fares but offered full service— free checked bags and free hot meals. The People Express marketing approach was an unmitigated disaster for Frontier, alienating loyal customers who abandoned Frontier and flew United or Continental. Frontier hemorrhaged dollars unmercifully.[162] Load factors remained high, but yields plummeted. Meanwhile, Continental expanded service at Denver and signed marketing pacts with Rocky Mountain Airways and Trans-Colorado Airlines to provide feeder service at Stapleton.[163]

Frontier stopped offering advance seat assignments either by travel agents or its own reservations department.[164] Needless to say, travel agents had little enthusiasm for the difficulty they were encountering in satisfying customers' needs, the absence of sufficient phone lines, the overbooking, the decision to charge customers for food and drink, or the low commissions offered by the discount fares.[165] They began to dissuade their customers from using Frontier.[166]

But economic problems were rippling throughout the industry.[167] *Aviation Daily* summarized the difficulties caused by the fare wars and the dampening of industry profitability in 1986:

[Robert] Joedicke [of Shearson Lehman/American Express] believes the stage was set for accelerated yield deterioration when American last year introduced ultimate super savers, restricted fares matching or undercutting tariffs by low-cost new entrants. Even greater pricing competition was spurred at the end of 1985 when People Express bought Frontier. For some time, Denver has been "a hotbed" of fare cutting. "The slugfest between United and Continental for supremacy at this key hub squeezed Frontier into an unviable and distant third position. Now, the impact of ever fiercer competition with a reconstituted Frontier has spread to several other markets in a classic example of the "domino theory". . . .

Joedicke characterized the current pricing fracas as "a knock-down battle" between arch competitors Frank Lorenzo of Texas Air and Donald Burr of People Express, "with many other participants being caught in the crossfire." This form of competitive attack can be dangerous, he said, citing as an example United possibly reacting with "a nationwide array of giveaway fares that would maul its weaker competitors in a financial bloodbath."[168]

In chess, checkers, war, and business, those with greater resources can best withstand a war of attrition.[169] People Express had so highly leveraged itself and Frontier that neither could long endure the bloodbath Burr had started.

As noted above, Frontier had been consistently profitable before the last quarter of 1982 when United and Continental began to dump seats in

Denver. Then Frontier earned operating profits during the third quarter of 1983 ($6 million), the third quarter of 1984 ($12 million), and the second quarter of 1985 ($10.9 million). But after People Express acquired it, Frontier failed to turn an operating profit during any quarter.[170] Between September 1, 1985, and July 31, 1986, Frontier experienced 11 straight months of losses, totaling $47 million, by People's calculations.[171]

By the spring of 1986, People Express had realized its tactical mistake and reinstituted Frontier's full-service operations.[172] Donald Burr confessed, "It was a mess—it was a disaster."[173] People Express announced that Frontier's ticket prices would be increased between $20 and $40 and that lower fares would require a 14-day advance purchase.[174] It reintroduced advance seat assignments and boarding cards.[175] By summer 1986, Frontier had borrowed $50 million, secured by eight 737-200s, to provide operating capital.[176] It also sold off its nonairline subsidiaries to raise cash.[177]

THE SINKING SHIP OF PEOPLE EXPRESS

People Express was hemorrhaging dollars as well. The economic problems that had preceded the acquisition of Frontier continued. In 1983 and 1984, People Express paid $10 million and $37 million in interest, respectively, on debt of $247 million and $326 million, respectively.[178] By 1985, People's half-billion-dollar debt burdened it with $60.5 million in interest payments, wiping out its operating revenue and leaving it with a net loss of nearly $28 million.[179]

By 1986, People Express had become the nation's fifth-largest carrier, but it was in serious trouble. In the first quarter alone, People Express suffered a $47.4-million operating loss and a net loss of $58 million (more than twice its losses for the entire preceding year).[180] In the second quarter, it suffered a $57-million operating loss and a $74-million net loss.[181] During the first six months of 1986, its operating expenses had increased by $66 million while its operating revenue remained stagnant over the same period in 1985.[182] By spring, People Express was near bankruptcy, down to a paltry $9 million in cash.[183] It was saddled with a crushing $560.5-million debt—2.8 times the company's equity.[184]

It was becoming clear that Burr's unorthodox organization and management style was ill suited for a company its size and that its operations and marketing efforts were a flop. Although the company had two hubs (Newark and Denver) and low labor costs, the other essential ingredients of survival in the deregulated airline environment were woefully lacking: People Express had failed to install yield and capacity management systems, a computer reservations system, or a process of measured and timely expansion.[185] As one analyst observed, "They've organized themselves with a very lean organization with no fat in it, but not much muscle either."[186] Another noted: "They still don't have secretaries. They were quite proud of the fact, but you try calling them on the phone. They just got someone as chief financial officer, but it's too late now."[187] The *New York Times* observed, "People Express, according to analysts, has long-term problems that might be solved only by a change of ownership or radical replanning."[188]

And service was miserable. As noted in a 1986 article in the *Wall Street Journal:*

In the past year, [People Express] operations have deteriorated badly, leaving many travelers in the lurch. Flights are being overbooked by huge margins, bags are being lost by the thousands, passengers complain that bargain fares are boosted unexpectedly, and planes are chronically late.

As a result, People Express has shot to the top of the charts in passenger complaints filed about major carriers.[189]

People Express was overbooking many of its flights by more than 100 percent. Its 185-seat Boeing 727s were being booked for 400 passengers; its 481-seat 747s were being booked for as many as 1,000 passengers.[190] During its final year, People Express cancelled 4.4 percent of its flights, more than seven times that of the industry's leader, Delta.[191] The result was thousands of bitter, stranded passengers and, in at least one instance, a near riot at the airport.[192] Frequent flyers began to refer to the company as "People's Distress." By May 1986, it was ranked worst in terms of number of complaints per 100,000 passengers.[193] The business editor of the *Rocky Mountain News* had these observations about People's service:

These People don't know what they're doing.
They don't know the airline business.
They have absolutely no concept of customer service.
"You get what you pay for," my mother-in-law bellowed. Agreed, but you don't pay for incompetence and deception.
In the future, I might buy my kids tickets on People Express if they turn into miserable teenagers. . . .
People Express doesn't deserve to share hangars with a classy airline like Frontier, and the only upside to the merger would be the hope that Frontier's professionalism might rub off.
A great deal of praise has been heaped on People Express founder Donald Burr for his "new wave" concept of corporate management. He encourages his employees to buy stock in their company and preaches that such ownership motivates his people.
Phooey.
What Burr does is hire kids (average age, 27) who, in their youthful idealism, believe the line he hands them. He gives them minimal training, a foot in the door of the airline business, stock in a company Wall Street loves, better-than-average pay and the opportunity for cheap travel with airline passes.[194]

People Express was hungry for cash. In 1985, the company offered $125 million in secured equipment trust certificates to repay debt and acquire aircraft and for operating expenses.[195] Most of it appeared to go to finance the purchase of Frontier Airlines.[196] It also issued an additional million shares of preferred stock.[197] In 1986, People Express announced the sale of $115 million in equipment certificates secured by fourteen 737-100s and seven 727-200s, $93 million of which went to refinance debt owed to nine banks and other creditors.[198] But even that wasn't enough. An internal Texas Air memorandum to Frank Lorenzo summarized the sad state of People Express:

But for $130 million in cash-on-hand, it would not be difficult to argue that People Express, Inc. is bankrupt. The company's stated equity amounts to $152.7 million at March 31, 1986 included redeemable (company elective) preferreds. Fair market value equity amounts to a negative $7.2 million including the preferreds.[199]

In June 1986, People Express announced it was exploring the possibility of selling all or parts of the company.[200] As one analyst noted: "Needless to say, People Express Inc. is facing a cash squeeze. The company has sold almost all its planes and has reduced cash flow from depreciation."[201]

By mid-1986, People's financial position was precarious. Burr tenaciously insisted: "I think we're here to stay. The reports of our death are greatly exaggerated."[202] But with more than $600 million in debt, People's interest payments and preferred dividends consumed nearly 8 percent of its revenues, the most of any major airline.[203]

But again, these problems preceded People's acquisition of Frontier. Julius Maldutis, an analyst for Salomon Brothers, observed: "[People Express] has been unable to generate a satisfactory level of earnings [since 1982]. . . . The company has funded its aggressive capital expenditures program and losses through debt, equity or equity-type financing."[204] As *Aviation Daily* noted, "People Express' current problems, according to some assessments, date to mid-1984."[205]

As the summer of 1986 dawned, People Express had $103 million in cash (the proceeds from the sale of equipment-secured debt obligations); but by midsummer, it had barely half that.[206] People Express was losing approximately $4 million a week.[207] Many of its suppliers complained that they were having serious problems collecting on debts owed by the airline.[208] Both Merrill Lynch and Salomon Brothers projected People Express to have the largest per-share losses of all airlines.[209] A power struggle ensued on the People Express board of directors, in which the board "clipped Burr's wings" and began restructuring the company.[210]

A business plan adopted in the summer of 1986 noted, "The Company must undergo a fundamental transformation if it is to survive the competitive pressures of a rapidly consolidating industry environment."[211] People

Express began installing a much-needed computer reservations system and planned for a move into new and badly needed terminal facilities at its Newark hub.[212] It also announced that it was abandoning its "no-frills" approach (catering to vacationers and backpackers) and becoming a more traditional full-service airline (targeting all types of flyers, including business travelers).[213] It would be adding first- and business-class sections, leather seats and plush carpeting throughout its aircraft, more leg room, and prices "bundled" with new perks, including "premium coach" (in which meals and baggage handling would be free), and a more generous frequent-flyer program than those of its rivals.[214] It was dropping service to Dayton, Greensboro, Columbia, Montreal, and Nashville and replacing jet service to Albany, Providence, and Melbourne with turboprop service from PBA's fleet.[215] It was reducing its fall schedule by 110 departures and downsizing its fleet by two 747s and eight 727s.[216]

The entire package was about as strong an admission of failure as a corporation could make. People Express had blundered badly by following Donald Burr's autocratic and unconventional rule, and it was drowning in a sea of red ink. But the changes came too late to avoid collapse.[217]

In July 1986, rejecting a bid by Texas Air to purchase People Express for $237 million, People agreed to sell Frontier to United for $146 million.[218] At that time, United dominated Denver, with nearly 40 percent of its domestic passengers (followed by Continental, with 28 percent, and Frontier, with 18 percent).[219] United needed Frontier's assets "to keep competing carriers out of Denver."[220] The deal would allow United to consume Frontier's 42 aircraft, 15 of its 18 gates at Denver's Stapleton International Airport, and 4,600 employees.[221]

But People Express was desperate for cash. Hence, the agreement was structured so that United could buy Frontier assets immediately while infusing People Express with cash. This allowed United to buy several of Frontier's valuable properties, including five takeoff and landing slots at Chicago O'Hare, three gates at Dallas/Ft. Worth, two hangars at Denver, contracts to acquire two MD-80 aircraft, and six gates at Denver.[222] These asset sales, totaling $43.2 million, were not contingent on consummation of the stock transaction and indeed were paid to People rather than Frontier, despite agreements that payments were to be made to Frontier.[223] Thus, Frontier's assets were being cannibalized in order to save ailing People Express.

People Express was losing money so fast it needed the $43 million infusion merely to stay aloft and was counting on the remaining $100 million from United to give it "time to complete its new main terminal in Newark and realign as a low-cost full-service carrier."[224] People Express begged the DOT to approve the United acquisition under the failing-company doctrine, saying that it "urgently" needed the cash to stay in business.[225] In the third quarter of 1986, People Express suffered a $27-million operating loss and a $112.9-million net loss.[226] People Express was a very sick bird.

As Bankruptcy Judge Charles Matheson noted: "People Express got the net cash . . . and, perhaps, willingly abandoned the remaining shell of its failing subsidiary, Frontier. The shutdown of Frontier gave United the opportunity to gain part or all of Frontier's share of the Denver air carrier market utilizing the assets it had acquired."[227]

Meanwhile, Continental was causing trouble, trying to tie up the transaction with threats of antitrust litigation. Said a former Frontier official: "People Express has about $50 million in cash and that is about all. Texas Air [TAC] is not about to be accommodating and let the deal go through particularly since it has twice been beaten out for control of Frontier. TAC wants gates at Denver. If they can tie up this merger in court and have People Express bleed financially at the same time, that will not bother them."[228]

The United-People agreement also included a proviso that labor agreements "satisfactory to United" be concluded by August 31, 1986, with the unions of United and Frontier and that United "use its best efforts to obtain such agreements." The machinists union, which had experienced some acrimony with Frontier's management under People Express, anticipated a harmonious relationship with United, saying, "United was one of our top choices."[229]

But United met only with the United members of the Air Line Pilots Association (ALPA), with United demanding that the labor contract be amended to keep the low wages Frontier pilots had accepted when People Express consumed them (in effect, creating a new "C"-tier pay scale at United, at wages 40 percent below those of United "A"-scale pilots).[230] At the time, a Frontier 737 captain earned an average $68,000 a year, whereas the equivalent United captain earned $115,000.[231] One analyst described United's posture in these terms: "United was in no mood to be overly generous, mainly because the Frontier people are between a rock and a hard place, between working and not working."[232] But the Frontier pilots (indeed, all Frontier unions) were locked out of the negotiations and had no say. And both United and the United pilot Master Executive Council (MEC) were dragging their feet in the labor negotiations. It was at this point that United was offered all of People Express for nothing if United would just complete the Frontier acquisition. But United had no interest in People Express, even for free.

On August 24, 1986, People Express shut down Frontier, grounding its 42 Boeing 737s, putting its 4,700 workers out of work, cancelling 325 flights, and stranding 17,000 passengers in 55 cities and airports.[233] People Express claimed that it had to shut down Frontier because it was out of cash and its liabilities exceeded its assets by $60 million.[234] People announced, "[Frontier] is out of funds and, in the absence of assurances that the sale to United will take place, People Express is unwilling to commit any more of its funds to Frontier."[235] The truth was that People Express was nearly broke. By now, it was losing about $7 million a week (up from $4 million a week in midsummer) and was largely being kept alive on the $43.2-million cash infusion from United for Frontier's assets.[236]

Frontier became the first airline to go out of business because of an attempted merger. Texas Air aptly summarized the reasons for Frontier's demise: "During 1986, Frontier suffered significant losses as a result of the fare wars in Denver which at times depressed yields below costs, and the loss of passenger traffic as a result of its adoption of the no-frills marketing strategy of People Express Airlines."[237] In four decades of operation, Frontier had never suffered a service disruption due to labor strife and had earned the industry's best safety record.

Having stripped Frontier of the assets it wanted and having eliminated Frontier as a hub competitor at Denver, United dusted its hands off and walked away from the deal, leaving Frontier and People Express dangling in the wind. United announced that it would not purchase the rest of Frontier because the grounded airline had been damaged beyond repair. Said a United spokesman, "The airline we attempted to purchase does not exist anymore."[238]

In fact, United walked away with everything it wanted. As one source noted, "The collapse of Frontier is seen as a major blow for People Express and as a triumph for arch-rival United."[239] Another observed: "United will benefit from eliminating the instability of Denver's three-carrier hub. This will translate into higher fares and better returns and will ensure that another carrier does not attempt to build a presence in Denver."[240] Yet another said, "Airline executives and observers concurred that the elimination of a below-cost fare competitor, Frontier, should bring some revenue relief to both United and Continental, the two other major carriers serving Denver."[241]

United had the best of both worlds—the essential assets it wanted from Frontier without paying the full purchase price and without absorbing Frontier's employees and, better yet, the demise of Frontier at the Denver hub.[242] The United-Continental-Frontier oligopoly had at long last become the United-Continental duopoly.

Frontier formally entered bankruptcy on August 28, 1986. (Ironically, Continental emerged from bankruptcy one week later.) At that time People Express was losing money at the rate of $35 million a month.[243] But People's economic position had been precarious long before the acquisition of Frontier. As Paul R. Schlesinger, an analyst at DLJ Securities, noted, "People Express made a big splash, but they've never made any money."[244] Table 5.1 reveals that People Express was never a bastion of profitability.

As People Express stockholders were later to charge, in misleading financial statements, reports, proxy statements, and other documents, Peo-

Table 5.1
People Express Revenue and Profit (Loss)
(in Millions of Dollars)

Year	Revenues	Profit
1981	38.4	(9.2)
1982	138.9	0.5
1983	286.6	6.5
1984	586.8	1.7
1985	977.9	(27.5)
1986	(1250.0)*	(345.0)*

* estimates

Source: CHRISTIAN SCIENCE MONITOR, Jan. 26, 1987, at 16.

ple Express, its officers and financial institutions "sold a misinformed public millions of dollars of various types of People Express securities, while concealing the fact that the company and its subsidiaries were heading towards economic disaster."[245] People Express, its directors, and investment bankers allegedly engaged in fraud, deceit, and negligent misrepresentation in connection with the public offering of securities and the filing of reports and documents required under the securities laws. These gave the public "a false and misleadingly and unduly optimistic picture" of the business operations of People Express.[246] The stockholders ultimately recovered $10.5 million in settlement of their suit alleging deception of the financial condition of People Express.[247]

FRANK LORENZO ACQUIRES PEOPLE EXPRESS AND FRONTIER AIRLINES

In September 1986, Frank Lorenzo offered $138.4 million for People Express and $176 million for Frontier—an offer that Burr quickly accepted.[248] (Recall that People Express had bought Frontier for $307 million the year before and that Texas Air had offered $237 million for People Express and its subsidiaries only two months earlier.)

Texas Air promised that if 75 percent of Frontier's pilots, flight attendants, agents, and dispatchers dropped their claims against Frontier and People Express, Continental would hire them.[249] Of course, the employees were hardly in an equal bargaining position, for they were then out of work because of the Frontier shutdown. They had little choice but to sign away their legal rights to work for Continental. Those who went to work for Continental were promised seniority under a "fair and equitable formula."[250]

People Express was desperate to consummate the transaction. It begged the DOT to approve its acquisition by Texas Air under the failing-company doctrine, saying that without it, People Express would be unable to meet upcoming interest payments of $13.4 million and $31 million.[251] People owed $309 million in equipment certificates, secured on 53 of its 59 aircraft, and $165 million in unsecured notes.[252] Texas Air's debt load would reach $5.5 billion with the People Express purchase.[253]

Lorenzo had offered $237 million for People Express the year before but had been rebuffed by Burr.[254] After the Frontier sale to United collapsed, Lorenzo offered $138.4 million for People Express, less than he offered for Frontier.[255] By December 1986, Lorenzo had lowered his offer to $122 million, then to $113.7 million, citing People's deteriorating financial condition.[256] Most of the deal involved an exchange of People Express stock for Texas Air stock, with the number of shares, their value, and the dividend rate reduced by Lorenzo in several steps.[257] Burr would also have to work with his debt holders to reduce annual interest payments by $12.6 million.[258] It was "take it or leave it," and Burr had no choice but to take it, for People Express had no realistic prospects of continuing operations and retaining substantial assets during bankruptcy or of emerging successfully from reorganization.[259] People Express sustained a net loss of $199 million in 1986, and its liabilities exceeded its assets by $135 million.[260] With its subsidiaries, People Express sustained a net loss of approximately $344 million in 1986, of which about $36 million (10 percent) was attributable to the operations of Frontier Airlines.[261] People Express was in such a pathetic economic condition that Lorenzo was able to acquire the company primarily on a stock swap, without a significant outlay of capital, and to unilaterally dictate a lower selling price than originally offered.

Continental sought to merge these carriers and New York Air in a messy overnight consummation on February 1, 1987. These acquisitions, as well as that of Eastern Air Lines, made Texas Air the nation's largest airline, growing from 160 aircraft in 1985 to 636 planes in 1987.[262]

Texas Air bought Frontier's stock for $9.5 million in April 1987. Continental subsequently bought Frontier's stock for $10 million from Texas Air and became the debtor in possession. Frontier transferred 44 aircraft and certain airport and related properties, including gates, to New York Air, a Texas Air subsidiary, for $64 million and an assumption of $49 million of Frontier's indebtedness.[263] Forty-one of these aircraft were then folded into Continental's fleet.

At Continental's direction, Frontier sued United for fraud and return of assets transferred under People's patronage. With a settlement reached with United, Continental possessed most of Frontier's aircraft, as well as three hangars and two concourses (concourses C and D) in Denver. The acqui-

Figure 5.2
Stapleton International Airport Domestic Market Shares, 1985–1990

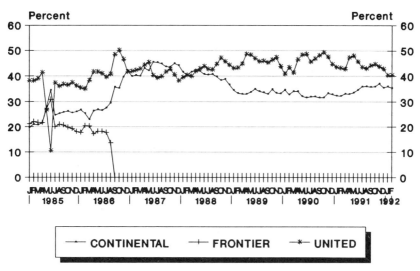

Frontier ceased operations 8/24/86
Continental bought People Express 3/87

sition would give Continental 236 daily departures from Denver by the end of 1986 (including service to eight additional cities formerly served by Frontier), compared with United's 218 departures.[264] As Texas Air was proud to inform its stockholders, "The Company acquired most all of the assets of Frontier Airlines, Inc., resulting in a strategic strengthening of Continental's Denver hub."[265] For the first time, in December 1986, Continental became the largest carrier at Denver Stapleton, which, as figure 5.2 reveals, is a title it held from June 1987 until May 1988.

After Frontier's demise, United added 51 flights at Denver, significantly increasing its market share from nearly 40 percent when it had agreed to buy Frontier from People in July 1986 to a record 50 percent two months after Frontier's disintegration.[266] Continental's market share skyrocketed from 28 percent to nearly 40 percent during the same period.[267] Denver has become a duopoly—or, perhaps better termed, a "shared monopoly."

In January 1987, Frank Lorenzo called a news conference, speaking from a podium at which he had $5 million in a five-foot-tall pile of cash. This was the amount Continental had allegedly saved consumers a day, or about $2 billion a year. Lorenzo proclaimed that although People Express was no longer, "To borrow from Mark Twain, reports of the death of low fares are exaggerated."[268] Lorenzo said that his commitment to low fares was "not based on altruistic motives" but reflected "Continental's business strategy—to maintain low costs and to use the low-cost structure to keep

Figure 5.3
Stapleton International Airport Average Roundtrip Ticket Price Analysis, 1986–
1991

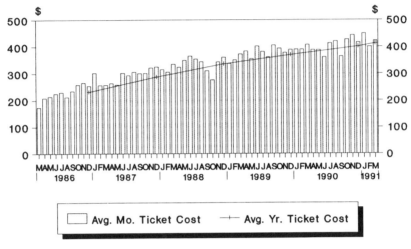

9/88 - New Fares Set
11/88 - Restrictions Increased

fares down."[269] But ticket prices at Denver, which had fallen during 1985 and 1986, began to climb sharply—17.6 percent in 1987 and a record 39.2 percent in 1988.[270] Thus, the demise of Frontier allowed United and Continental to raise ticket prices robustly at Denver. (See figure 5.3.) The destruction of Frontier was essential to that result.

Recent years have been difficult ones for Texas Air.[271] In 1988 it posted the worst losses in the history of the industry—$885.6 million, on revenues of $6.7 billion—surpassing its 1987 record of $718.6 million.[272] In 1989, Eastern went on strike, and Texas Air put the company into bankruptcy. Scandinavian Airline Systems (SAS) purchased a significant equity interest in the airline. To improve its public image, Texas Air was renamed Continental Airline Holdings, and the controversial chairman, Frank Lorenzo, resigned, leaving the company saddled with more than $5 billion in debt.[273] As this book goes to press, Continental is in bankruptcy once again (some call it Chapter 22 bankruptcy).

POSTMORTEM ON PEOPLE EXPRESS

Several experts have commented on the disintegration of People Express, which dragged Frontier into the abyss of bankruptcy. These observations provide some insight into the strategic mistakes made by Donald Burr and into the predatory behavior of United and Continental.

John W. Teets, chairman and chief executive officer of the Greyhound Corporation, observed, "People Express expanded too fast."[274] Gail H. Ruderman, co-owner of Revere Travel, Inc., provided the travel agents' perspective:

People has alienated travel agents. . . . It also has been almost impossible to get through to People's reservations center. Even when we can, we cannot count on that reservation being honored when the customer arrives at the airport. Most major carriers oversell flights, but this has been a greater problem with People.[275]

Professor D. Quinn Mills of Harvard, an early admirer of Burr's, once described People Express as "the most interesting company in America today."[276] But he had this to say about the demise of People Express: "Some people are absolutely convinced the management system was at fault. It was unusual, and it looked chaotic."[277]

Alfred Kahn, the former chairman of the Civil Aeronautics Board and architect of deregulation, had once taken credit for the success of People Express, saying: "People Express is clearly the archetypical deregulation success story and the most spectacular of my babies. It is the case that makes me the proudest."[278] His pride was not to last long. When People Express was going down the tubes, Kahn had a different view:

United, American and other major, full service airlines . . . developed very sophisticated, computerized scheduling programs, which have enabled them to determine how many seats on each flight are likely to go unsold at normal fares but might be filled if they offer a steep discount. . . .

Thus, it appears the People Express model of offering uniform, low fares with no restrictions cannot survive. And I am pessimistic about new companies being able to compete successfully by offering uniform low fares to everybody without discriminating among different travelers and without imposing such restrictions as advance purchase or cancellation penalties.[279]

And finally, from Donald Burr himself, several years after People Express collapsed: "I practiced a number of things that were ideological but not practical. If I were doing it over, I would hire from the outside and amend our compensation terms to attract good people. You couldn't get a chief financial officer for a $1 billion company for $75,000 a year—Burr's salary."[280]

Burr's People Express was like a rocket on the Fourth of July—ascending up and up at a frantic pace, then exploding, and quickly disappearing. It was a wondrous thing to behold. But it was doomed to self-destruction. Although People Express initially enjoyed meteoric growth, there were major flaws in Donald Burr's highly unusual approach.

- The company was grossly undercapitalized, all assets heavily laden with debt. The debt burden crushed what operating profits there were. Like a house of cards, People Express rested on a foundation that was highly delicate.

- People Express lacked measured, timely expansion. It pursued frantic growth without adequate planning or market assessment. Its fleet grew from about 20 aircraft in 1983 to 80 in 1986. It then acquired Frontier, Britt, and PBA.

- People Express had a primitive yield and capacity management system, whereas its competitors were highly sophisticated in managing yield and capacity so as to lure away the low-fare passengers to whom People Express catered. As People Express began to compete in markets dominated by established airlines, the incumbents managed yield by lowering fares on some seats, causing People Express to suffer terribly. Although People Express began the fare wars by offering no-frills discount service, it had insufficient capital to withstand the competitive response.

- A cultlike corporate culture might work in the short term or in small enterprises, but as employee enthusiasm begins to wane or management becomes more impersonal, such a culture becomes difficult to sustain. Twenty-four-hour-a-day industries like airlines often have difficulty keeping labor-management relations on a friendly course. The unusual management philosophy invented by Donald Burr was not adaptable to a company the size that People Express ultimately became.

- Since wages were low, and tasks and hours taxing, employee morale was tied heavily to the success of People Express stock, which they were forced to buy and which soared and plummeted in tandem.

- Many key positions were held by inexperienced managers. Requiring employees to double in other positions sacrificed efficiency. Burr recruited few executives from other airlines and paid corporate officers poorly. They were denied such fundamental productivity-enhancing staff as secretaries. The managerial system was, in a word, chaotic.

- Poor and undependable service alienated passengers and travel agents. Travel agents (on whom most airlines depend for bookings) were appalled by People's no-seat reservations policy, its deliberate overbooking, the insufficient phone lines, its practice of charging customers for meals and checked luggage, and the low commissions. Imposing this no-frills service on Frontier was a marketing disaster and alienated its loyal western customer base.

NOTES

1. Ennis, *Sky King,* BUS. MONTH, Sept. 1988, at 27, 32.

2. The "Wine of the Week" was one of President Michael Levine's contributions to travel in the Washington–New York–Boston corridor.

3. *At the Controls: Frank Lorenzo Will Pilot a Troubled Continental,* WALL ST. J., July 24, 1987, at 18.

4. R. SERLING, MAVERICK: THE STORY OF ROBERT SIX AND CONTINENTAL AIRLINES 18–19 (1974).

5. *Id.* at 30–33; R. DAVIES, AIRLINES OF THE UNITED STATES SINCE 1914 198 (1972).

6. R. SERLING, *supra* note 4, at 36.

7. *Id.* at 37.

8. *Id.* at 81–82.

9. *Id.* at 95, 151–53.

10. *Id.* at 59.

11. *Id.* at 60–61.

12. *Id.* at 98–99.

13. *Id.* at 108.

14. *Id.* at 131.

15. *Id.* at 172.

16. *Id.* at 204–8.

17. *Id.* at 203.

18. *Id.* at 230.

19. M. MURPHY, THE AIRLINE THAT PRIDE ALMOST BOUGHT: THE STRUGGLE TO TAKE OVER CONTINENTAL AIRLINES 5 (1986).

20. *Id.* at 5.

21. *Id.* at 6–7.

22. *Id.* at 8.

23. *Id.* at 8.

24. *Id.* at 9.

25. *Id.* at 11.

26. *Id.* at 11–12.

27. *Id.* at 32–33, 54.

28. Letter from A. L. Feldman to Francisco A. Lorenzo (April 3, 1981).

29. H. LAWS & R. LOSEE, FRONTIER 2 (monograph, December 2, 1986).

30. *Id.*

31. See J. CASH, JR., F. MCFARLAN, J. MCKENNEY, & M. VITALE, CORPORATE INFORMATION SYSTEMS MANAGEMENT 18 (1988) [hereinafter cited as CORPORATE].

32. See *id.* at 17.

33. H. LAWS & R. LOSEE, *supra* note 29.

34. *Id.*

35. See CORPORATE, *supra* note 31, at 17.

36. Salpukas, *Frontier Files for Bankruptcy,* N.Y. TIMES, Aug. 29, 1986, at D-1, col. 6.

37. H. LAWS & R. LOSEE, *supra* note 29.

38. See CORPORATE, *supra* note 31, at 17.

39. H. LAWS & R. LOSEE, *supra* note 29.

40. *Market Shares of Top Three Carriers at 30 Busiest Airports,* AVIATION DAILY, May 24, 1984, at 138; *CAB Report Details Capacity Changes at Major Hubs,* AVIATION DAILY, June 21, 1984, at 291.

41. *Texas Air Makes Bid for Frontier,* AVIATION DAILY, Apr. 5, 1985, at 201.

42. *Frontier Loses $13.8 Million—First Annual Loss Since 1971,* AVIATION DAILY, Feb. 2, 1984, at 179. In 1982, Frontier had earned a net profit of $17 million. *Id.*

43. Salpukas, *supra* note 36, at D-1, col. 6. *Frontier Negotiates Tentative Contract Concessions,* AVIATION DAILY, Mar. 30, 1984, at 171.

44. Stapleton International Airport, *Average Airline Fares for Selected U.S. Airports* (1988).

45. *Intelligence,* AVIATION DAILY, Apr. 16, 1984, at 257.

46. *CAB Gives Frontier Horizon Go-Ahead to Begin Service Today,* AVIATION DAILY, Jan. 9, 1984, at 34.

47. *Frontier Says Competition at Denver Responsible for Losses,* AVIATION DAILY, Apr. 30, 1984, at 341.

48. *Record of Certificated Air Carrier Reporting Performance to CAB,* AVIATION DAILY, Oct. 24, 1984, at 287.

49. *Frontier Auditors Qualify Annual Report,* AVIATION DAILY, Apr. 3, 1985, at 185.

50. H. LAWS & R. LOSEE, *supra* note 29.

51. Amended Complaint of Frontier Airlines, Inc., In re Frontier (Reorganization Case N. 86-B-8021E, U.S. Bankruptcy Court for the District of Colorado, Jan. 2, 1987) [hereinafter Frontier Bankruptcy Complaint]; *Frontier Begins Down-Sizing Operations,* AVIATION DAILY, May 15, 1985, at 83. Frontier also sold nine Convair 580s to Metro for $12 million. *Frontier Directors Approve Employee Buyout Plan,* AVIATION DAILY, July 18, 1985, at 97.

52. *Frontier Begins Down-Sizing Operations,* AVIATION DAILY, May 15, 1985, at 83.

53. Levere, *United to Buy Frontier Air from People,* TRAVEL WEEKLY, July 17, 1986, at 1.

54. *Frontier to Terminate Service to Five Cities in September,* AVIATION DAILY, July 8, 1985, at 35.

55. *Frontier Directors Approve, supra* note 51, at 97; *Frontier Employees Would Fund Buyout from Several Sources,* AVIATION DAILY, Aug. 8, 1985, at 217.

56. See Salpukas, *supra* note 36, at D-1, col. 6.

57. *Texas Air to Make Tender Offer for Frontier,* AVIATION DAILY, Sept. 20, 1985, at 105.

58. *Id.* at 105, 106.

59. *Texas Air Bid for Frontier Called No-Lose Situation,* AVIATION DAILY, Sept. 23, 1985, at 114.

60. *Id.*

61. *Texas Air, supra* note 57, at 105, 106.

62. "Lorenzo's bid to acquire Frontier had run into opposition by labor because he was viewed as being a union buster." Csongos, *People Express, Frontier Airlines Merger Approved,* United Press International Wire Report (Nov. 21, 1985). Ironically, when Texas Air sold its 800,000 shares to People Express, it made $5.7 million in arbitrage. *Texas Air Withdraws Offer for Frontier,* AVIATION DAILY, Oct. 11, 1985, at 226.

63. *Texas Air Makes Bid for Frontier,* AVIATION DAILY, Apr. 5, 1985, at 201, 202.

64. *Id.* at 201; *Texas Air Takeover of Frontier Faces Numerous Obstacles,* AVIATION DAILY, Sept. 23, 1985, at 114, 115.

65. *Consumer Complaints Filed at DOT,* AVIATION DAILY, Feb. 3, 1986.

66. As of April 1985, Frontier had 18 unencumbered 737s out of 33 it owned. *Texas Air Makes Bid, supra* note 63, at 201; *Texas Air Bid, supra* note 59, at 114.

67. *A Yankee Preacher in the Pilot's Seat,* TIME, Jan. 13, 1986, at 46.

68. *Id.*

69. Byrne, *Donald Burr May Be Ready to Take to the Skies Again*, Bus. WEEK, Jan. 16, 1989, at 74.

70. Burns, *Rapid Expansion Meant Collapse for People Express*, United Press International Wire Report (Dec. 29, 1986).

71. *Yankee Preacher, supra* note 67, at 46.

72. Clayton, *The People at People Express Say 'Bye to a Dream*, CHRISTIAN SCIENCE MONITOR, Jan. 26, 1987, at 16.

73. *Super Savings in the Skies*, TIME, Jan. 13, 1986, at 40, 43.

74. *Yankee Preacher, supra* note 67, at 46.

75. Clayton, *supra* note 72, at 16.

76. *Super Savings, supra* note 73, at 40, 42.

77. Clayton, *supra* note 72, at 16.

78. Thomas, *Again, Burr and Lorenzo Part Company as People's Founder Leaves Texas Air*, WALL ST. J., July 24, 1987, at 1.

79. *Bitter Victories*, INC. (Aug. 1985), at 25.

80. Byrne, *supra* note 69.

81. *Id.*

82. *Super Savings, supra* note 73, at 40, 42.

83. Deposition of Charles W. Thomson In Re Frontier Airlines, Inc., U.S. Bankruptcy Court (D-Colo) No. 86B8021E (Dec. 10, 1986), at 14.

84. Bitter Victories, *supra* note 79.

85. Byrne, *supra* note 69.

86. Clayton, *supra* note 72; Byrne, *supra* note 69.

87. *Super Savings, supra* note 73, at 40, 42.

88. *Yankee Preacher, supra* note 67.

89. *Id.*

90. *Id.*

91. Clayton, *supra* note 72.

92. *Id.*

93. Prokesch, *People Express: A Case Study*, N.Y. TIMES, July 6, 1986, at 3-1, col. 3.

94. Form 10-K Filed by People Express Airlines, Inc., for the Fiscal Year Ending Dec. 31, 1986, with the U.S. Securities and Exchange Commission (Mar. 31, 1987), at 10.

95. *Super Savings, supra* note 73, at 40, 44.

96. See *Bitter Victories, supra* note 79, at 25, 28.

97. Labich, *How Long Can Quilting-Bee Management Work?*, TIME, Nov. 25, 1985, at 136.

98. Deposition of Charles W. Thomson In Re Frontier Airlines, Inc., U.S. Bankruptcy Court (D-Colo) No. 86B8021E (Dec. 10, 1986), at 16.

99. *Bitter Victories, supra* note 79, at 25, 34; *People Express Contracts Out Entire Computer Network*, AVIATION DAILY, Sept. 4, 1985, at 13.

100. Clayton, *supra* note 72.

101. *Id.*

102. *Yankee Preacher, supra* note 67.

103. *Super Savings, supra* note 73, at 40, 45.

104. Clayton, *supra* note 72; Form 10-K Filed by People Express with the U.S. Securities and Exchange Commission, Dec. 31, 1985, at 25.

105. Form 10-K, Filed by Texas Air Corporation with the U.S. Securities and Exchange Commission 35 (for fiscal year ending Dec. 31, 1986); Carley, *Many Travelers Gripe about People Express, Citing Overbooking*, WALL ST. J., May 19, 1986, at 1; Form 10-K Filed by People Express with the U.S. Securities and Exchange Commission, Dec. 31, 1985, at 25.

106. People Express, Inc., Business Plan-Preliminary Draft 5 (Aug. 7, 1986).

107. *Intelligence*, AVIATION DAILY, June 3, 1985, at 177.

108. *Bitter Victories, supra* note 79, at 25, 26.

109. Burns, *supra* note 70.

110. *Super Savings, supra* note 73, at 40, 44; *People Express Seeks Certificate Rights to Six European Countries*, AVIATION DAILY, Apr. 19, 1985, at 282.

111. *Super Savings, supra* note 73, at 40, 45.

112. Salpukas, *The Woes at People Express*, N.Y. TIMES, June 25, 1986, at D-1, col. 3.

113. Clayton, *supra* note 72.

114. *Id.*

115. People Express, Inc., Business Plan-Preliminary Draft 4-5 (Aug. 7, 1986).

116. Salpukas, *supra* note 112.

117. Byrne, *supra* note 69.

118. Clayton, *supra* note 72.

119. *Id.*; Prokesch, *People Express: A Case Study*, N.Y. TIMES, July 6, 1986, at 3-1, col. 3.

120. *Super Savings, supra* note 73, at 40, 44.

121. Burns, *supra* note 70.

122. Byrne, *supra* note 69.

123. *Super Savings, supra* note 73, at 40, 44.

124. *Bitter Victories, supra* note 79, at 25, 32.

125. *Super Savings, supra* note 73, at 40, 44.

126. Britt had enabled him to boost the number of flights from Chicago. *People Express, Inc., Requests Exemption to Acquire Britt*, AVIATION DAILY, Jan. 14, 1986, at 69; *People Express to Acquire Provincetown-Boston Airline*, AVIATION DAILY, Feb. 3, 1986, at 179. Burr's acquisition of Frontier was described by a People Express board member as perhaps "a bridge too far for him." *Yankee Preacher, supra* note 67.

127. *Super Savings, supra* note 73, at 40.

128. *PBA Reviewing People Express Buyout Offer, Seeking Alternatives*, AVIATION DAILY, Mar. 19, 1986, at 426.

129. Prokesch, *Two Dreamers Awake to a Dogfight*, N.Y. TIMES, Mar. 9, 1986, at 3-1.

130. Clayton, *supra* note 72.

131. Form S-2 Filed by People Express with the U.S. Securities and Exchange Commission, Mar. 13, 1986, at 5; see also Form 10-K Filed by People Express with the U.S. Securities and Exchange Commission, Dec. 31, 1985, at 7-8.

132. Based on revenue passenger miles for 1985. *Industry Market Share*, AVIATION DAILY, Feb. 5, 1986.

133. *Frontier Shareholders, DOT Approve People Express-Frontier Purchase*, AVIATION DAILY, Nov. 22, 1985, at 113; Csongos, *supra* note 62.

134. *People Express 1985 Expenses Increase 68.4 Percent,* AVIATION DAILY, Feb. 20, 1986, at 276.

135. *National Carriers Financial,* AVIATION DAILY, Mar. 21, 1986, at 449; *Industry Market Share,* AVIATION DAILY, Mar. 27, 1986, at 480.

136. Labich, *supra* note 97.

137. McCormick, *People Chief Is a Burr in Many Saddles,* ROCKY MOUNTAIN NEWS, Oct. 9, 1985.

138. *Denver Hub Possible for People Express,* AVIATION DAILY, Oct. 11, 1985, at 227.

139. Henry, *People on the Move,* TIME, Oct. 21, 1985, at 64.

140. See CORPORATE, *supra* note 31, at 17. Frontier was then Denver's third-largest airline, with 19 percent of the market (compared with United's 37 percent and Continental's 26 percent). Data for 1985 taken from Stapleton International Airport, Domestic Market Shares (Nov. 1986).

141. *Frontier to Revamp Fare Structure, Marketing Strategy,* AVIATION DAILY, Feb. 11, 1986, at 225.

142. *Super Savings, supra* note 73, at 40, 44.

143. Summary of Testimony of Expert Witness Thomas R. Sandler, In re Frontier Airlines, Inc. (U.S. Bankruptcy Court, District of Colorado, Case No. 86B-8021-E, Dec. 19, 1986), at 2; Frailey, *Retreat from Denver,* U.S. NEWS & WORLD REP., July 21, 1986, at 31; *People Express, Frontier Sign Final Merger Agreement,* AVIATION DAILY, Nov. 26, 1985, at 132.

144. Summary of Testimony of Expert Witness Thomas R. Sandler, In re Frontier Airlines, Inc. (U.S. Bankruptcy Court, District of Colorado, Case No. 86B-8021-E, Dec. 19, 1986), at 2.

145. *Id.*

146. *Domestic News,* United Press International Wire Report (July 10, 1986).

147. Form 10-Q Filed by People Express Inc. with the U.S. Securities and Exchange Commission for the Quarter Ending Sept. 30, 1985 (Nov. 14, 1985), at 29.

148. Deposition of Charles W. Thomson In Re Frontier Airlines, Inc., U.S. Bankruptcy Court (D-Colo) No. 86B8021E (Dec. 10, 1986), at 13 [hereinafter cited as Thomson].

149. Dubroff, *Burr Aims for Frontier 5-Cent Mile,* Denver Post, Jan. 30, 1986.

150. Thomson, *supra* note 148, at 26; *Frontier President Replaced with People Express Appointee,* AVIATION DAILY, Jan. 31, 1986.

151. *United-Frontier Merger Likely to Reduce Denver Flights,* AVIATION DAILY, July 15, 1986, at 141.

152. *Frontier to Revamp, supra* note 141.

153. *Id.*

154. Thomson, *supra* note 148, at 27.

155. *Id.* at 28.

156. See *Frontier Marketing VP Thomas Volz Resigns,* AVIATION DAILY, June 12, 1986, at 412.

157. Thomson, *supra* note 148, at 28.

158. *Frontier to Revamp, supra* note 141; *People Express/Frontier Put High Priority on Cutting Costs,* AVIATION DAILY, Mar. 5, 1986, at 345; *Frontier Predicts Sharp Increase in Summer Traffic,* AVIATION DAILY, May 19, 1986, at 276.

159. *Frontier Predicts, supra* note 158.

160. H. LAWS & R. LOSEE, *supra* note 29.

161. *Frontier Service Changes Bring Carrier into Line with People Express,* AVIATION DAILY, Mar. 20, 1986, at 438.

162. See Salpukas, *supra* note 112.

163. *Continental Expands Denver Service, Signs Marketing Pact with Regionals,* AVIATION DAILY, June 12, 1986, at 410.

164. *Frontier Inflight Service Changes Result in Layoffs,* AVIATION DAILY, Mar. 25, 1986, at 461.

165. See Dubroff, *Frontier's Low Fares Score with Travelers,* DENVER POST, Feb. 13, 1986, at 1-D.

166. Weber, *Travel Agents Air Gripes about Frontier's No-Frills,* ROCKY MOUNTAIN NEWS, Mar. 4, 1986, at 1-B.

167. The industry suffered more than half a billion dollars in operating losses during the first quarter of 1986. *Industry Market Share,* AVIATION DAILY, June 2, 1986.

168. *Analyst Says First Quarter Losses May Top $625 Million,* AVIATION DAILY, Apr. 10, 1986, at 63.

169. H. LAWS & R. LOSEE, FRONTIER 11 (monograph, December 2, 1986).

170. *Frontier Operations Shut Down, Carrier's Future Uncertain,* AVIATION DAILY, Aug. 26, 1986, at 313.

171. Frontier Bankruptcy Complaint, *supra* note 143, at 3.

172. Levere, *United to Buy Frontier Air from People,* TRAVEL WEEKLY, July 17, 1986, at 1.

173. Stevenson, *People Express Says It Might Have to Sell Part or All of Airline,* N.Y. TIMES, June 24, 1986, at A-1, col. 1.

174. Salpukas, *Frontier Files for Bankruptcy,* N.Y. TIMES, Aug. 29, 1986, at D-1, col. 6.

175. *Frontier Unveils New Premium Coach Service,* AVIATION DAILY, May 28, 1986, at 325.

176. See *Frontier Gets $50 Million Loan, New Entrant Air Puerto Rico $5.2 Million,* AVIATION DAILY, June 3, 1986, at 353.

177. *Frontier Holdings Sells Off Last Non-Airline Subsidiary,* AVIATION DAILY, June 17, 1986, at 437.

178. Form S-2 Filed by People Express with the U.S. Securities and Exchange Commission, Mar. 13, 1986, at 9. There may be some inconsistency in the reports filed by People Express with the Securities and Exchange Commission. For example, People elsewhere indicated that its long-term debt for 1984 was $289 million and for 1985 was $479 million. Form 10-K Filed by People Express Inc. with the U.S. Securities and Exchange Commission for the 1985 Fiscal Year (Mar. 13, 1986), at 39.

179. Burns, *supra* note 70; *National Carriers Financial,* AVIATION DAILY, Mar. 21, 1986, at 488; Form 10-K Filed by People Express with the U.S. Securities and Exchange Commission, Dec. 31, 1985, at 32.

180. *People Express Posts $47.4 Million Quarterly Operating Loss,* AVIATION DAILY, May 2, 1986, at 186; Prokesch, *supra* note 119. However, in data files at the Securities and Exchange Commission, People Express reported only a $20-million operating loss and a $35-million net loss for the quarter. Form 10-Q Filed

by People Express Airlines, Inc. with the U.S. Securities and Exchange Commission (May 15, 1986), at 4. Nonetheless, even People's numbers show a rapidly deteriorating condition. Revenue had increased only 12 percent while maintenance expenses had risen 23 percent and operating losses had soared over 300 percent for the same period in 1985. *Id.*

181. *People Express Loses $57 Million in Second Quarter,* AVIATION DAILY, Aug. 14, 1986, at 251.

182. Form 10-Q Filed by People Express Airlines Inc., for the Quarter Ending June 30, 1986, with the U.S. Securities and Exchange Commission (Aug. 15, 1986), at 4. People Express reported a less severe economic picture to the SEC than did media accounts. It said that during the first half of 1986, People Express suffered a net loss of only $80 million and that its working capital fell by $89 million. *Id.* at 6, 12.

183. Salpukas, *supra* note 112.

184. Prokesch, *supra* note 129.

185. Burns, *supra* note 70.

186. *Id.*

187. *Id.*

188. Stevenson, *supra* note 173.

189. Carley, *Many Travelers Gripe about People Express, Citing Overbooking,* WALL ST. J., May 19, 1986, at 1.

190. *Id.* at 1, 12.

191. Byrne, *supra* note 69.

192. Carley, *supra* note 189, at 1, 12.

193. *Consumer Complaints Filed at DOT, Per 100,000 Passengers,* AVIATION DAILY, June 10, 1986.

194. McCormick, *Ordeal in Frustration Makes People the Airline to Avoid,* ROCKY MOUNTAIN NEWS.

195. *People Express,* AVIATION DAILY, June 14, 1985, at 251.

196. Form 10-Q Filed by People Express Inc. with the U.S. Securities and Exchange Commission for the Quarter Ending Sept. 30, 1985 (Nov. 14, 1985), at 22.

197. *People Express Seeks More Financing, Predicts First Quarter Loss,* AVIATION DAILY, Mar. 15, 1985, at 83.

198. *People Express Restructuring Debt Load,* AVIATION DAILY, Mar. 21, 1986, at 451; *People Express Launches Sale of $115 Million in Equipment Certificates,* AVIATION DAILY, Apr. 21, 1986, at 115.

199. Memorandum from R. R. Ferguson to F. Lorenzo (May 5, 1986).

200. *People Express Studying Sale of Part or All of Company,* AVIATION DAILY, June 24, 1986, at 474.

201. *Id.*

202. Levere, *Pilots' Dispute Imperils Plan for United to Buy Frontier,* TRAVEL WEEKLY, Aug. 11, 1986, at 45.

203. Prokesch, *supra* note 119.

204. *People Express,* AVIATION DAILY, June 25, 1986, at 482.

205. *Id.*

206. Levere, *United to Buy Frontier Air from People,* TRAVEL WEEKLY, July 17, 1986, at 1.

207. Russell, *Cliff Hanger; People Sells Off Frontier,* TIME, July 21, 1986, at 49.

208. Levere, *supra* note 202.

209. *Airlines 1986 Earnings Estimate Per Share Projected by Analysts,* AVIATION DAILY, Apr. 23, 1986.

210. Russell, *supra* note 207.

211. People Express, Inc., Business Plan-Preliminary Draft 1 (Aug. 7, 1986).

212. Burns, *supra* note 70.

213. Moran & Levin, *People Planning to Put Frills Back,* ADVERTISING AGE, July 14, 1986, at 74.

214. *Id.*

215. People Express, Inc., Business Plan-Preliminary Draft 5 (Aug. 7, 1986).

216. People Express, Inc., Business Plan-Preliminary Draft 6-7 (Aug. 7, 1986); Form 10-Q Filed by People Express Airlines Inc., for the Quarter Ending June 30, 1986, with the U.S. Securities and Exchange Commission (Aug. 15, 1986), at 13.

217. Burns, *supra* note 70.

218. *United to Buy Frontier from People Express,* AVIATION DAILY, July 11, 1986, at 58.

219. Stapleton International Airport, Domestic Market Shares (July 1986).

220. *People Express Likely to Sell Frontier Assets to Ensure Survival,* AVIATION DAILY, June 25, 1986, at 482.

221. Levere, *supra* note 206.

222. In re Frontier Airlines, Inc. (memorandum opinion and order on motion to approve settlement, Case No. 86 B 8021 E, Mar. 23, 1987), at 4.

223. *Id.* at 5. Most of these assets were ultimately transferred back under an agreement reached in settlement of litigation by Frontier against United. *Id.* at 6.

224. Frailey, *supra* note 143.

225. *United-Frontier Merger to Focus on Failing Company Doctrine,* AVIATION DAILY, July 25, 1986, at 138.

226. *People Express Posts $112.9 Million Quarterly Net Loss,* AVIATION DAILY, Nov. 12, 1986, at 228. Oddly, People's filings with the Securities and Exchange Commission show a less gloomy picture. For the third quarter of 1986, People Express reported a net loss of only $40 million (interest expenses were more than half of that) to the SEC. Form 10-Q Filed by People Express Inc. with the U.S. Securities and Exchange Commission for the Quarter Ending Sept. 30, 1986 (Nov. 14, 1986), at 4. For the first nine months of 1986, People Express reported a net loss of $120 million and a decrease in working capital of $158 million. *Id.* at 5.

227. In re Frontier Airlines, Inc. (memorandum opinion and order on motion to approve settlement, Case No. 86 B 8021 E, Mar. 23, 1987), at 12.

228. *United-Frontier Merger Likely to Reduce Denver Flights,* AVIATION DAILY, July 15, 1986, at 141.

229. *United Directors Approve Frontier Acquisition,* AVIATION DAILY, July 14, 1986, at 66.

230. *Domestic News,* Reuters Wire Service (Aug. 29, 1986).

231. Sing, *Frontier Halts Flights, Cites Large Losses,* LOS ANGELES TIMES, Aug. 25, 1986, at 1-1, col. 6.

232. *United-Frontier Merger, supra* note 228.

233. Salpukas, *supra* note 174; Sing, *supra* note 231.

234. Sing, *supra* note 231; Lassiter & Dubroff, *Petition for Bankruptcy Reveals People Subsidiary Had No Cash*, TRAVEL WEEKLY, Sept. 8, 1986, at 1.

235. *Frontier Operations Shut Down, Carrier's Future Uncertain*, AVIATION DAILY, Aug. 26, 1986, at 313, 314.

236. Pierce, *The Perils of Competition*, TIME, Sept. 8, 1986, at 50.

237. Form 10-K, Filed by Texas Air Corporation with the U.S. Securities and Exchange Commission 35 (for fiscal year ending Dec. 31, 1986).

238. Pierce, *supra* note 236.

239. Lassiter & Dubroff, *supra* note 234.

240. *Reporters, Analysts Applaud United's Frontier Purchase*, FRIENDLY TIMES, Aug. 1986, at 2.

241. *Id.*

242. Lassiter & Dubroff, *supra* note 234.

243. TEXAS AIR, INC., 1987 ANNUAL REPORT 2 (1988).

244. Stevenson, *supra* note 173.

245. Burns, *Judge OKs Settlement of People Express Suit*, United Press International Wire Report (Apr. 25, 1989); *People Express Investors File Suit against Carrier*, AVIATION DAILY, Sept. 25, 1986, at 484.

246. Form 10-K, Filed by Texas Air Corporation with the U.S. Securities and Exchange Commission 22 (for fiscal year ending Dec. 31, 1986).

247. In re Frontier Airlines, Inc. (memorandum opinion and order on motion to approve settlement, Case No. 86 B 8021 E, Mar. 23, 1987), at 4.

248. *Texas Air Lowers People Express Purchase Price by about $25 Million*, AVIATION DAILY, Nov. 4, 1986, at 185. Some sources report the original offer at $122.1 million for People Express. See, e.g., *Texas Air Corp. Seeks to Purchase People Express, Frontier Assets*, AVIATION DAILY, Sept. 16, 1986, at 425.

249. *Texas Air Corp.*, *supra* note 248.

250. *Frontier Employees to Vote on Continental Work Offer*, AVIATION DAILY, Sept. 18, 1986, at 441, 442.

251. *People Express Using Failing Company Doctrine to Speed Merger Okay*, AVIATION DAILY, Oct. 9, 1986, at 50.

252. *Texas Air Debt Load Reaches $5.5 Billion with People Express Purchase*, AVIATION DAILY, Oct. 16, 1986, at 82.

253. *Id.*

254. Moran & Levin, *supra* note 213.

255. *Texas Air Lowers*, *supra* note 248.

256. Salpukas, *Company News; People's Acquisition Is Voted*, N.Y. TIMES, Dec. 30, 1986, at 4, col. 1; *Texas Air Lowers*, *supra* note 248; *Texas Air Again Reduces Value of People Express Takeover Offer*, AVIATION DAILY, Dec. 2, 1986.

257. *Texas Air Again Reduces*, *supra* note 256, at 324.

258. *Texas Air Negotiating to Lower Price for People Express*, AVIATION DAILY, Nov. 3, 1986, at 179.

259. *People Express Requests Confidentiality for Board Proceedings*, AVIATION DAILY, Oct. 17, 1986, at 90.

260. Form 10-K Filed by People Express Airlines, Inc., for the Fiscal Year Ending Dec. 31, 1986, with the U.S. Securities and Exchange Commission (Mar. 31, 1987), at F-1.

261. Form 10-K Filed by People Express Airlines, Inc., for the Fiscal Year End-

ing Dec. 31, 1986, with the U.S. Securities and Exchange Commission (Mar. 31, 1987), at F-12.

262. TEXAS AIR CORPORATION, 1986 ANNUAL REPORT 15 (1987).

263. Form 10-K, Filed by Texas Air Corporation with the U.S. Securities and Exchange Commission 3 (for fiscal year ending Dec. 31, 1986).

264. *Continental Expects to Operate 250 Daily Denver Departures by Yearend,* AVIATION DAILY, Nov. 3, 1986, at 182.

265. TEXAS AIR CORPORATION, 1986 ANNUAL REPORT 2 (1987).

266. Stapleton International Airport, Domestic Market Shares (July & Oct. 1986).

267. *Id.* See Pasxtor & Carley, *People Express Bid by Texas Air Clears a Hurdle,* WALL ST. J., Oct. 2, 1986, at 2, col. 2.

268. Krauss, *Financial,* United Press International Wire Report, Jan. 13, 1987.

269. *Id.*

270. Stapleton International Airport, Average Airline Fares for Selected U.S. Airports (1988).

271. Lorenzo made the helm of Continental a revolving door of executives. For example, Lorenzo ousted Continental President Thomas Plaskett (who had previously served as an executive at American Airlines for 12 years) after only eight months on the job. Thomas, *Frank Lorenzo, Builder of Airlines, Now Faces Task of Running One,* WALL ST. J., July 24, 1987, at 1.

272. Salpukas, *Airlines' Big Gamble on Expansion,* N.Y. TIMES, Feb. 20, 1990, at C1, C5.

273. *Id.* at C5.

274. Prokesch, *supra* note 119.

275. *Id.*

276. *Super Savings, supra* note 73, at 40.

277. Byrne, *supra* note 69.

278. *Super Savings, supra* note 73, at 40, 41.

279. Prokesch, *supra* note 119.

280. Byrne, *supra* note 69.

6

DELTA AIR LINES

Hubs: Atlanta, Salt Lake City, Cincinnati
Mini-hubs: Dallas/Ft. Worth, Los Angeles, Orlando
Post-deregulation Merger: Western (1986)
Computer Reservations System: DATAS II, a part of WORLDSPAN
Rank and Market Share: 1978—fifth, 10.3%; 1990—third, 13.0%

Delta Air Lines began modestly, as a crop-dusting outfit in Monroe, Louisiana, in 1928—the first professional crop duster in the nation.[1] Delta was born and nurtured by C. E. Woolman, who headed the company for 38 years, until his death in 1966.[2] He moved Delta's headquarters to Atlanta in the early 1940s, and that hub became the heart of its operations and, ultimately, the source of most of its management.

Delta vigorously opposed deregulation. During the late 1970s, it was the most litigious of the bunch, taking the CAB to court on nearly every one of the deregulatory initiatives of Alfred Kahn, who was then chairman. Indeed, the litigation branch of the CAB's Office of General Counsel came to be known as the "Delta Wing."

Nonetheless, Delta entered deregulation with a number of strengths. By growing, it had elbowed its way into the "big five." Delta had expanded significantly in the Southeast by acquiring Citizens & Southern Airlines in 1953.[3] It expanded north with its acquisition of Northeast in 1972.[4] And in 1986, Delta joined the stampede to merge by acquiring Western Air Lines, hubbed in Salt Lake City, for $900 million (see figure 6.1).[5]

Because Delta paid its workers well and had never laid any off, it enjoyed relatively amicable labor relations and had few union contracts. That enabled it to enjoy high productivity, excellent service, and high worker

Figure 6.1
Airline Market Share at Salt Lake City

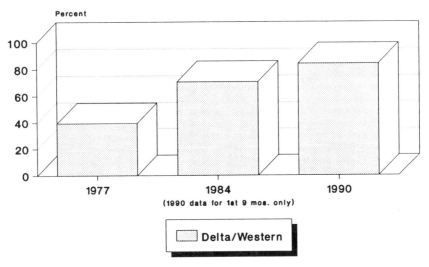

Sources: AVIATION DAILY, Apr. 19, 1985, at 28; Feb. 1, 1990, at 230; Apr. 29, 1990, at 628; Mar. 29, 1991, at 590; and CONSUMER REPORTS.

morale with little turnover.[6] Delta's greatest asset of all was its people.[7] While deregulation has brought the industry tremendous labor strife, labor-management relations are relatively peachy at Delta's Atlanta headquarters. In 1986, Delta's workers dug into their pockets and bought their company a jet.

Delta is a conservative company with little debt. Its debt-to-equity ratio in 1988 was 33 percent, compared with an industry median of more than 80 percent.[8] But Delta's salary expenditures are high. In 1988, Continental's average employee earned $29,700, whereas Delta's earned $51,400.[9] When it acquired Western in 1986, Delta immediately raised the salary of all Western employees from their average of $36,500 to Delta levels, without laying off a single employee.[10] As a former Delta executive noted, "A job with Delta is security for life, providing you don't get caught sleeping with one of your employees."[11] Hence, Delta is a bit heavy with employees and is saddled with the largest labor expenditures in the industry.

Nonetheless, Delta has been blessed with profitability. Before 1990, the only year since the Great Depression in which Delta failed to post a profit was 1983, when it lost $16.1 million.[12] In part, Delta's success lies in its competition.

Its slower southern cousin, Eastern, was traditionally a poor competitor at their common Atlanta hub. Eastern's bankruptcy experience under Frank

Figure 6.2
Airline Market Share at Atlanta

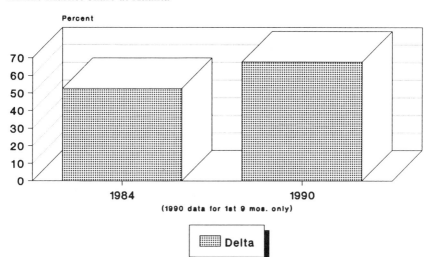

Percent

(1990 data for 1st 9 mos. only)

[░░░] Delta

Sources: AVIATION DAILY, Apr. 19, 1985, at 28; Feb. 1, 1990, at 230; Apr. 29, 1990, at 628; Mar. 29, 1991, at 590; and CONSUMER REPORTS.

Lorenzo ultimately benefited Delta. Because of Delta's superior service, most businesspeople prefer to fly Delta. (See figure 6.2.)

Delta is, however, a sluggish price innovator.[13] Higher prices tend to cover its higher costs. With its shared monopoly in Atlanta, passengers who board or depart there recently paid 22¢ a mile, whereas other Delta passengers paid 13¢ a mile.[14]

Above all, Delta is conservative. Tradition is the bedrock of its management philosophy.[15] Delta still holds its annual stockholders' meeting in Monroe, Louisiana, the city in which it was founded.

Delta's headquarters at Atlanta's Hartsfield International Airport was traditionally austere. Although newer facilities have been built at Hartsfield, Delta still has linoleum floors in all offices except those of the assistant vice presidents and higher officers. Indeed, if a lower-ranking officer takes over a carpeted office, the carpet is ripped out and replaced with linoleum.[16]

One analyst noted, "Tradition . . . still shapes the basic approach of Delta's management."[17] Conservative management has led it to follow the pack, so to speak, in buying aircraft, in offering pricing discounts, in adopting two-tier wages, in developing a frequent-flyer program and computer reservations system, and in acquiring other carriers. Delta acquired Western Airlines when it saw the industry charging toward acquisitions. It jumped into the computer reservations system (CRS) business by buying DATAS II for $130 million three years after American pioneered the industry with SABRE.[18] By 1989, DATAS II had only about 5 percent of the national CRS market.[19] A Delta critic charged, "The corporate culture at Delta does not honor innovation; it honors loyalty to tradition."[20]

Since the death of C. E. Woolman in 1966, it has been said that Delta's management has been dominated by "good old boys," mostly Georgia Tech graduates. Promotion from within has created a bureaucratic culture at Delta, one with inbred management.[21] However, Delta picked up 4 of Western's 16 officers in the merger, which strengthened its marketing team. Delta also stole away the "Official Airline of Walt Disney World" title from Eastern and expanded operations in Orlando.[22] If indeed the good old boys run Delta, they know what they are doing.

"Buster Tom" Beebe served as chairman of Delta between 1970 and 1980.[23] His protégé, Ronald W. Allen, became president and chief operating officer in 1983.[24] Allen is clean-cut, tall, and athletic.[25] One critic described him as lacking "the breadth of background and the experience to be a successful competitor in this environment where airlines continue to fumble through the chaos created by the Airline Deregulation Act of 1978."[26] Nonetheless, the disintegration of Eastern has been a tremendous boon for Delta, and Delta is among the three airlines most analysts agree will survive the market Darwinism unleashed by deregulation. When asked about Delta's good-old-boy management style, Allen said, "Being nice to our workers, paying them well, that's not good ol' boy, it's modern and enlightened."[27]

Delta's weaknesses include "higher labor costs, a weaker computer reservations system and a slower start on international service than its major rivals."[28] Nonetheless, it has a dedicated labor force, a relatively high level of service, and loyal customers. With its acquisition of Western, it became an airline with a national presence. Delta's development of a hub at Cincinnati has been profitable. It is also enthusiastically expanding in the transatlantic market. It acquired most of Pan American's transatlantic and European operations in 1991. And with 517 aircraft on order or option, Delta is poised to grow if the economy is strong.[29]

NOTES

1. S. DAVIS, DELTA AIR LINES: DEBUNKING THE MYTH 134 (1988); R. DAVIES, AIRLINES OF THE UNITED STATES SINCE 1914 117 (1972).

2. S. DAVIS, *supra* note 1, at 2, 134.

3. *Id.* at 131.

4. R. DAVIES, *supra* note 1, at 566.

5. S. DAVIS, *supra* note 1, at 153.

6. *Id.* at 19, 75, 81, 99.

7. *Id.* at 39.

8. Banks, *Is Delta Too Nice for Its Own Good?*, FORBES, Nov. 28, 1988, at 91, 94.

9. *Id.* at 94.

10. S. DAVIS, *supra* note 1, at 19.

11. *Id.* at 81.

12. *Id.* at 43.

13. *Id.* at 17.

14. *Id.* at 119.

15. Salpukas, *Delta a Favorite in Airline Stocks,* N.Y. TIMES, Oct. 11, 1988.

16. S. DAVIS, *supra* note 1, at 2.

17. Salpukas, *supra* note 15.

18. S. DAVIS, *supra* note 1, at 17, 47.

19. Salpukas, *supra* note 15.

20. S. DAVIS, *supra* note 1, at 32.

21. *Id.* at 43, 167, 173.

22. Banks, *supra* note 8, at 94.

23. S. DAVIS, *supra* note 1, at 3, 45, 140.

24. *Id.* at 148, 154.

25. *See id.* at 133.

26. *Id.* at 3.

27. Banks, *supra* note 8, at 94.

28. Salpukas, *Airlines' Big Gamble on Expansion,* N.Y. TIMES, Feb. 20, 1990, at C1, C5.

29. *See id.* at C5.

7

EASTERN AIRLINES

Hub: Atlanta

Post-deregulation Acquisition: Braniff's Latin American Routes (1983)

Computer Reservations System: System One (now owned by Texas Air, renamed Continental Airline Holdings)

Rank and Market Share: 1978—fourth, 11.1%; 1990—ninth, 4.5% (ceased operations in 1991)

Eastern Airlines, founded in 1928, was the first major carrier to fly the north-south routes along the East Coast. This gave it strong presence in Atlanta and Miami and beyond to the Caribbean. Eastern pioneered the Washington–New York–Boston shuttle, a major innovation.[1]

Before Lorenzo devoured it in 1986, Eastern was the pilots' airline, a company headed by a long line of aviation heros. "Captain Eddie" Rickenbacker, the World War I flying ace (only Sergeant Alvin York earned as much public adoration),[2] reputedly knew all its fliers by name.[3] Rickenbacker became general manager of Eastern in 1935. In 1938, he and several friends bought the airline for $3.5 million from General Motors.[4] That year, Rickenbacker took it off federal subsidy.[5]

Almost single-handedly, Rickenbacker made Eastern one of the "big four," despite the absence of a transcontinental or transoceanic route.[6] He was a colorful and dominant leader. The Eastern "family" was a close one, cultivated by management to take pride in the airline. It was christened "The Wings of Man."

Although among the most profitable of airlines in the late 1940s, Eastern endured a roller-coaster ride that placed the company near bankruptcy by the early 1960s.[7] Employee morale during this period has been compared to that of "a football team with a twenty-five game losing streak."[8] Competing head to head in the Atlanta hub with healthier Delta was an uphill battle. Delta could charge more, pay labor more, and still turn a profit with largely nonunionized workers while Eastern paid labor less, had poor labor-management relations, and was anemic in the 1980s.

But Eastern was growing nonetheless. In 1969, Eastern carried 21.6 million passengers, had a fleet of slightly more than 200 aircraft, and employed 31,900 employees. By 1985, the company carried 41.7 million passengers—more than any other airline in the free world—with a fleet of 289 aircraft and 41,100 employees.[9] Indeed, Eastern ranked first or second in the number of passengers from 1976 to 1985.[10]

Colonel Frank Borman, the Apollo astronaut, was faced with a number of challenges when he took the reigns of Eastern in 1975.[11] Borman also took a gamble with the first U.S. airline purchase of the new Airbus aircraft, manufactured in Europe by a consortium of European governments. Airbus was eager to penetrate the U.S. market and offered Eastern a sweetheart deal with generous financing. For his part, Borman was interested in refurbishing his fleet with modern, fuel-efficient equipment. Had fuel prices continued to ascend, the major purchase might have given Eastern a competitive advantage over its rivals. But in the 1980s, they fell.

The 1970s were, on balance, profitable years for Eastern. Over the decade it enjoyed an operating profit of more than half a billion dollars and earned $58.8 million in net income.[12] The 1980s were a bit worse. Between 1980 and 1985, Eastern earned $244.3 billion in operating profit but lost $380 million in net income, although later years showed some improvement.[13] But interest expenses, levied to finance Eastern's new planes and its tenacious debt, were crushing its profits.

The ailing bottom line required a confrontation with labor over wages. Eastern workers, for their part, could not understand why Eastern didn't make a profit when rival Delta, with higher labor costs, turned a profit in all years except 1983. Surely, Eastern's inferior management was to blame. Nevertheless, Eastern's employees surrendered $1.14 billion in concessions between 1979 and 1988.[14] Eastern enjoyed the lowest labor costs in the industry, with only nonunion Continental having lower wages.[15]

Borman traded wage concessions for equity in the company, and the unions soon sat on Eastern's board of directors, holding four seats and 30 percent of the stock. Charlie Bryan, the tenacious leader of the machinists union, took a seat on Eastern's board and badgered poor Borman relentlessly.[16]

In 1985, Borman informed the unions that he intended unilaterally to impose a continuation of the 1984 wage cuts. The unions voted to strike in February 1986, by margins of over 90 percent. Borman threatened that if the unions did not accede to his demands, he would sell or liquidate the company. All the unions except the machinists capitulated. But on the evening of February 24, 1986, when Borman threatened to sell Eastern to the dreaded Frank Lorenzo, the destroyer of unions, the machinists too agreed to a wide range of concessions, with one condition: that the board of directors replace Borman with a chairman "acceptable to the employee groups."[17] Borman was outraged. Within hours, he had persuaded the board to sell the company to Lorenzo's Texas Air.

Texas Air paid $615 million, or $10 a share, for Eastern. Eastern paid a $20-million inducement to Texas Air to make an offer to buy the airline; Eastern financed Texas Air's purchase with $110 million of its cash and $230 million in preferred stock. Thus, of the $615 million Texas Air ostensibly paid for Eastern, $374 million (or about 61 percent) was paid for by Eastern itself.[18] Five weeks before the deal, Eastern's top management had given itself "golden parachutes" totaling $7.3 million. Colonel Borman's fee for assisting the Texas Air purchase exceeded $1 million.[19] Borman became a director and vice chairman of Texas Air.[20]

The pilots summarized their view of working for Frank Lorenzo in these terms:

Texas Air is the best example of the worst abuses of deregulation. Under the guise of airline building, Texas Air has gotten control of ten airlines and driven the two survivors, Continental and Eastern, into bankruptcy. This manipulation of the intent of deregulation has protected Lorenzo's financial empire and forced his repressive work ethic upon the remaining employees. Texas Air has a record of reduced competition, reduced jobs, reduced compensation, reduced benefits, reduced service, reduced maintenance, reduced safety margins, reduced pilot quality, and reduced communities served. The major increases Texas Air are responsible for are compensation for its top executives (Lorenzo's 1988 salary was reportedly over $1.25 million) and consumer complaints at a level unequaled in the airline industry.[21]

The pilots were not alone in their disdain for Texas Air. *Fortune* magazine ranked the company as one of the five least-admired corporations in the United States, the least-admired transportation firm in the nation, and one of the ten worst U.S. corporations at providing rewards for shareholders.[22] Between 1971 and 1989, under Lorenzo's control, Texas Air's four airlines (Texas International, Continental, New York Air, and Eastern) suffered net income losses totaling nearly $2.3 billion.[23]

Under Lorenzo's tutelage, between 1986 and 1988, Eastern lost $85.5 million in operating profit and nearly $648 million in net income.[24] Eastern was radically downsized. Between 1986 and 1988, the work force was reduced from 42,000 to 29,000, and its newly established Kansas City hub was abandoned. Eastern, which had been the airline that served the most passengers in the free world, was now the sixth largest. Still, Eastern was the pilots' airline, one traditionally run by men with an aviation background, and it remained proud.

Lorenzo changed all that. His stripping the company of assets and demanding wage reductions created "an intense, highly personal anger, a feeling that [employees'] self-respect, honor and dignity" had been "undercut by a management obsessed with pinching pennies and controlling every aspect of their working lives."[25] As the psychologist Linda Little said:

I sat with Eastern management and talked to them about the process of bonding with an infant, that within the first hours and weeks the mother makes contact and the infant knows that this is a safe place to be. But Eastern management came in and rather than saying to its employees, "We want your advice, we want your help, we want to work with you, to bond with you," they came in and said, "You've got to change, you are bad, you are wrong." They did everything a bad mother would do.[26]

On taking over Eastern, Lorenzo put into place much of the same management team that had led Continental into bankruptcy.

Continental	Eastern
IAM Strike: Aug. 13, 1983	*IAM Strike:* Mar. 4, 1989
Bankruptcy: Sept. 24, 1983	*Bankruptcy:* Mar. 9, 1989
Frank Lorenzo: Board Chairman, President	*Frank Lorenzo:* Board Chairman
Phil Bakes: Director, Exec. VP	*Phil Bakes:* Director, President, CEO
Tom Matthews: Sr. VP Employee Relations, Chief Labor Negotiator	*Tom Matthews:* Sr. VP Human Resources, Chief Labor Neg.
Barry P. Simon: VP, Gen. Counsel, Secretary	*Barry P. Simon:* Sr. VP Legal Affairs, Gen. Counsel, Sec'y
Gary H. Lanter: Various positions	*Gary H. Lanter,* VP Properties and Purchasing
Mickey Foret: VP, Treasurer	*Mickey Foret:* Director, Past VP Finance, CFO
Lindsay E. Fox: Past Chairman Int'l Advisory Board	*Lindsay E. Fox:* Director
James W. Wilson: Director	*James W. Wilson:* Director
Carl R. Pohlad: Director	*Carl R. Pohlad:* Director
John Adams: Sr VP Personnel	*John Adams:* Past VP Human Res.
Donald Breeding: VP Flight Ops	*Donald Breeding:* Past VP Flight Operations
Robert D. Snedeker: Director, Sec./Treas. (Jet Cap), Director, Sr. VP, Treasurer (TAC)	*Robert D. Snedeker:* Director (EAL), Sec./Treas. (Jet Cap), Director, Sr. VP, Treasurer (TAC)[27]

These interlocking directorates allowed Texas Air and its subsidiaries to strip Eastern of its assets for only a fraction of their worth. As the bankruptcy examiner Shapiro concluded:

The history uncovered in the course of the investigation reveals a host of transactions where Eastern apparently suffered from conflicts of interest on the part of Texas Air and the interlocking officers and directors it has put in place. Indeed, there are indications that from time to time Texas Air's officers entertained the notion to "Cherry pick" Eastern assets to the benefit of Continental.[28]

Examiner Shapiro found that such self-dealing resulted in the following amounts owed by Texas Air to Eastern for the following reasons:

Texas Air's Acquisition of Eastern $61 million

Eastern paid Texas Air a $20 million inducement fee and funded approximately $109 million of the total cash paid by Texas Air to Eastern's shareholders. Texas Air in turn paid Eastern approximately $68 million to purchase previously unissued Eastern stock. Thus, the acquisition of itself by Eastern cost Eastern a net amount of $61 million in cash. This payment of cash by Eastern directly benefited Texas Air by reducing the cash it had to pay for the acquisition, and it is doubtful that Eastern received any benefit from making this payment; . . . there is no clear evidence even that the acquisition by Texas Air was anything other than harmful to Eastern economically.[29]

Eastern's $16 Million Receivable from Texas Air $5 million

Apparently in order to improve the appearance of Eastern's financial statements, approximately $16 million in expenses paid by Eastern in connection with the acquisition were recorded as a receivable from Texas Air based upon its undertaking to reimburse Eastern. . . . [N]o effort was made to collect until the matter became the subject of adverse attention in litigation. . . . At that time . . . Texas Air paid the receivable without interest. At the rate being paid by Eastern to borrow money at the time, this two-year forbearance cost Eastern approximately $5 million. . . . [A] controlling stockholder's failure to pay a substantial receivable (and officers' and directors' failure to make any collection effort) in a time of financial need would appear to be an appropriate subject for a claim of breach of fiduciary duty.[30]

Management Fees Paid by Eastern to Texas Air $1.4 million

Eastern paid Texas Air a management fee of $500,000 per month from January 1, 1987 until filing for bankruptcy in March 1989, when the fee was reduced to $250,000 per month. . . .

The evidence raises questions . . . as to whether the value of the services received by Eastern, or the cost incurred by Texas Air in providing those services, are equivalent to the fee paid by Eastern. Further, although Eastern initially was charged a greater fee than Continental because it was a larger carrier, Eastern continued to pay more after Continental had surpassed it in size in the latter part of 1987, and there is no clearly sufficient justification for its having been required to do so.[31]

Continental-Eastern Sales, Inc. $1 to $3 million

Continental-Eastern Sales, Inc., ("CESI") was formed in December 1986 to combine the sales forces and City Ticket Office ("CTO") operations of Continental and Eastern. . . . Eastern contributed substantially greater numbers of experienced sales personnel, sales offices and CTO's without an equivalent corresponding contribution from Continental. . . . CESI maintained excessive accounts receivable from Continental which improved Continental's cash flow position, at times at Eastern's direct expense.[32]

Aircraft Transactions by Eastern with Continental   **$5.7 to $11.5 million**

Eastern did not receive fair value for its lease and eventual sale of [six] A300s to Continental. . . . Further, the restructuring of the lease agreement appears to have remedied Continental's cash-flow problems to Eastern's detriment.[33]

People Express Notes Purchased by Eastern from   **$8 to $11 million**
Texas Air

In March 1987, Eastern acquired unsecured People Express notes due 2001 with a face value of $30 million from Texas Air for approximately $26 million in cash. This acquisition was approved by the Board of Directors of Eastern under circumstances warranting an inference that Eastern's management and the common directors made material omissions and misrepresentations. In addition, the price that Eastern paid appears excessive.[34]

Sale of Newark Gates Lease from Eastern to   **$10 to $14 million**
Continental

Eastern assigned its lease of sixteen gates at Newark International Airport to Continental in exchange for an $11 million note and a sublease of five gates. . . . Before the transaction Eastern's Newark gates were valued [at between $1.875 and $2.03] per gate. . . . The evidence concerning the transaction suggests something less than arm's-length negotiations.[35]

Texas Air Fuel Management Arrangement with   **$18.8 million**
Eastern

[Eastern paid a one-cent-per-gallon fee to a Texas Air subsidiary, TAC FMI, for aviation fuel.] The cost savings Eastern has realized from TAC FMI's activities appear to be less than the fees Eastern has paid.[36]

Eastern's Sale of System One to Texas Air   **$150 to $250 million**

In February 1986, Merrill Lynch valued [Eastern's computer reservation system] SODA at $200–$250 million. . . .

On March 3, 1987 Eastern's Board unanimously approved the sale [of SODA to Texas Air] for a $100 million, 6 percent convertible subordinated Texas Air note due in 2012. . . .

[It is reasonable to contend that Eastern sold SODA . . . for less than fair value, by an amount in the range between $150 to $250 million.][37]

Bar Harbor Airways Transactions   **$10.6 to $12.4 million**

The evidence suggests that Eastern provided disproportionate financial support for a commuter that was virtually controlled and operated by Continental and from which Continental received substantial benefits, and that it did so under unfavorable financial terms.[38]

| Eastern's Deposit of Cash Collateral for Texas Air's Guarantee in Connection with $200 million Private Placement | $5 to $7 million |

Texas air treated Eastern less favorable in connection with this guarantee than it had treated Continental in an analogous situation.[39]

| Eastern's Payment of Cash Dividends on Preferred Stock Guaranteed by Texas Air | $8 million |

[T]he last cash dividend payment of approximately $6.7 million on the merger preferred occurred just days before Eastern declared bankruptcy, at a time when Eastern's borrowing cost was practically infinite because of its poor financial condition, and therefore should not have not been made.[40]

| Total | $284.5 to $403.1 million[41] |

Lorenzo demanded concessions from the machinists unions. But labor insisted that wages were not the cause of Eastern's economic problems; the rape of Eastern's assets by Texas Air was. A machinists strike in March 1989, followed by the decision of the pilots and flight attendants to honor the picket line, shut down Eastern. Lorenzo responded by throwing the company into Chapter 11 bankruptcy reorganization and selling off its most promising assets: the Boston–New York–Washington shuttle (landing slots and 21 aircraft) to Donald Trump for $365 million; the Philadelphia mini-hub, slots at LaGuardia and Washington National, and Philadelphia-Canada routes to Midway Airlines for $207 million; and the Latin American and Miami-Europe routes to American Airlines for $349 million.[42] Sixty-one aircraft were sold, and 23 aircraft were leased. Asset sales totaled nearly $1.5 billion.[43] Under Lorenzo, Eastern's assets were depleted by more than $1 billion; its capacity was reduced by more than 15 percent; and its employees were cut by more than 30 percent.[44] A much demoralized and downsized Eastern resumed operations from its Atlanta hub in 1989. In 1990, the creditors in Eastern's bankruptcy proceeding insisted that Frank Lorenzo and his lieutenant, Phil Bakes, be removed from the controls of Eastern, and the court appointed Martin Shugrue, a former Pan Am and Continental executive, as trustee to operate Eastern. Lorenzo and Bakes are the only people in history to have bankrupted two airlines (Continental and Eastern).

Chapter 11 bankruptcy was much more difficult than anticipated. As one source noted:

While its pilots and flight attendants have called off their strike [but the machinists have not], the airline's passenger loads have been low and, under pressure from unsecured creditors, Eastern's management has scaled back its plan to rebuild the carrier.

The goal now is to be about two-thirds its former size, compared with more than 80 percent [in the fall of 1989]. Eastern lost $852.3 million [in 1989] and expects to lose $155 million in the first six months of 1990.[45]

Examiner Shapiro's findings of self-dealing, quoted above, forced Texas Air to promise a contribution of $280 million to settle possible claims that it had underpaid for Eastern's assets, a promise it later withdrew.[46]

Lorenzo was the central figure who galvanized the unions. One commentator summarized his impact on the airline industry and the American corporate scene:

> The ruthlessness of Lorenzo, although it would have been enough to set the old muckrakers scavenging to expose him, allowing him no place of hiding, is not really the moral of the abysmal story. It is more important that he represents in a particularly comprehensive way the atmosphere of greed and reckless manipulation, of sham financial transactions, that has grown in the past eight years of Reaganism, at the expense of productive commercial and industrial leadership. The reason Lorenzo has not been called to account, except by the union, is that he represents practices that are so common and tolerated.
>
> The destruction of a great airline is in part the destruction of the pride of its work force at all levels, and its loyalty to the company. As Robert Reich . . . said . . . "If companies are run on the principle of financial management, reducing wages, going deeply into debt, moving pieces of the company around like they were pieces of a Monopoly game, workers are not going to feel loyal. . . ." But disloyal to partners and his own management, Lorenzo does not prize loyalty in his own employees. He relies on discipline by terror.[47]

For a time, there was speculation that Eastern's flight operations and aircraft would be folded into nonunion Continental.[48] In early 1991, Eastern Airlines ceased operations, and its properties were placed on the auction block for liquidation. American Airlines purchased the Latin American routes Eastern had earlier bought from Braniff.

NOTES

1. Stockton, *Tearing Apart the Airlines*, N.Y. TIMES MAGAZINE, Nov. 6, 1988, at 36, 37.

2. R. SERLING, FROM THE CAPTAIN TO THE COLONEL 92 (1980).

3. Solis and de Cordoba, *Eastern Strike Bares Long-Festering Anger over Sense of Betrayal*, WALL ST. J., Mar. 17, 1989, at 1.

4. R. DAVIES, AIRLINES OF THE UNITED STATES SINCE 1914 196 (1972); R. SERLING, *supra* note 2, at 116, 143.

5. R. SERLING, *supra* note 2, at 151.

6. R DAVIES, *supra* note 4, at 533.

7. *Id.* at 393.

8. *Id.* at 393.

9. EASTERN PILOTS, EASTERN AIR LINES: AN AIRLINE IN CRISIS 3.2 (1990).

10. *Id.* at 3.3.
11. R. DAVIES, *supra* note 4, at 473.
12. See EASTERN PILOTS, *supra* note 9, 3.2.
13. *Id.*
14. *Id.* at 2.94.
15. *Id.* at 2.93.
16. *See* S. DAVIS, DELTA AIR LINES: DEBUNKING THE MYTH 12 (1988).
17. Fairlie, *Air Sickness,* NEW REPUBLIC, June 5, 1989.
18. EASTERN PILOTS, *supra* note 9, 4.22.
19. *Id.*
20. *Id.*
21. *Id.* at 10.
22. *America's Most Admired Corporations,* FORTUNE, Jan. 29, 1990.
23. EASTERN PILOTS, *supra* note 9, 2.91.
24. See *id.* at 3.2.
25. Solis & de Cordoba, *supra* note 3.
26. Quoted in Boyer, *The Double Life of Frank Lorenzo,* VANITY FAIR, Dec. 1989.
27. EASTERN PILOTS, *supra* note 9, 5.2.
28. In the Matter of Ionosphere Clubs, Inc, U.S. District Court for the Southern District of New York, Chapter 11 Case Nos. 89 B 10448 and 89 B 10449.
29. *Id.* at 80.
30. *Id.* at 81, 82.
31. *Id.* at 82–83.
32. *Id.* at 83–84.
33. *Id.* at 84–85.
34. *Id.* at 85.
35. *Id.* at 86.
36. *Id.* at 88.
37. *Id.* at 88–91.
38. *Id.* at 92.
39. *Id.* at 94.
40. *Id.* at 96.
41. *Id.* at 97.
42. EASTERN PILOTS, *supra* note 9, 7.15.
43. *Id.*
44. *Id.* at 3.
45. Salpukas, *Airlines' Big Gamble on Expansion,* N.Y. TIMES, Feb. 20, 1990, at C1, C5.
46. Lowenstein, *Eastern Veers into Texas Air's Flight Path,* WALL ST. J., Mar. 9, 1990, at C1.
47. Fairlie, *supra* note 17.
48. Lowenstein, *supra* note 46, at C2.

8

NORTHWEST AIRLINES

Hubs: Minneapolis/St. Paul, Detroit, and Memphis

Mini-hubs: Milwaukee, Tokyo, Seoul

Post-deregulation Merger: Republic (1986), which is the merged product of North Central, Southern, and Hughes Airwest

Computer Reservations System: an interest in PARS, renamed WORLDSPAN

Rank and Market Share: 1978—eighth, 3.1%; 1990—fourth, 11.6%

Born in 1926, Northwest "Orient" broke into international aviation with North Pacific route service on July 15, 1947, the first airline to serve the market.[1] For many years, it carried more passengers in the North Pacific than any other U.S. airline.[2] More recently, it acquired transatlantic authority, now funneling much of that through Boston.[3] With the acquisition of Republic in 1986, Northwest gained a death grip on its Minneapolis/St. Paul hub and a stranglehold on Detroit, as well as domestic feed for its international operations. It intends to keep its prominent position in the Pacific (the fastest-growing and most lucrative market in the world), where its traffic is growing 19 percent compounded annually.[4] However, many new carriers are entering the transpacific market. (See figures 8.1 and 8.2.)

The annual report lists several "firsts" for Northwest:

- The first airline to provide regularly scheduled service between Japan and the United States.
- The first airline to serve Seoul from the United States and the first to Shanghai and Manila on a North Pacific routing.
- The first U.S. airline to serve Osaka, Okinawa, and Taipei.

Figure 8.1
Airline Market Share at Detroit

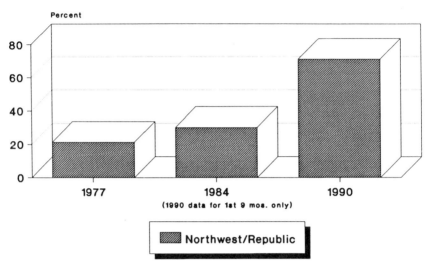

Sources: AVIATION DAILY, Apr. 19, 1985, at 28; Feb. 1, 1990, at 230; Apr. 29, 1990, at 628; Mar. 29, 1991, at 590; and CONSUMER REPORTS.

Figure 8.2
Airline Market Share at Minneapolis/St. Paul

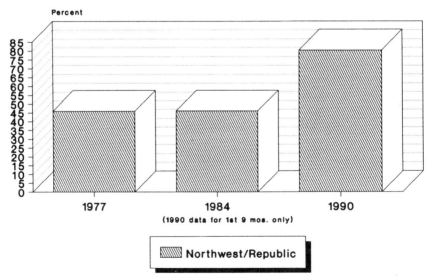

Sources: AVIATION DAILY, Apr. 19, 1985, at 28; Feb. 1, 1990, at 230; Apr. 29, 1990, at 628; Mar. 29, 1991, at 590; and CONSUMER REPORTS.

- The first U.S. airline to hire Japanese employees.
- The first airline to operate the new-technology Boeing 747-400 and the first to fly it on the 6,800-mile New York–Tokyo route.[5]

Northwest is a blend of several former local service airlines that merged after deregulation to form Republic—North Central, Southern, and Hughes Airwest. North Central merged with Southern to form Republic in 1979, then acquired Hughes Airwest the following year for $45 million.[6] With its $884-million acquisition of Republic, Northwest became an amalgamation of aircraft, hubs, labor, and management styles. The merger caused its problems:

Merging two companies of similar size but totally different management styles which previously competed tooth and nail has been fraught with practical difficulties. Those problems have resulted in one of the worst on-time performance records in the industry, and thousands of complaints from irate passengers suffering from delayed or cancelled flights, and mishandled baggage.[7]

Service problems in the 1980s led many of its passengers to dub it "Northworst."[8] The merger also festered labor sores as 16,400 Republic workers joined 17,500 Northwest employees in a long and difficult blending of seniority lists.[9] As Chairman and CEO Stephen Rothmeier said: "The resulting culture is neither Northwest nor Republic. Clearly, that doesn't come easily."[10] But Republic gave Northwest several significant assets: small aircraft to provide domestic feed for its international routes and, again, a stranglehold on several domestic hubs.

With the leveraged buyout in 1989 by Alfred Checchi, Northwest has been saddled with more than $3 billion in debt to pay for his acquisition. Checchi and his partners invested only $40 million of their own, for which they received half the common stock of Wings Holdings, Inc., established as the holding company. In the process, Northwest's debt-to-equity ratio rose from 0.42/1 to 5.85/1. Much of the LBO was financed by KLM Royal Dutch Airlines, which began pooling operations with Northwest from Minneapolis to Amsterdam. In an unprecedented move by the U.S. Department of Transportation in 1991, KLM was allowed to retain a 49 percent equity and debt interest in Northwest (traditionally, foreign ownership has been limited to 25 percent, the statutory standard for voting control).

Significant aircraft orders (229 aircraft will be delivered during the 1990s) have also laden the company with debt. This, of course, would make the airline vulnerable during a long economic downturn.[11] Because the airline is privately held, its true financial condition has been obfuscated.

Neither Checchi nor the man he installed as president, Frederic V. Malek, have airline experience. Both spent much of their careers in the Mariott Hotel chain.[12]

Although not well loved by many passengers, for whom Northwest is their only choice and who grumble about the ticket prices and domestic service, before 1990 Northwest had been consistently profitable each year after 1949.[13]

NOTES

1. R. Davies, Airlines of the United States since 1914 555 (1972). Northwest Airlines incorporated in 1934 to replace Northwest Airways. That year, it also purchased the Chicago-Fargo route from Hanford's Tri-State Airlines. *Id.* at 197.

2. NWA Inc., Annual Report 1 (1988).

3. *Northwest Shaping Strategy,* Airline Bus., June 1988, at 16.

4. *Id.; Northwest Changes Direction,* Flight International, June 11, 1988, at 33.

5. NWA Inc., Annual Report 8 (1988).

6. P. Dempsey, Law & Foreign Policy in International Aviation 146 (1987).

7. *Northwest Labour Pains,* Airline Bus., June 1988, at 19.

8. Reiff, *Being Down So Long It Looks Like Up to Me,* Forbes, June 13, 1988, at 96.

9. Salpukas, *Northwest Air Turns the Corner,* N.Y. Times, Jan. 7, 1989, at 36, 37.

10. Donoghue, *Northwest: Rough Ride to a Competitive Fortress,* Air Transport World, Oct. 1988, at 28, 31.

11. *Id.* at C5.

12. *Id.*

13. *See* NWA Inc, Annual Report 4 (1988.

PAN AMERICAN WORLD AIRWAYS

Hub: New York Kennedy

Mini-hub: Miami

Post-deregulation Merger: National (1979); Texas Air, Boston–New York–Washington Shuttle (1986)

Computer Reservations System: None

Rank and Market Share: 1978—fifth, 9.3%; 1990—eighth, 6.9% (ceased operations in 1991)

In the mid-1920s, Dr. Peter Paul Von Bauer, the managing director of Sociedad Colombo-Alemana de Transportes Aereos (SCADTA), flew a delegation from Barranquilla to Key West via Central America and Cuba in two flying boats, supplied by Kondor Synidkat of Berlin. Von Bauer went to Washington to negotiate landing rights in Florida and the Canal Zone. The idea of a German-sponsored airline based only a few hours' flight from the United States caused concern in Washington. Von Bauer's request was denied by the State Department,[1] which would give full support only to a U.S.-controlled airline.[2]

The following year, a group of "gentlemen flier" Yale classmates led by Juan Trippe, who had operated Long Island Airways in 1923, exhibited an interest in an air route in the Caribbean.[3] After losing out on a New York–Boston airmail contract to Colonial Air Transport, they formed a temporary alignment with their rival but soon became frustrated at the reluctance of the New England company to extend its ambitions south of New York. So Trippe and his friends looked southward at the broad horizons and the potential of an airline market in Latin America.[4] Trippe's group formed the Aviation Corporation of America on June 2, 1927, and six days later, Trippe formed Southern Air Lines Inc., later to become New York Airways.[5] On July 16, 1927, J. K. Montgomery, who was earlier influenced by Von Bauer and his group, Pan American, won the coveted mail contract for the Key West–Havana route.[6] On October 11, 1927, the Reed Chambers-Ed Hoyt group reincorporated Southeastern Air Lines Inc. as Atlantic, Gulf and Caribbean Airways.[7] These three separate financial groups competed for the high stakes of Latin American air routes.[8]

In 1927, President Calvin Coolidge recommended a comprehensive system of airmail services to South America. That year, Congress passed the Foreign Air Mail Act, and the postmaster general advertised for bids on a wide-ranging network of mail routes throughout Latin America.[9]

The three groups—Trippe's New York Airways, Montgomery's Pan American, and the Chambers-Hoyt group—merged on June 23, 1928, forming the Aviation Corporation of the Americas.[10] Forty percent of the new company's stock was held each by Trippe's and Hoyt's groups and 20 percent by Montgomery's group.[11] On June 27, 1928, the Aviation Corporation of the Americas formed Pan American Airways Inc. as the operating subsidiary.[12]

The new Pan American Airways Inc. was awarded every foreign airmail route for which bids were invited. Pan American precisely fit the concept of a U.S. "chosen instrument" for overseas airmail service.[13]

With Charles Lindbergh as its technical adviser, Pan American began passenger service in January 1928.[14] By 1930, after taking on W. R. Grace and Co., an international trading company, as a partner, Pan American quickly swallowed the air routes of Latin America.[15] However, none of this could have occurred without a special relationship with the State Department.[16]

Pan American merged with its competition—New York, Rio & Buenos Aires Line (NYRBA)—in South America.[17] The acquisition of NYRBA sealed Pan American's monopoly of the foreign air transportation in Latin America. All that was left to complete the circuit of Latin America was the Venezuelan overhang.[18]

In 1931, General Gomez granted rights to open service between Maracaibo and Port of Spain, thus enabling the airline to fly a ring around the Caribbean. Pan American was now the world's largest international airline, operating 21,000 miles of routes through 29 countries in the Western Hemisphere. In contrast, the entire domestic route system of the United States, operated by more than a dozen carriers, was 30,451 miles.[19]

During the depression, Pan American continued to expand.[20] As Pan American looked forward to spanning the globe, the fears of engine failure over jungles and oceans haunted both the passengers and the pilots flying the aircraft. To alleviate those fears, Pan American constantly improved the navigation and safety of its aircraft.[21]

In 1931, Pan American carried 820,000 pounds of mail and cargo and

45,000 passengers and flew 12,479,000 passenger miles. That year it showed its first profit of $105,000 on revenues of $7.9 million, after a loss of $700,000 in its first two and a half years of operation.[22] Postmaster General Brown convinced Trippe to give up the domestic field, so Trippe did as instructed and sold New York Airways to Eastern Air Transport.[23]

In the mid-1930s, retired Postmaster General Brown was charged with conspiracy and collusion in the awarding of airmail contracts.[24] The postmaster general canceled the domestic airmail contracts, and the army was directed to fly the mail. A spate of crashes revealed that the army was incompetent to carry the mail; consequently, the domestic mail routes were reopened for bidding.[25]

Even though Pan American was not a part of the domestic mail route system, it did catch heat in Washington for possible illegalities involved in the awards of Pan American's foreign airmail route contracts.[26] Fortunately for Pan American, foreign communication links were a necessity, and the U.S. military could not realistically hope to cross foreign territories on a daily basis. Thus, Pan American kept its foreign airmail contracts, but some had to be modified.[27]

On the completion of the first passenger flight across the Pacific, in October 1936, Trippe sent the following message to President Franklin Roosevelt from the cockpit of the *China Clipper:* "We are glad to inform you that this first flight was made during your administration by an American company with [an] aircraft built in [the] U.S. and in charge [an] American captain and his five flight officers."[28] Trippe was a consummate diplomat, stroking domestic politicians and negotiating his own "treaties" with other governments.[29] Pan Am was sometimes referred to as "the other State Department."[30] However, Japan's movements suggested that a war was brewing, and Japan opposed an American airline flying over or too near to its territory.[31]

While Pan American was conquering the Pacific, it also had its eyes on conquering the Atlantic.[32] In the late 1930s, Pan American and Imperial Airlines obtained a 15-year permit to operate over the Atlantic. Trippe obtained this authority with the recommendation of the British director of civil aviation, without actually being chosen by his own government.[33] This reciprocal agreement gave both airlines the authority to fly between the United States and Britain via Newfoundland and Bermuda.[34]

As of January 1, 1939, Pan American served 54,072 route miles in 47 countries with 126 aircraft, 145 ground radio stations, and 5,000 employees. That year, Pan Am began the world's first regularly scheduled transatlantic airplane passenger service. Pan American and its subsidiaries constituted, by far, the largest airline operation in the world.[35] At this time, however, the company was in poor financial shape, losing heavily in the Pacific and investing heavily in the Atlantic with no return as yet. Large loans had been needed, resulting in a board heavy with bankers.[36]

World War II broke out, and Pan American's passenger load boomed.[37] In the first two weeks of the war, 10,000 Americans in Europe wanted to fly home. Mail loads increased to a staggering amount. Profits soared, and Pan American purchased six more Boeings, making twelve in all.[38]

During the war, over half of Pan American's system was converted to military use.[39] Trippe was offered a job as the general of the U.S. Army Air Corps, but declined.[40]

Before World War II, the international air commerce of the United States was almost exclusively the domain of Pan American and its affiliate, Pan American–Grace Airways. The authority for these carriers to fly to Europe, Asia, the Caribbean, and South America had been granted by private agreements between the airlines and the governments of the foreign nations to which they flew. As the war wound down, the CAB announced that treaties henceforth would be negotiated by the CAB and the U.S. Department of State and not by individual airlines.[41]

By 1945, Pan American had 13 terminal cities: Boston, New York, Philadelphia, Baltimore-Washington, Miami, Seattle, San Francisco, Los Angeles, Detroit, Chicago, New Orleans, and Houston. Trippe proposed to link them with high-speed, nonstop service.[42] The only real competition Trippe feared was foreign competition.[43] In 1947, Pan Am began the first round-the-world service but was prohibited from providing the New York-San Francisco domestic link.

Pan American's proposal for routes inside the United States was opposed by all 17 domestic airlines.[44] Pan American upgraded its fleet with larger and more powerful aircraft in anticipation that the CAB might grant Pan American its domestic express route applications.[45] The CAB, however, did not allow Pan American to provide domestic U.S. operations due to the harm anticipated to the other domestic airlines.[46]

Pan American eventually purchased American Overseas Airlines from its parent, American Airlines. The CAB denied the merger, but the CAB's decision was reversed by President Harry Truman. As a result, Pan American secured access to Paris and Rome.[47]

Before the first jets flew,[48] a civil antitrust suit brought by the Department of Justice complained that Panagra, Pan American, and W. R. Grace and Co. were engaged in a conspiracy to monopolize air commerce between the United States and South America. In 1956, Pan American served 111 cities while TWA served only 26. In 1959, Pan Am became the first airline to offer scheduled turbojet service. Although 14 U.S. airlines now operated overseas, Pan American flew 54.4 percent of all overseas route mileage flown by U.S. airlines and carried about 59 percent of the passengers and 60 percent of the air freight. Pan American's capital investment amounted to 67 percent of the international industry total, while TWA's capital investment was only 11 percent.[49] Pan American was the largest

U.S. company engaged in air transportation in terms of assets. No statistics adequately reflected Pan American's economic power, for it owned or controlled 49 affiliated companies, of which 12 were airlines.[50] Trippe was sitting on top of a billion-dollar conglomerate in the 1960s with hotels, missile ranges, business jets, a midtown Manhattan office building, and the mightiest international airline in the world.[51]

By 1964, Pan American was flying to Europe 214 times a week, up from 170 flights in 1958 in propeller planes carrying half as many people. In 1965, there were 258 flights a week to Europe and 152 across the Pacific. By 1966, traffic had risen another 25 percent. The planes now touched down in 118 cities, and the net profit was $132 million.[52]

By now Pan American was so vast that Trippe had become a difficult individual to see. Trippe's subordinates would sometimes wait days, even weeks, before seeing him. Trippe ruled absolutely. Trippe was a hard worker, who worked long hours straight through, and he expected the same out of his executives. When he took a lunch, it was not a social occasion. A social occasion to Trippe was his annual Christmas party, for one social event was enough.[53] Trippe lacked a sense of time, a consideration of others and their schedules.[54] Pan American employees grumbled about low pay, except for the pilots.[55] Pan American was like a government service. The company appealed to those who liked to give public service and to those who were attracted by the glamor of the airline business and who believed that they were building and serving.[56]

Trippe boldly placed orders for 25 DC-8s and 20 707s, ushering in the jet age in international aviation.[57] By the late 1960s, Trippe was looking forward to supersonic air transport; but more so, he was looking toward the development of a new aircraft by Boeing—the 747.[58] Pan Am had always challenged the manufacturers to produce long-range equipment.[59] Indeed, it was Trippe's vision that persuaded Boeing to launch the B-747; he placed an order for 25 in 1966.[60] Thus came the dawn of widebodied aircraft.

At a stockholders' meeting in 1968, Trippe announced his successor, Harold Gray. By then, Pan American had 40,000 employees. Its fleet numbered 143 multi-engined jets, with a billion and a half dollars' worth of planes on order. Pan Am still had hotels, business jets, airports, and a midtown skyscraper.[61]

As Trippe departed, Pan American began to lose money. In 1969, Pan American lost $26 million, and in 1970, $49 million. Trippe shouldered part of the blame as the press claimed he had bailed out when he foresaw these losses coming.[62] Throughout the 1960s, overseas air traffic had increased at nearly 25 percent per year. Market projections pointed to continued growth.[63] Unfortunately, traffic stagnated. A two-year recession hit at the beginning of the 1970s, which led to acres of empty seats. Thirteen domestic and foreign airlines competed across the Pacific, 23 in Central and South America, and 29 across the North Atlantic. Pan American's share of overseas traffic dropped sharply.[64]

The ruinous competition Trippe had predicted had at last emerged. Unanticipated increases in labor and maintenance expenses and higher fuel costs also contributed to Pan American's financial woes.[65]

A few domestic routes might have offset all this. Though Pan American continued to apply for domestic routes from time to time, none were ever awarded. On top of the recession, the increased competition, and the unfavorable CAB decisions rode Pan American's monumental debt due to the purchase of new 747s and the cost of a new Kennedy Airport terminal building, which amounted to nearly a billion dollars borrowed at then near-usurious rates of 11 percent.[66] As a result of all this, Gray laid off 2,000 employees.[67] Gray had the extremely orderly mind typical of an engineer, but he was raw in public relations. Gray lasted only 18 months and was succeeded by Najeeb Halaby.[68]

Pan American continued to lose money through the 1970s. Halaby, in panic, began firing executives and bringing in new executives who were inexperienced in the airline business.[69]

The president and chief operating officer under Halaby, William Seawell, was promoted to chairman and replaced Halaby. Seawell began to reduce the tariff structure, to abandon unprofitable routes such as the Caribbean, and to cut the personnel roster almost in half. Contradicting Seawell's moves was the Mideast oil embargo, which caused fuel prices to skyrocket. Seawell traded off routes with TWA, abandoning Paris and Rome but gaining exclusivity in Germany and the Pacific.[70] Pan American continued to lose money. At last, Seawell considered declaring bankruptcy.[71]

Pan American lost money for eight consecutive years, until 1977, when it finally showed a profit. Seawell's business tactics had turned the situation around, and he temporarily saved the airline.[72]

In 1979, President Jimmy Carter approved Pan American's $308-million acquisition of National Airlines, which served the entire U.S. East Coast and the West Coast via a southern transcontinental route. Frank Lorenzo had secretly begun buying the stock at $26 a share and ultimately walked off with $47 million in arbitrage.[73] Pan Am offered $41 a share, but after Eastern's Frank Borman entered the bidding war, the price was bid up to $50 a share, which is what Pan Am paid.[74] By the 1980s, Pan American at long last had a domestic artery to pump traffic into its international system, but the prize was too late in coming.

After the Airline Deregulation Act of 1978, Pan American faced greater competition, greater than any Trippe had conceived in his wildest dreams, when domestic and foreign carriers began to fly the routes that Pan American had once had all to itself.[75] And National's domestic feed was not all Pan Am had hoped. Pan Am's economic problems continued throughout the 1980s, causing it to cannibalize its assets—the Manhattan headquarters building (sold to Metropolitan Life Insurance Company for $400 million in July 1980, then the largest real estate transaction for a single building in history),[76] Intercontinental Hotels (sold to Grand Metropolitan of London for $500 million), its transpacific routes and corresponding aircraft including all the 747SPs (sold to United Airlines in 1985 for $750 million), queues for new Airbus aircraft (16 A320s and options for 34 more sold to Braniff in 1988 for $115 million), and its contract services unit, Pan Am World Services (sold to Johnson Controls in 1989 for $165 million).[77] Yet the losses continued. During the 1980s, Pan Am had two barely profitable years.[78] But it lost $73 million in 1988 and $337 million in 1989.[79] In the 1990s, it sold its intra-German Berlin operations to Lufthansa and its London Heathrow and beyond rights to United Airlines.

Pan Am was able to pick up from Texas Air valuable landing slots and gates in Boston, New York, and Washington to run a shuttle in competition with Eastern. The Justice Department had insisted on divestiture as a condition of approval of the Texas Air acquisition of Eastern.[80] Pan Am sold the shuttle and its remaining European routes to Delta in 1991.

Pan Am's problems were summarized as follows:

Pan American World Airways has survived on money raised by selling assets, but there is general agreement that the carrier will end up slowly liquidating itself unless it can link up with another major airline. . . .

Thomas G. Plaskett, chairman and chief executive, acknowledges that the carrier has no future unless it can link up with a strong airline. Talks between Pan Am management and each of the major carriers have taken place from time to time. Pan Am's international routes, particularly its operations out of Heathrow Airport in London, are especially attractive. But no carrier has been willing to take on the enormous task of renewing Pan Am's aging fleet and integrating its unionized workers with its own employees.[81]

Pan Am entered Chapter 11 bankruptcy in early 1991, and ceased operations late that year. United purchased its Latin American routes in a bankruptcy auction for $315 million. The loss of this pioneer of global aviation was a profound national tragedy. The demise of Pan Am is of the same magnitude as if Ford Motor Company produced its last car, for Henry Ford had the same impact on automotive manufacturing that Juan Trippe had on international aviation. Both were brilliant innovators in their respective industries. As R. E. G. Davies eloquently wrote:

During the 60 years of its brilliant history, [Pan Am] pioneered transocean and intercontinental air routes, it sponsored airplane types which were in the van of technical progress, and as the Chosen Instrument of commercial aviation policy overseas, it became a powerful political force. Without Pan American the course of air transport, even some nations' destinies, would have been different.[82]

NOTES

1. R. Davies, Airlines of the United States since 1914 210–11 (1972).
2. *Id.* at 211.
3. *Id.* at 211; Carlson, *Juan Trippe Took World Travel into Skies,* Wall St. J., June 13, 1989, at B2.
4. R. Davies, *supra* note 1, at 211. Another investment group was also looking for an airline route in Latin America. This group, led by Eddie Rickenbacker and Reed Chambers, had organized Florida Airways early in 1926 and operated a route between Atlanta and Miami with the objective of obtaining a Havana route. Because it did not have a good northern connection, Florida Airways fell into bankruptcy. *Id.* at 211.
5. *Id.* at 212.
6. *Id.* at 212.
7. *Id.* at 212.
8. *Id.* at 212.
9. *Id.* at 213.
10. *Id.* at 213.
11. *Id.* at 213.
12. *Id.* at 214.
13. *Id.* at 214.
14. M. Bender & S. Altschul, The Chosen Instrument: The Rise and Fall of an American Entrepreneur 102 (1982); Carlson, *supra* note 3, at B2.
15. Braniff acquired Pan American–Grace Airways in 1967. R. Davies, *supra* note 1, at 562.
16. M. Bender & S. Altschul, *supra* note 14, at 123. By the summer of 1929, Pan American owned a dozen and a half planes, mostly Sikorsky S-38 amphibians and Fokker-10 land planes. *Id.* at 170.
17. *Id.* at 174–75.
18. *Id.* at 176.
19. *Id.* at 176.
20. *Id.* at 181.
21. *Id.* at 182–83.
22. *Id.* at 190.
23. *Id.* at 191.
24. *Id.* at 215.
25. *Id.* at 219–20.
26. *Id.* at 221–23.
27. *Id.* at 221–23.
28. Carlson, *supra* note 3, at B2.
29. *Id.* at B2. P. Dempsey, Law & Foreign Policy in International Aviation 18, 48 (1987); A. Sampson, Empires of the Sky 44, 46 (1984).
30. Carlson, *supra* note 3, at B2.
31. M. Bender & S. Altschul, *supra* note 14, at 230.
32. *Id.* at 258.
33. *Id.* at 261.
34. *Id.* at 260–61. However, actual commercial flights on these routes over the Atlantic were delayed. During the next year and a half, transatlantic bargaining

took on a vicious character. The contract between Pan American and Imperial was signed on January 25, 1936. The French and the German air officials saw the Pan American and Imperial agreement as an effort to freeze their airlines out of the North Atlantic. *Id.* at 262. So Trippe signed an agreement with Air France in February 1936 for operation of a North Atlantic service between New York and Paris. *Id.* at 262. But because of Hitler's decision to rearm and unleash his anti-Semitic campaign, the United States acceded only to permits for experimental flights of German aircraft. *Id.* at 263. In June 1937, commercial air service between Port Washington and Hamilton, Bermuda, was inaugurated at last. *Id.* at 265.

35. R. DALY, AN AMERICAN SAGA: JUAN TRIPPE AND HIS PAN AM EMPIRE 231 (1980).

36. *Id.* at 232.

37. In 1939, Sonny Whitney temporarily deposed Trippe. *Id.* at 238. Whitney soon tired of the chief executive role and was also aware that he lacked both the knowledge and the dynamism to run a company the size of Pan American. Whitney voluntarily resigned his position back over to Trippe in January 1940. *Id.* at 253. Trippe held a few grudges and promoted those who had previously supported him and demoted those who had not. *Id.* at 253–55.

38. *Id.* at 252.

39. *Id.* at 335.

40. *Id.* at 336.

41. P. DEMPSEY, *supra* note 29, 18–19.

42. R. DALY, *supra* note 35, at 358.

43. His principal postwar job was to reestablish his company's "chosen instrument" status, no matter what the cost. *Id.* at 346.

44. *Id.* at 359.

45. *Id.* at 359. Pan American sometimes engaged in unethical competition with the other airlines in some foreign ports, such as extinguishing the landing lights so that the competitor aircraft could not land. Pan American would also charge its competitors exorbitant landing fees or demand that permission to land be obtained from Pan American's division headquarters in Miami far in advance of the scheduled flights, so far in advance that it became practically impossible for competitors to provide flexible schedules or to change them to meet passenger demand. *Id.* at 372. Trippe denied responsibility for these actions and shifted the possible blame to his divisions, which were run as personal fiefdoms by the men in charge. *Id.* at 372. The late 1940s was a time of depression in the airline business, which brought about merger fever. *Id.* at 384–85. Trippe and Howard Hughes met to discuss the possible merger of TWA with Pan American. This merger never took place. *Id.* at 384–85.

46. *Id.* at 386.

47. *Id.* at 385–86.

48. By 1955, airlines were lining up to buy turboprops like the Lockheed Electras and British Viscounts. Pan American avoided this stepping-stone in converting to jets. *Id.* at 401. Once delivered in 1958, the jets were an immediate success. During the first quarter of 1959, 33,400 passengers were carried on Pan American's jets, with a 90.8 percent seat occupancy, an all-time high. *Id.* at 413–14. During the first five years of the jet age, overseas traffic doubled. *Id.* at 414.

49. *Id.* at 420–21.

50. See P. DEMPSEY, *supra* note 29, at 180; R. DALY, *supra* note 35, at 421. No antitrust proceedings were ever initiated as a result of these hearings, but Pan American bore the scars on its reputation for some time. *Id.* at 421. Pan American purchased a hotel chain, Intercontinental Hotels, sprinkled across 37 countries on five continents. *Id.* at 426–27. Intercontinental had been established in 1946 at the insistence of the U.S. government. R. DAVIES, *supra* note 1, at 562.

51. R. DALY, *supra* note 35.

52. *Id.* at 427.

53. *Id.* at 428.

54. M. BENDER & S. ALTSCHUL, *supra* note 14, at 463.

55. *Id.* at 463.

56. *Id.* at 464.

57. R. DAVIES, PAN AM 78 (1987).

58. R. DALY, *supra* note 35, at 430–32.

59. R. DAVIES, *supra* note 1, at 358.

60. R. DAVIES, *supra* note 57, at 79.

61. R. DALY, *supra* note 35, at 441.

62. *Id.* at 442.

63. *Id.* at 443.

64. *Id.* at 443–44.

65. *Id.* at 444.

66. *Id.* at 444.

67. *Id.* at 444.

68. *Id.* at 444–45.

69. *Id.* at 446–47.

70. P. DEMPSEY, *supra* note 29, at 20–21.

71. R. DALY, *supra* note 35, at 447.

72. *Id.* at 447.

73. P. DEMPSEY, *supra* note 29, at 146.

74. *Aviation Daily Reports on the Decade of the Eighties,* AVIATION DAILY, Jan. 2, 1990, at five.

75. M. BENDER & S. ALTSCHUL, *supra* note 14, at 524.

76. *Id.*

77. *Id.* Nomani, *Ailing Pan Am Still Relying on Asset Sales,* WALL ST. J., Mar. 12, 1990, at A4, col. 2; R. DAVIES, *supra* note 57, at 84.

78. Reed, *Pan Am Hopes Miami Strategy Will End Era of Crisis,* CHICAGO TRIBUNE, Dec. 3, 1989, at 7-22C, col. 5.

79. *Id.*

80. P. DEMPSEY, *supra* note 29, at 147.

81. Salpukas, *Airlines' Big Gamble On Expansion,* N.Y. TIMES, Feb. 20, 1990, at C1, C5.

82. R. DAVIES, *supra* note 57, at 1.

10

TRANS WORLD AIRLINES

Hubs: St. Louis, New York Kennedy

Post-deregulation Merger: Ozark (1986)

Computer Reservations System: an interest in PARS, renamed WORLDSPAN

Rank and Market Share: 1978—third, 11.9%; 1990—seventh, 7.1%

Trans World Airlines (TWA) was formed in 1930 with the merger of Transcontinental Air Transport (TAT) and part of the Western Air Express System (WAE) into what was originally named Transcontinental and Western Air.[1] TWA flew the central mail route from New York to Los Angeles via St. Louis and Kansas City.

By 1938, TWA had slightly more than 1,100 employees, more than half of whom were based in Kansas City, the core of the airline's technical operations. Jack Frye, TWA's president, continuously focused TWA's energies on customer service and technological development in airline safety and navigation.

Even though the DC-3 was the state-of-the-art plane in the late 1930s, Frye realized that the DC-3's limitations in altitude and speed capabilities restricted airline growth.[2] The development of the 307 Stratoliner by Boeing attracted Frye's attention, but Frye was outvoted on buying the aircraft by John Hertz and the Hertz-dominated board of directors. The Hertz-Frye feud threatened to erupt into a full-scale battle for control of TWA's destiny, and it was a war in which Frye was sadly outmanned. Frye was essentially the heart and soul of TWA, but John Hertz was TWA's financial source. Frye's position was weakened in 1937 because for all his service innovations and commitment to technical excellence, the airline had suffered significant losses.

At a directors' meeting at Hertz's home early in 1939, an argument broke out over the type of propellers to be installed on all TWA aircraft. Hertz opposed the propeller installation, and the meeting was quickly adjourned so that Hertz and his friends could attend the horse races. Frye left the meeting furious at Hertz's lack of business concern, and he decided to do something about it. Flying to Los Angeles to meet with Howard Hughes, Frye was to shape TWA's destiny for the next 20 years.

Frye and Hughes had known each other for quite some time. The two were not the best of friends, but aviation was a common interest. Frye explained that he wanted to quit TWA and go to work for Hughes when Hughes bought the airline Pacific Air Transport. Frye told Hughes that he simply would not work for Hertz any more.

Hughes then offered to buy TWA. By the end of January 1939, Hughes had acquired about the same amount of TWA stock as Hertz and the Lehman brothers. Hertz and the Lehman brothers soon sold out to Hughes, and shortly thereafter, Hughes owned around 78 percent of the TWA stock. With controlling interest in TWA, Hughes provided the company with nearly unlimited financial resources.[3]

Howard Hughes had inherited his father's Hughes Tool Company at an early age. It provided the revenue for his diverse ventures. In 1925, Hughes moved to Hollywood. After producing a series of movies, he began production of an epic on the First World War airmen, a movie entitled *Hell's Angels.* Hughes had always been fascinated by airplanes, and by 1928, he was an expert pilot. Hughes observed with awe as the classic planes soared through their aerial scenes for his silent film.[4]

In 1932, after the success of his latest movie, *Scarface,* Hughes thought of making another aerial film. But before that movie idea developed into reality, he had another grand idea. It was typical of Hughes throughout his life to focus all his attention and energy in one field only to abandon it suddenly in favor of another.

Hughes's new goal and obsession was to purchase a small airplane and to remodel and make it the fastest plane in the world. Hiring a crew of airplane mechanics and designers, Hughes established the Hughes Aircraft Company, which would later become one of the nation's largest and most powerful defense contractors.

For years, airplanes were Hughes's hobby. By the summer of 1932, aviation had become the focus of his mental and financial energies. During this period, American aircraft manufacturers were obsessed with conquering time and distance by designing more powerful and faster airplanes. Hughes was caught up in this craze and entered all sorts of flying shows and contests where he broke records in speed and distances traveled, including a transcontinental flight followed by an around-the-world flight.

At TWA, Jack Frye and Howard Hughes worked as a team. The two men were united at first by a piece of machinery, the Boeing 307 Stratoliner, an aircraft that was to remain a significant force in the airline's development. Although TWA had some financial troubles, it was achieving a reputation for technical excellence, safety, and good service—characteristics that dominated the airline until the 1980s.[5]

In 1944, the airline applied to the CAB for a round-the-world route serving more than 20 countries. This postwar expansion plan was larger than that of any other carrier. Frye's optimism was both his own and that of Howard Hughes, who had a keen interest in global flight. During the war, Hughes suggested to Frye that the airline change its name while retaining its initials. In 1945, TWA registered the name Trans World Airlines, but it was not until 1950 that the name change became official.[6]

Coming out of the war, TWA sponsored the development of the Constellation, Lockheed's challenge to Douglas's supremacy. The Constellation's pressurized cabin was more advanced than that of the Boeing 307, and the plane was 80 mph faster than the unpressurized Douglas DC-4. The new Constellations, with their four engines, improved comfort, speed, and reliability and were more profitable with their longer ranges between stops.[7]

Throughout the 1930s, Pan American Airways had held a monopoly over all U.S. international air routes. On June 1, 1945, the CAB announced that both American Export and TWA could compete with Pan American in the North Atlantic.[8] American Export was authorized to serve all of Europe north of the 50th parallel. The rest of Europe was divided between TWA and Pan American.[9]

In 1946, the pilots went on strike and Jack Frye retired, leaving Hughes to run the show for a while.[10] Hughes's participation was successful in major policy decision making during the presidency of Ralph Damon, who had joined Hughes from American Airlines in 1949.[11] Like Jack Frye, Damon was Hughes's working companion and TWA's president until January 1956. Damon's untimely death in 1956 hurt TWA when the vital decisions that were required in ordering the big jet airliners, which were to be the backbone of TWA's fleet for the next 10 years, had to be made without his guidance. In addition to the loss of Damon's help, the industry was beginning to suffer from a minor economic recession in 1958–59.[12]

During this period, Hughes became more of a recluse, seeking privacy to the extent that the next TWA president, Carter Burgess, never met Hughes.[13] Hughes lost control of TWA in 1960.[14] But in 1969 he acquired Airwest, renaming it Hughes Airwest (subsequently acquired by Republic, which itself was absorbed by Northwest).

With the departure of Hughes, Charles Tillinghast directed the airline's fortunes for nearly 16 years, first as president, then as chief executive officer and board chairman.[15] In 1967, in an effort to balance the seasonal fluctuations of air travel, TWA diversified vertically and merged with Hilton International Hotels.[16]

Following Pan American's lead in placing the first Boeing 747 into flight on January 22, 1970, TWA opened transcontinental Boeing 747 service from New York to Los Angeles on February 25, 1970.[17] TWA and Pan Am tried to merge twice in the 1970s, but these negotiations wound up instead in an exchange of routes.[18] TWA surrendered its Pacific routes, thus getting rid of the money-losing around-the-world service it had begun in 1968. TWA also abandoned service at Frankfurt, Hong Kong, Bombay, and Bangkok. Pan Am agreed to pull out of Paris, Barcelona, Nice, Vienna, and Casablanca, in addition to dropping its London service out of Chicago, Los Angeles, and Philadelphia. The CAB authorized the route swap for a two-year period.[19] When the agreement expired, TWA restored service only to Frankfurt. TWA purchased the Barcelona route from Pan Am because it linked up well with its Madrid service. However, TWA eventually abandoned service at the African points of Kenya, Tanzania, and Uganda because it proved economically unfeasible. During this period, TWA was contracting while struggling with heavy equipment commitments.[20]

The 45-day strike of flight attendants in 1973 was a tragedy, showing the gradual deterioration in company and attendant relations, although TWA had been one of the first carriers to employ married women as flight attendants in 1938.[21] By 1975, TWA's domestic operations were suffering from built-in deficiencies. Not having enough of the right airplanes was one of them. TWA's domestic system was the most embattled in the industry because it had no monopoly markets. TWA enjoyed a booming international market. Unfortunately, its inherent domestic vulnerability was a problem.

The resolution was simple but long term. TWA established a long-range schedule-planning department. The plan was twofold. First, TWA wanted to make St. Louis the hub around which most of its domestic system revolved. Second, TWA wanted to develop New York Kennedy Airport as a second major hub, one that would link the domestic system with the international routes. The latter would become a reality several years later, when Pan Am acquired National and sold TWA its JFK terminal directly adjacent to TWA's international terminal.

TWA fought deregulation when it was first proposed, as did most of the airlines.[22] After deregulation, TWA became a leaner carrier by reducing the number of employees and by trimming its fleet.[23] Labor relations at TWA deteriorated due to the resulting employee layoffs. Labor-management relations at TWA have generally been acrimonious since.

Not only did the airline shrink, but the company shed itself of nonairline operations as well. In the late 1970s, Ed Smart was chairman of TWA, which owned the profitable Hilton International and Canteen corporations. They had been acquired to provide a more balanced revenue flow to offset the highly seasonal fluctuating peaks and valleys of the airline's operations. Smart decided to form a holding company, Trans World Corporation, with the airline and the other two companies as subsidiaries. In part, this may have been motivated by the desire to get the Hilton and Canteen profits off TWA's books so that labor would concede more in wage cuts. TWA was given notes for its Hilton and Canteen investments for a fraction of their worth. Rather than expanding the airline's operations, Trans World Corporation acquired other nonairline ventures, including Century 21 real estate and Spartan Foods. So rather than grow, TWA lost market share to the other airlines.

During this period, the board of directors of both TWA and Trans World Corporation were substantially identical. Smart made sure they received generous stock options, even when their main subsidiary, TWA, was losing money and when its employees were surrendering wages and benefits. Ultimately, to fend off corporate raids, one of which was led by Donald Trump, Trans World Corporation spun off its subsidiaries (Hilton International, Century 21, Canteen Corporation, and Trans World Services), leaving TWA, which had provided the capital for the acquisition of the other subsidiaries, dangling in the wind. But soon TWA was to be taken by a corporate raider extraordinaire—Carl Icahn.

In the mid-1980s, TWA became a target of Frank Lorenzo and then Carl Icahn. The pilots surrendered generous wage concessions to Icahn to avoid the dreaded union-buster Lorenzo. The TWA board of directors seemed most concerned that they retain their lifetime term passes on TWA.

After acquiring TWA, Icahn did three major things. First, he executed the existing plan to acquire Ozark, which gave TWA market dominance in St. Louis. As figure 10.1 reveals, TWA's market share at Lambert International Airport soared.

Second, Icahn effectively crushed the flight attendants union when it struck. In demanding wage concessions, he told the flight attendants that whereas the machinists are "bread winners . . . you girls are second incomes."[24] Finally, he took the company private and so highly leveraged the airline (with $2.6 billion in debt) to pursue leveraged buyouts of nonairline ventures (including Texaco Inc. and USX Corporation) that TWA had a negative net worth, had the oldest fleet of aircraft in the industry, and was not reinvesting its profits. Prodded by labor, Icahn placed $3 billion in orders for new aircraft in the late 1980s, although the first planes were not scheduled to roll off the assembly line until 1994.[25]

In 1990, Icahn offered to buy a number of narrowbodied aircraft if the pilots would surrender another $80 million in concessions. In the shadow of his sale and then lease of eight L-1011s and three B-747s, Icahn threatened to cannibalize more aircraft if the pilots didn't succumb to his demands. He had already sold the Chicago-London route, gates, and landing slots to American.

The pilots, whose contract did not expire until 1992, estimated the cuts

Figure 10.1
Airline Market Share at St. Louis

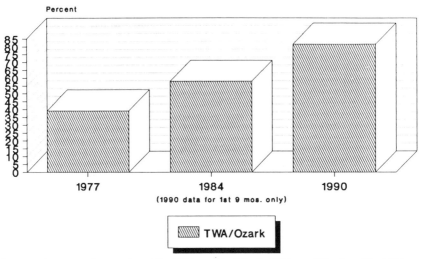

Percent

1977   1984   1990
(1990 data for 1st 9 mos. only)

▨ TWA/Ozark

Sources: AVIATION DAILY, Apr. 19, 1985, at 28; Feb. 1, 1990, at 230; Apr. 29, 1990, at
628; Mar. 29, 1991, at 590; and CONSUMER REPORTS.

at between 30 percent and 40 percent of their salaries, which would put
them near the bottom of the industry, slightly above nonunion Continental. Kent Scott, head of the TWA pilots union, said: "There is only one
word that describes the tactics being used here, extortion. You and I are
being asked to pay protection money to keep our airline intact and keep
it from being scattered into the winds."[26] Scott said that Icahn chose "to
squander the $1.5 billion in total concessions that TWA's labor group
gave him by crippling the company with debt in his privitization transaction of 1988."[27] In response, TWA announced that it was selling two B-
747s and selling and leasing back six B-767s.[28] The fleet was shrinking
rapidly. In 1991, TWA sold to American Airlines several of its primary
transatlantic routes to London Heathrow Airport.

Thus, with Hughes, Frye, and Tillinghast, TWA has consistently been
bled of its capital and resources to finance nonairline ventures, and this
once magnificent international airline has consequently lost market share
to its rivals. All the while, employee morale has disintegrated. As a senior
TWA pilot recently observed:

The entire physical plant of what is still TWA is deteriorating at a rate never even
approached in the past. I'm not talking about the aircraft . . . everyone knows we
have the oldest fleet in the industry . . . I'm speaking about the rest of the company. Have you walked through the terminals in New York or St. Louis lately?
The latest Business Week mentions that the president of US Air inspects and makes

notes of the carpet and upholstery as he walks through his terminals when he travels. If he were with us he would need a full time stenographer to keep track of the notes! . . . Employee morale and pride is non-existent and most of us have been here a lot longer than some of the management and can remember when we were proud to say we worked for TWA. Now when asked if I fly for an airline, I usually respond in my N.Y. smart alec way, "no, I'm with TWA." . . .

Management should be thankful that the ontime performance does not include mechanical delays or we would be the joke of the industry in yet another category! Doesn't anyone back there know what is going on and try to tell who ever is supposedly steering this drifting carcass of what had once been a decent airline that we are sinking rapidly and that we need some help? We used to have excellent maintenance and I really believe that the people are still outstanding for the most part, but we need parts and planes if we are going to survive. . . .

Doesn't anyone at [TWA's headquarters at] Mt. Kisco understand that there are a lot of good people out here, wanting and trying to do a decent job but are fed up with all the lies and misdirection that is going on. Make no mistake about it . . . despite what comes out of the office of the president, TWA is not close to being a premier airline and will not be a survivor the way we are going. . . .

Have you looked at the old planes; have you looked at the paint jobs on most of the Ozark fleet? Even Earl Schieb would be embarrassed. . . . I close with a simple question that has been going around the airline: "Do you know why people are still flying TWA? Because they are born faster than we can piss them off."[29]

The pilot was referring to the low esteem in which TWA is held by its passengers. Recently, TWA has ranked among the major airlines with the highest level of consumer complaints per 100,000 passengers.[30] As this book goes to press, TWA entered Chapter 11 bankruptcy.

NOTES

1. R. DAVIES, AIRLINES OF THE UNITED STATES SINCE 1914 89 (1972).
2. *Id.* at 93. TWA's original transcontinental mail route stretched from New York to Los Angeles, connecting 12 intermediate cities: Philadelphia, Harrisburg, Pittsburgh, Columbus, Dayton, Indianapolis, St. Louis, Kansas City, Wichita, Amarillo, Albuquerque, and Winslow.
3. *Id.* Howard Robard Hughes was born on Christmas Eve of 1905. Hughes's early life was transient and unsettled as his father followed the oil rush from one strike to the next. In 1909, Howard Hughes, Sr., designed and patented a revolutionary drill bit, which was later to make Howard Hughes, Sr., and Howard Hughes, Jr., incredibly wealthy by forming what was to become the financial foundation for the Hughes Tool Company, later to be renamed Toolco.
4. *Hell's Angels* was modified by adding sound and was a box office hit. Hughes became a popular man in the glamorous society of Hollywood. However, no matter how popular Hughes became, he still remained shy, uncomfortable around strangers, and eccentric. And what was worse, the phobias instilled by his mother were maturing within, compounded by the compulsive striving for perfection in all his projects.

5. *Id.* at 368.

6. R. SERLING, FROM THE CAPTAIN TO THE COLONEL at 115 (1980).

7. R. DAVIES, *supra* note 1, at 291. Up until 1940, the DC-3 was the standard aircraft. On March 1, 1946, TWA placed the Lockheed Constellation on the transcontinental route from New York to Los Angeles, the first civil airline to challenge Douglas's supremacy. *Id.* at 283–29. TWA developed its Super Constellation on September 10, 1952. *Id.* at 332–33.

8. P. DEMPSEY, LAW & FOREIGN POLICY IN INTERNATIONAL AVIATION 19 (1987).

9. R. DAVIES, *supra* note 1, at 366–67. Both TWA and Pan Am were granted onward rights to India. TWA was granted a southern route via Cairo to Bombay. Pan Am was granted the route via Istanbul to Karachi and Calcutta. American Export and Pan Am were given the right to fly to London. TWA was given the right to fly to Paris. P. DEMPSEY, *supra* note 8, 18–23.

10. R. DAVIES, *supra* note 1, at 368. Hughes emerged from World War II with a magnificent route award that gave TWA the distinction of being the only airline with both transcontinental and transatlantic traffic rights. *Id.* at 535.

11. *Id.* at 535.

12. *Id.* at 536.

13. *Id.* at 536.

14. *Id.* at 537. After Howard Hughes departed, TWA appointed two new leaders, Ernest Breech and Charles Tillinghast. One of Breech and Tillinghast's first decisions was to file an antitrust suit against Hughes Tool Company, Toolco, on June 30, 1961. *Id.* at 537.

15. R. SERLING, *supra* note 6, at 311.

16. *Id.* at 316; R. DAVIES, *supra* note 1, at 563.

17. R. DAVIES, *supra* note 1, at 573; R. SERLING, *supra* note 6, at 311.

18. P. DEMPSEY, *supra* note 8, 20–21.

19. CAB Order 75-1-133 (1975).

20. R. SERLING, *supra* note 6, at 318.

21. *Id.* at 320.

22. *Id.* at 323.

23. *Id.* at 323.

24. *Supreme Court Upholds Rulings for Management over Unions,* AVIATION DAILY, Jan. 17, 1990, at 115.

25. Salpukas, *Airlines' Big Gamble on Expansion,* N.Y. TIMES, Feb. 20, 1990, at C1, C5.

26. See *TWA Chairman Offers Pilots New Aircraft for Concessions,* AVIATION DAILY, Feb. 16, 1990, at 334.

27. *TWA Pilots Call Aircraft Sale a Major Step toward Liquidation,* AVIATION DAILY, Feb. 27, 1990, at 400.

28. *Id.*

29. For obvious reasons, the identity of the pilot will not be disclosed.

30. See *Rankings of U.S. Carriers Consumer Complaints Per 100,000 Passengers,* AVIATION DAILY, Mar. 8, 1990, at 474; and *Consumer Complaints against U.S. Carriers Reported to DOT, By Category,* AVIATION DAILY, Jan. 10, 1990, at 70.

11

UNITED AIRLINES

Hubs: Chicago, Denver, San Francisco, Washington Dulles

Post-deregulation Acquisition: Pan American's Transpacific Routes (1985); Pan American's Transatlantic authority to London Heathrow and beyond (1991)

Computer Reservations System: Apollo

Rank and Market Share: 1978—first, 17.4%; 1990—second, 16.0%

United Airlines has always been a massive enterprise. It was organized in 1931 as "the World's Largest Air Transport System."[1] Boeing Air Transport was a predecessor of both United and today's Boeing aircraft manufacturing giant.[2]

The intercorporate structure of the United companies was based on the consolidation of several corporations under United Aircraft & Transport Corporation, a holding company formed on October 30, 1928, under the name of Boeing Airplane & Transport Corporation.[3] In general, United's system in 1931 consisted of airmail routes linking New York and Chicago, Chicago and San Francisco, Chicago and Dallas, Salt Lake City and Seattle and Spokane, and Seattle and San Diego.[4]

The Air Mail Act of 1934 contained prohibitions against holding companies owning stock in airmail contractors and against airmail contractors holding stocks of companies engaged in aviation except those operating airports and other ground facilities. As a consequence, United Aircraft & Transport Corporation was put into dissolution that year, and its holdings were transferred to three new companies it had previously formed. Its air-transport holdings were transferred to United Air Lines Transport Corporation, one of the three new companies. On December 28, 1934, United

Air Lines Transport Corporation, the former United Air Lines, Inc., and two of the old transport companies were merged and consolidated into United Air Lines Transport Corporation, which thereby became the operator of the United system. In 1943, the name of the company was changed to United Airlines, Inc.[5]

In 1933, shortly after W. A. Patterson assumed the vice presidency of the United Airlines management company, the directors decided to merge Varney Air Lines, purchased three years before, into Boeing Air Transport, of which Patterson was also president.[6] Patterson became president of United Airlines in 1934.[7] One of Patterson's axioms was that he learned the airline business from the people running the airline.[8] Patterson spent more than half of his first year as president and over half of the six years that followed talking with people in the airline system. These people included everybody from janitors to superintendents, and they had plenty of ideas. Many of the better suggestions flowered into standard airline procedures— safety practices such as guaranteed base pay for pilots whether they flew or not, special fares to induce wives to fly with husbands, and flying around or over the weather, to mention only a few.

As late as 1945, Patterson made it a practice to try to see every United employee at least once a year. Thereafter, this program overwhelmed him. The United family had grown too large. Patterson negotiated strikes by listening to his employees' grievances and then solving the problem with a solution agreeable to both sides—the employees and the company.[9] Patterson was an apostle of pensions, sick benefit pay, and guaranteed income for pilots, and he gave an all-around "best friend" treatment to employees.[10]

The business was greatly enlarged during the later 1930s and the 1940s, largely as a result of the passage of the Civil Aeronautics Act of 1938 and World War II. First, the Civil Aeronautics Act gave rise to an era of route expansion, and second, war production and wartime mobilization created unprecedented and virtually insatiable demands for air transportation.[11]

As early as 1943, plans had been drawn up for a postwar route-expansion program, and as of January 1, 1946, the company had on file with the Civil Aeronautics Board various applications designed to add 59 cities to its route system, exclusive of Mexican points served by LAMSA, United's foreign subsidiary. At that time, the company's route system covered 6,341 unduplicated miles, compared with the 5,313 miles acquired by the corporation on August 22, 1938, when certificates of "public convenience and necessity" were first issued by the Civil Aeronautics Board.[12]

After World War II, United's operation policies were summed up in what the company called its "Rule of Three"—Safety, Passenger Comfort, and Schedule Dependability. In December 1952, this motto was expanded to the "Rule of Five," with the addition of Honesty and Sincerity.[13]

In 1947, United acquired a Los Angeles–Denver route from Western Air

Lines, thereby gaining direct entry from the East into Southern California.[14] In the same year, United placed the pressurized cabin DC-6 into coast-to-coast service and began San Francisco–Honolulu service. In 1948, an operations hub opened at Denver, setting a new pattern for centralized control of airline activities.[15]

United was the first domestic airline to order jetliners, 30 Douglas DC-8s, costing $175 million. United then completed the largest single airline-financing program in history—about $150 million over a period of five years of expansion into the jet age.[16] In 1956, United retired the last of its DC-3s from its fleet.

As the 1960s began, United completed the construction of a new $14.5-million passenger terminal at New York International Airport, and it also placed DC-7A Cargoliners in service. In 1961, United acquired Capitol Airlines, making United the nation's largest airline. United also placed the new Instamatic reservations system into operation, which was the largest interconnected electronic data-processing facility ever built for business use.[17]

The old-timers who had led the airlines to the heights were beginning to retire, and Patterson was no exception. He left United in 1966.[18]

In the mid-1960s, there was substantial competition between the carriers. Initially, this took the form of equipment competition, as new jet aircraft and, later, widebodied jet aircraft lowered operating costs and appealed to passengers. Toward the end of the 1960s, the airlines shifted their competition toward schedule and capacity contests and obtaining favorable regulatory decisions. Long-haul routes were particularly desirable because, although fares tapered with distance, the operating costs of large, widebodied jets tapered even more sharply. Thus, a carrier with a substantial market share on such routes could achieve high load factors and high profits. The most desirable long-distance routes at the time stretched across the Pacific.[19]

The CAB's *Transpacific Route Investigation* in the 1960s was a central event in this competition. Traditionally, in its route policies, the CAB attempted to strengthen the weaker carriers by giving them access to the more lucrative routes. The CAB would also give routes to the carriers that could best develop a route system. Therefore, the CAB's search for evidence of route-development ability focused on such factors as a carrier's ability to offer low prices, ability to provide accommodations, and knowledge of the local market. Possession of a fleet of low-cost widebodied jet aircraft was requisite for offering low prices; travel-related diversification, such as hotel ownership, was seen by the carriers as necessary to develop the market.

In the original 1961 transpacific award, the CAB had allowed United, Pan American, and Western to serve the U.S. mainland–Hawaii route. A fourth applicant, Hawaiian Airlines, was denied certification because it had no jet equipment. This case was reopened in 1966 because the CAB was disturbed about high fares, high load factors, and high traffic growth. By this time, all carriers had jets, and almost every domestic airline was a party to the case; a total of 18 carriers applied for nearly every major Pacific route. Many undertook diversification as part of this effort. During the mid to late 1960s, all of the large carriers and two of the smaller ones cited hotels and other travel-related facilities as part of their route-development efforts. United developed hotels in an effort to secure or protect coveted routes.[20]

The transpacific route awards led to disastrous consequences for United and Pan American, which had earlier dominated the Hawaiian market. All new carriers assigned their new widebodied, long-distance aircraft to the prime Hawaiian routes. This created severe overcapacity as competition ratcheted up schedule frequencies. In the year of the route awards, United, which was the dominant carrier on the Hawaiian routes, saw its earnings on these routes plunge from a $19-million profit in 1969 to a $17-million loss in 1970.

As a response to these losses, United diversified even further.[21] The major reason for United's diversification was that the airline perceived itself as severely hurt by regulation. Rexford Bruno, United's former senior vice president for finance and administration, believed that regulation was stifling United's growth. Because United was the largest airline, the CAB gave it no new routes, instead giving new routes to the smaller carriers. Bruno cited the CAB decision in the transpacific route case as particularly damaging to United because the CAB let five new competitors into United's very lucrative Hawaiian routes. The CAB also ruled that because of its size, United should not be considered for any new routes to the Far East. United acted to alleviate its regulatory problems by pushing for deregulation and diversifying.[22]

Although some diversification had taken place in the air carrier industry, until the late 1960s, the air carriers were parent companies to the diversified subsidiaries, so the entire companies remained subject to CAB regulation. This changed, however, in 1970, when United Air Lines reorganized into a subsidiary of a newly formed holding company, UAL, Inc., and acquired Western International Hotels, a sizable chain that has since been renamed Westin.[23]

With United's vigorous support, Congress passed the Airline Deregulation Act of 1978, which dismantled the system of economic regulation under which U.S. airlines had operated, explicitly leaving the industry to the forces of the marketplace. Regulation of safety matters was ostensibly retained by the U.S. Federal Aviation Administration.

The elimination of entry, pricing, route, and other restrictions that had prevented or impaired competition in U.S. domestic airline markets transformed the U.S. air transportation industry. Freed of CAB control, airlines

Figure 11.1
United vs. American, 1979–1989 (Operating Earnings/Loss)

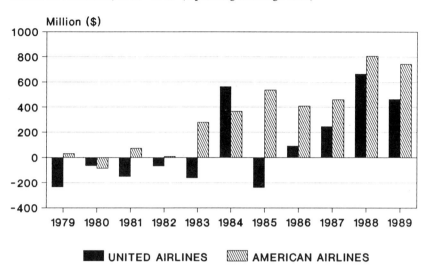

■ UNITED AIRLINES   ▨ AMERICAN AIRLINES

radically realigned their route structures, explored new pricing strategies, adjusted their aircraft fleets, developed a range of new competitive weapons including computerized reservations systems and frequent-flier programs, and consummated a rash of mergers and acquisitions.[24] Led by United and American Airlines, U.S. airlines introduced increasingly sophisticated computerized reservations systems.[25]

The airlines have had very mixed results in their transition into deregulation. United, for one, has been somewhat successful.[26] American and United were large carriers that developed wide-market, quality-service differentiation strategies, drawing on their corporate diversification, as needed, for resources. Both entered deregulation with relatively costly operations and large amounts of debt. Both airlines ultimately achieved success under deregulation by developing strong, multiple-hub route systems.[27] But as figures 11.1 and 11.2 reveal, American has generally outperformed United in profitability and in growth rates, overtaking United as the industry leader in the late 1980s.

United's early deregulation strategy was problematic, but after a mid-course correction, the airline emerged from the transition period with a successful wide-market, quality-service competitive strategy resting on multiple, strong hubs. Initially, United drove toward a route structure that would have been ideal under regulation but was imprudent in a competitive airline industry: United increased its long-haul routes and dropped many of its smaller, short-haul feeder markets. In support of this move, it

Figure 11.2
United vs. American, 1979–1989 (Net Earnings/Loss)

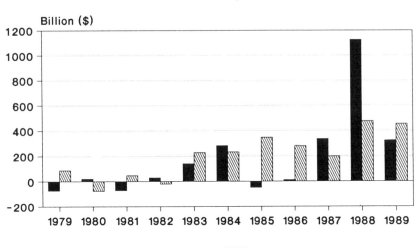

made plans to sell its short-haul B-737 airplanes and to move into longer-range aircraft. However, it retained large connecting hubs in Chicago and Denver.[28]

By 1981, United was facing mounting losses and seeing new competition move into its exposed long-haul routes. In response, United started the high-density, low-cost, point-to-point Friendship Express service. At this stage, United was pursuing several strategies at once, attempting to position itself as both a low-cost and a high-service carrier.

By 1983, faced with continuing poor performance, United shifted strongly toward a multiple-hub feeder-route system. United focused its route-development efforts on strengthening its primary hubs at Chicago, Denver, and San Francisco by adding spokes, increasing frequency, and adding one-stop flights. The carrier's marketing efforts largely paralleled those of American. It developed a ubiquitous travel-agency computer reservations system (Apollo) and strengthened its frequent-traveler program, Mileage Plus. Those provisions, along with overall economic prosperity, returned United to profitability in 1983. (See figures 11.3 and 11.4.)

As UAL focused on repositioning United, it slowed its diversification so that the airline's proportionate share of the corporation's revenues began increasing. Bruno, the former senior vice president of UAL, stated that deregulation had caused UAL to change its diversification strategy: it stopped searching for acquisitions because it saw many opportunities in the airline business as well as in its hotel and insurance businesses. However, UAL did not divest itself of its nonairline businesses for several reasons: United

Figure 11.3
Top Carrier Concentration at Major Airports

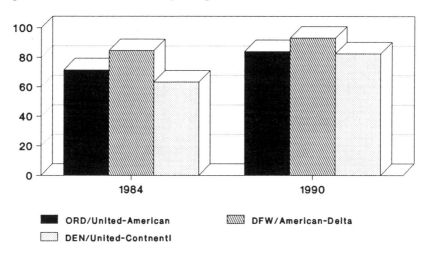

ORD--Chicago O'Hare
DFW--Dallas/Fort Worth
DEN--Denver

Figure 11.4
Airline Market Share at Washington Dulles

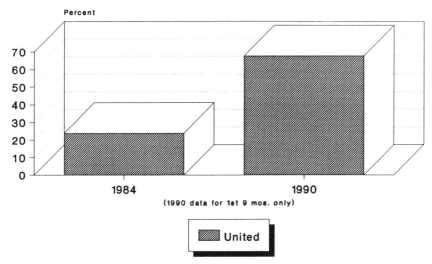

Sources: AVIATION DAILY, Apr. 19, 1985, at 28; Feb. 1, 1990, at 230; Apr. 29, 1990, at 628; Mar. 29, 1991, at 590; and CONSUMER REPORTS.

was so large that its management was concerned that antitrust or diseconomies of scale might curtail growth; the company had enough capital to reposition the airline without having to sell other subsidiaries; the corporation's top managers, Edward Carlson and Richard Ferris, had come from the hotel business and were loyal to it; and the hotel chain was generating substantial returns.[29]

In 1985, United acquired Pan American's transpacific routes, ground facilities, and aircraft for $750 billion. United already operated a Seattle-Tokyo route, but this was the first significant foray into the transpacific—the fastest-growing aviation market in the world. Rehabilitation of the aging Pan Am 747s was a lengthy and expensive process, but it ultimately paid off. By 1989, the Pacific system accounted for 20 percent of United's revenue and a third of its operating profit.[30]

Other than the transpacific purchase, Dick Ferris, chairman of United, diverted resources and attention from airline operations in order to become a travel conglomerate, sweeping Westin Hotels, Hilton International Hotels, and Hertz Rent-a-Car under the United umbrella. Ferris changed the holding company's name from UAL to Allegis, a move that failed to impress Wall Street. Diversification led the corporate raider vultures to circle, and Ferris got the boot as United shed itself of its $3.5-billion non-airline properties and the perplexing Allegis name.[31]

The diversion cost United market share. It lost its dominance as the nation's largest domestic carrier, first to Frank Lorenzo's Texas Air and then to Bob Crandall's American. Some analysts predicted that United would fall behind Delta as well.[32]

Under the towering Stephen Wolf, six feet six inches tall and United's current CEO, the airline has gotten back to the basics. Wolf is credited with turning things around at Republic and Flying Tigers before coming to United.

Always doing things in a big way, United in 1989 placed an order for 370 Boeing jets worth $15.7 billion—the largest order in the history of the industry.[33] In 1991, it purchased Pan Am's U.S.-London (Heathrow) and beyond authority. But it still is distracted by corporate machinations. As one source noted:

The nation's largest airline until American passed it [in 1989], United has attempted to regain the lead but has frequently been distracted by turmoil involving the ownership and holdings of the company.

Most recently, Coniston Partners, which owns 11.8% of United's parent, the UAL Corporation, has put pressure on the company's board to do more for shareholders. The stock price has fallen 56 percent, to $130.75 from $294 since a buyout deal fell apart in October [of 1989]. . . .

Whatever is done, United will almost certainly end up with much higher debt and less money to spend on new aircraft and expansion. Management worries that the airline will therefore be at a disadvantage against the likes of American. . . .

United grew by only 2.8 percent [in 1989] in terms of available seat miles, compared with 12.9 percent for American.[34]

NOTES

1. R. DAVIES, AIRLINES OF THE UNITED STATES SINCE 1914 79 (1972).

2. *Id.* at 195.

3. UNITED AIRLINES, INC., CORPORATE AND LEGAL HISTORY OF UNITED AIR LINES AND ITS PREDECESSORS AND SUBSIDIARIES: 1925–1945 I.3 (1953).

4. *Id.* at I.4. Effective February 19, 1934, all domestic airmail contracts were cancelled, and the air-transport operating companies were held to be disqualified by the Post Office Department from bidding for new contracts. The former United Air Lines, Inc., was not disqualified, however. It succeeded in obtaining new airmail contracts for substantially the same routes as those operated by the old companies, with the major exception of the Chicago-Dallas route. On May 1, 1934, it became the operator of the United system. *Id.* at I.4.

5. *Id.* at I.4.

6. F. TAYLOR, HIGH HORIZONS: DAREDEVIL FLYING POSTMEN TO MODERN MAGIC CARPET—THE UNITED AIR LINES STORY 155 (1962).

7. *Id.* at 93.

8. *Id.* at 157.

9. *Id.* at 158.

10. *Id.* at 164.

11. *Supra* note 3 at I.386. UNITED AIR LINES, INC., CORPORATE AND LEGAL HISTORY OF UNITED AIR LINES, INC., AND ITS SUBSIDIARIES: 1946–1955 II.3 (1965).

12. *Id.* at II.6.

13. *Id.* at II.800–801.

14. R. DAVIES, *supra* note 1, at 331–32. United had substituted Denver for Cheyenne as a staging point in 1937. *Id.* at 328.

15. F. TAYLOR, *supra* note 6, at 248.

16. *Id.* at 251.

17. *Id.* at 254.

18. C. SOLBERG, CONQUEST OF THE SKIES: A HISTORY OF COMMERCIAL AVIATION IN AMERICA 412 (1979).

19. J. BYRNES, DIVERSIFICATION STRATEGIES FOR REGULATED AND DEREGULATED INDUSTRIES: LESSONS FROM THE AIRLINES 31 (1985).

20. *Id.* at 33–34.

21. *Id.* at 38.

22. *Id.* at 45.

23. In 1975, United purchased GAB Business Services, an insurance adjustment firm. *Id.* at 47.

24. D. KASPER, DEREGULATION AND GLOBALIZATION: LIBERALIZING INTERNATIONAL TRADE IN AIR SERVICES 29 (1988).

25. *Id.* at 35.

26. J. BYRNES, *supra* note 19, at 63.

27. *Id.* at 64.

28. *Id.* at 67.

29. *Id.* at 67–68.

30. *United Swallows Hard and Goes for Growth*, Bus. Week, May 15, 1989, at 34.

31. Salpukas, *Steve Wolf's Big Test*, N.Y. Times, Jan. 8, 1989, at 23.

32. *Id.* at 24.

33. *United Swallows Hard, supra* note 30.

34. Salpukas, *Airlines' Big Gamble on Expansion*, N.Y. Times, Feb. 20, 1990, at C1, C5.

12

USAIR

Hubs: Charlotte, Pittsburgh

Mini-hubs: Baltimore, Dayton, Philadelphia

Post-deregulation Mergers: Pacific Southwest Airlines (1987), Piedmont (1988)

Computer Reservations System: A minority position in the Covia Partnership, which owns Apollo

Rank and Market Share: 1978—ninth, 1.8%; 1990—sixth, 7.4%

Originally begun as All American Aviation in the 1920s, USAir was certificated by the CAB after World War II as Allegheny Airlines, a local-service carrier.[1] In 1968, it acquired Lake Central.[2] In 1971, it merged with Mohawk.[3] In 1979, Allegheny changed its name to USAir, corresponding with its expanded geographic emphasis.[4]

Edward Colodny became president of USAir in 1975. A Harvard Law School graduate and former Civil Aeronautics Board attorney, he joined USAir (then Allegheny Airlines) in 1957, serving in most areas of the company's management before becoming president.[5]

He opposed deregulation in 1978 fearing, rightly, that the industry would eventually come to be dominated by a handful of megacarriers. He entered deregulation with a conservative growth-from-within strategy, concentrating his operations geographically in the East, strengthening his Pittsburgh hub into a fortress, and retaining Allegheny's spartan headquarters at Washington National Airport.[6] In 1979, the spokes beginning to grow longer out of the Pittsburgh hub, Colodny dropped the Allegheny name in favor of USAir.[7] Colodny kept the airline's costs down and reduced indebtedness.[8] As one observer noted, "Colodny has emphasized the significance of

Figure 12.1
Airline Market Share at Baltimore/Washington

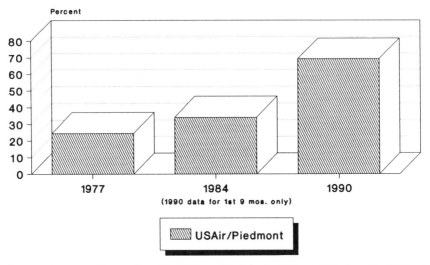

Sources: AVIATION DAILY, Apr. 19, 1985, at 28; Feb. 1, 1990, at 230; Apr. 29, 1990, at 628; Mar. 29, 1991, at 590; and CONSUMER REPORTS.

niching, or concentrating on providing specified services and amenities in a given geographic area and using equipment tailored to those requirements."[9] Like the tortoise chasing the hare, Colodny plodded slowly along, gaining ground in the race.[10] The approach paid off as USAir became a strong small airline.

The son of a Vermont grocer, Colodny has been described as "a conservative manager with strong views on how to run an airline" but willing to change his mind when circumstances demand it.[11] Even in his personal life, his boyhood dreams of becoming a professional violinist gave way to the reality that he was perfectly awful at playing the violin.[12]

Witnessing the unprecedented mergers that swept the industry in 1986, Colodny jumped on board, more than doubling the size of his airline by expanding westward with the acquisition in 1987 of Pacific Southwest Airlines for $400 million and southward in 1988 with his purchase of Piedmont Airlines for $1.56 billion.[13] USAir also gathered together a string of smaller carriers, including Henson, Jet Stream International, Pennsylvania Airlines, and Suburban Airlines.

Piedmont was a healthy little carrier with a reputation for excellent service. The combined carriers, known for their efficiency and profitability, would saturate the East like Sherman Williams covers the globe, with hubs in Pittsburgh, Philadelphia, Charlotte, Baltimore, and Dayton (see figures 12.1 through 12.4).[14] Colodny promised to merge the carriers slowly, in-

Figure 12.2
Airline Market Share at Charlotte

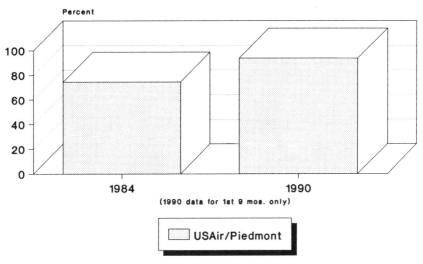

Sources: AVIATION DAILY, Apr. 19, 1985, at 28; Feb. 1, 1990, at 230; Apr. 29, 1990, at 628; Mar. 29, 1991, at 590; and CONSUMER REPORTS.

Figure 12.3
Airline Market Share at Philadelphia

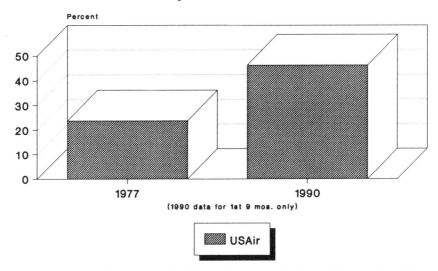

Sources: AVIATION DAILY, Apr. 19, 1985, at 28; Feb. 1, 1990, at 230; Apr. 29, 1990, at 628; Mar. 29, 1991, at 590; and CONSUMER REPORTS.

Figure 12.4
Airline Market Share at Pittsburgh

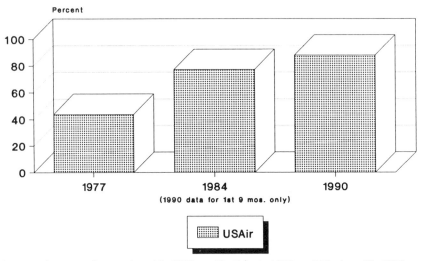

Sources: AVIATION DAILY, Apr. 19, 1985, at 28; Feb. 1, 1990, at 230; Apr. 29, 1990, at
628; Mar. 29, 1991, at 590; and CONSUMER REPORTS.

tegrating labor and operations with the least amount of disruption possible.

But its monopoly presence in the eastern United States allowed USAir to
raise prices even higher, much to the chagrin of passengers without a competitive alternative. The only bit of justice was that ticket prices out of
Ithaca, New York, and surrounding cities began to rise significantly. Ithaca
is the home of the Cornell economics professor and former CAB chairman
Alfred Kahn, the godfather of deregulation. Justice demands that airline
ticket prices should be as high as the heavens in Ithaca. For his part, Colodny now sings the Fred Kahn tune of deregulation, arguing that reregulation would be "the worst thing Congress could do."[15]

As noted above, USAir merged with Pacific Southwest Airlines in 1987
for $400 million and with Piedmont Aviation in the following year for
$1.56 billion.[16] Both were fine small airlines with good reputations for
service. Except for the latter purchases, USAir's corporate strategy has been
cautious and conservative, taking small steps to solidify its dominance of
its Pittsburgh hub and its lesser Northeast hubs, Philadelphia and Buffalo.[17] Only recently did USAir move out of its shabby headquarters at
Washington National Airport, and only then because its lease was up and
it badly needed space.[18] USAir is the largest U.S. airline not to have made
a major move into international markets. Its consolidation of Piedmont
and PSA into USAir was planned to be accomplished slowly and deliber-

ately, the former concluded in 1988 and the latter in 1989. In taking on Piedmont, USAir adopted Piedmont policies and thereby upgraded its operations in two ways: it changed its plane configurations from single-class to dual-class cabins, and it continued Piedmont's policy of offering passengers the whole can of Coke, instead of a single cup.[19]

Colodny sought to further strengthen USAir's grip on the Northeast by trying to buy eight gates at Philadelphia and two Canadian routes at the Eastern bankruptcy sale for $85 million.[20] The Pennsylvania attorney general threatened litigation if USAir proceeded, on the grounds that the firm was already charging monopoly rates at its Pittsburgh hub and that the transfer would allow USAir to dominate the state's two major cities. USAir dropped the effort when the U.S. Department of Justice threatened to intervene on antitrust grounds. But in 1991, it was able to pick up the assets in a fire sale by bankrupt Midway Airlines, which, only a couple of years earlier, had acquired them from Eastern Airlines.

Colodny maintained a tight grip on USAir as chairman, president, and chief executive officer of the airline until his retirement in 1991. Under his relatively conservative reign, USAir was consistently profitable each year between 1975 and 1989, when it suffered losses of $63.2 million in trying to digest Piedmont.[21] Service and on-time performance suffered during the transition.[22] In 1990, USAir lost half a billion dollars and began to retrench from the California markets it had acquired from PSA. USAir was projected to lose half a billion dollars in 1991 as well. Unless its performance improves significantly, it too may find itself in bankruptcy court.

NOTES

1. W. Lewis & W. Trimble, The Airway to Everywhere 3 (1988).
2. R. Davies, Airlines of the United States since 1914 410, 419 (1972).
3. *Id.* at 420.
4. W. Lewis & W. Trimble, *supra* note 1, at 190.
5. Rose, *USAir's Colodny Nurtures Crucial Merger,* Wall St. J., Nov. 2, 1987, at 34.
6. W. Lewis & W. Trimble, *supra* note 1, at 190.
7. *Id.* at 190.
8. *Id.* at 191.
9. *Id.* at 191.
10. Hamilton, *A Tale of Two Airlines: Texas Air, USAir Survive at Different Speeds,* Washington Post, May 22, 1988, at H1, H7.
11. Rose, *supra* note 5.
12. *Id.*
13. Rose & McGinley, *Merger of USAir, Piedmont Approved; Challenge to "Mega-Carriers" Expected,* Wall St. J., Nov. 2, 1987, at 23.
14. *See What's Standing between USAir and Piedmont,* Bus. Week, Oct. 5, 1987, at 40.

15. *USAir's Colodny Warns Congress against Re-Regulation*, AIRLINE FINAN- CIAL NEWS, Nov. 7, 1988.

16. W. LEWIS & W. TRIMBLE, *supra* note 1, at 192.

17. *Id.* at 191.

18. Feldman, *Blending the Elements of a Major Empire*, AIR TRANSPORT WORLD, June 1988, at 28, 30.

19. USAIR, ANNUAL REPORT 7 (1989).

20. O'Brian & Valente, *Eastern to Sell Assets to USAir for $85 Million*, WALL ST. J., Mar. 9, 1989, at A3.

21. *See* USAIR, ANNUAL REPORT 2 (1989), and Salpukas, *Airlines' Big Gamble on Expansion*, N.Y. TIMES, Feb. 20, 1990, at C1, C5.

22. Salpukas, *supra* note 21, at C5.

Part II

REGULATION AND DEREGULATION: THE METAMORPHOSIS IN AMERICAN PUBLIC POLICY

13

ORIGINS OF REGULATION: THE LEGISLATIVE HISTORY OF THE CIVIL AERONAUTICS ACT OF 1938

We begin with a review of the conditions leading to the creation of a regime of economic regulation over aviation. As we shall see in the ensuing chapters, there are many parallels between the economic environment preceding regulation and that following deregulation.

The legislative history of the Civil Aeronautics Act of 1938, the predecessor of the Federal Aviation Act of 1958, reveals that Congress recognized the air transport industry to be in its infancy and believed that the existing competitive environment could, in the absence of regulation, inhibit or impede its sound development. The existing airmail legislation was believed to have imposed certain undesirable influences on the industry. Moreover, to avoid the deleterious consequences of "cutthroat," "wasteful," "destructive," "excessive," and "unrestrained" competition, and the economic "chaos" that had so plagued the rail and motor carrier industries, Congress sought to establish a regulatory structure similar to that devised for those industries, which had also been perceived as "public utility" types of enterprises. Such a system, it was believed, would enhance economic stability and thereby contribute to the sound economic growth and development of air transportation, which was thought to be an industry of potentially vast significance to the economic development of the nation. It would insure service to small communities and the protection of smaller carriers. It would not be a system that would prohibit the entry of new carriers. The regulatory scheme would assure that the industry adhered to the highest standards of safety and satisfied the public interest, including the needs of commerce and the national defense.

AN INFANT INDUSTRY IN A HOSTILE ECONOMIC ENVIRONMENT

At the outset, the perspective from which Congress viewed the aviation industry before the Civil Aeronautics Act should be examined. As has been indicated, the air transportation industry was perceived to be in its infancy[1] and potentially of fundamental importance to the national economic growth.[2] The Senate Commerce Committee expressed serious concern over the "intensive," "extreme," and "destructive" competition in which all transport modes, air and surface, were engaged; such an economic environment was having injurious effects on the industry and its ability adequately to provide the service required to satisfy the needs of commerce and the national defense.[3] By establishing a system for the orderly development of air transportation analogous to that used in regulating public utilities and other modes of transportation (i.e., the railroads and motor and ocean carriers), Congress believed that these deleterious consequences could be avoided.

Among the difficulties faced by air carriers before 1938 was an inability to attract sufficient investment capital.[4] It was argued that the order and stability insured by public regulation would create an environment in which this difficulty would be diminished.[5]

THE REGULATORY ENVIRONMENT BEFORE 1938

Among the difficulties that the pending legislation sought to alleviate were those arising under the existing structure of airmail legislation. In 1918, airmail service was inaugurated by the army. The Kelly Act (Air Mail Act of 1925) established economically feasible commercial air transportation autonomous from the military by permitting the postmaster general to award contracts to private airlines for the movement of mail.[6] The Air Commerce Act of 1926 vested jurisdiction over safety and the maintenance of airways and navigation facilities in the secretary of commerce.[7] The McNary-Watres Act of 1930 established a formula for airmail payments based on the amount of mail transported.[8] Congressional discontent with the administration of this legislation by the postmaster general led to an investigation by a Senate Special Committee chaired by Senator (later U.S. Supreme Court Justice) Hugo Black.[9] The outrageous activities revealed by this investigation led President Franklin Roosevelt to respond by terminating all existing airmail contracts on the ground that there had been collusion between air carriers and the Post Office Department in route and rate establishment.[10]

THE SURFACE CARRIER AND PUBLIC UTILITY ANALOGY

The legislative history of the 1938 act also reveals a concern that the unfortunate economic experience of surface carriers might be repeated in the air transport industry. The Great Depression was, undoubtedly, the most intense economic calamity of the century. It was an era of economic upheaval and uncertainty during which the fatality level of businesses was robust. Certain industries were deemed so fundamental to the existence of a sound national economy that the federal government intervened to regulate competition, restore order, and diminish the uncertainty that prevailed. Among those industries perceived as essential to general economic recovery and therefore entitled to the benefits of "public utility" regulation was transportation. In 1935, Congress promulgated the Motor Carrier Act, which established federal regulation of motor carrier entry and rates and placed such jurisdiction in the Interstate Commerce Commission (ICC), which had held extensive regulatory authority over rail carriers since 1887.[11] Commissioner Joseph Eastman of the ICC stated:

Important forms of public transportation must be regulated by the government. That has been accepted as a sound principle in this country and . . . in practically every country in the world. . . .
Transportation is of such vital importance to the public welfare and the business is so affected with a public interest that some measure of government regulation is . . . necessary.[12]

Transportation was viewed by some to be on the order of a public utility for which regulation was deemed essential. For example, a representative of the National Association of Railroad and Utilities Commissioners (NARUC) testified:

Any important public-utility industry requires regulation in the public interest and will be regulated sooner or later. . . . [T]he full purpose of regulation can be accomplished only by regulation from the beginning of the development of the industry. . . .
[Congress must establish] such conditions that there may be an encouraged development of the aircraft business . . . and [create] conditions—and this is of paramount importance—which will avoid the wastes and losses which will be inevitable if the business is left to struggle to establish itself in open competition.[13]

AVOIDANCE OF EXCESSIVE COMPETITION

In light of the economically catastrophic experience of other transport modes, it was believed that regulation might insure that such consequences were avoided in the nascent air transport industry. Colonel Edgar Gorrell, the president of the Air Transport Association, stated: "Cutthroat competition is nowhere so dangerous as in transportation. And in no form of transportation would it be more disastrous . . . [than] in the case of air carriers."[14] The legislative history is replete with references to destructive competition.[15] For example, the Senate reports indicate that the air carriers were "engaged in intensive competition with each other and with the railroads and other carriers [which] is being carried to an extreme which tends to undermine the financial ability of the carriers and jeopardize the maintenance of transportation facilities and service appropriate to the needs of commerce and required in the public interest and the national defense." The Senate Committee on Interstate Commerce felt that the proposed legislation not only would "promote an orderly development of transportation in the United States" but also would "prevent the growth of bad practices and uneconomic capital structures resulting from a period of *destructive competition.*"[16] The House Committee on Interstate and Foreign Commerce, in its 1937 report recommending adoption of the proposed legislation, maintained, "The government cannot allow unrestrained competition by unregulated air carriers to capitalize on and jeopardize the investment which the government has made during the last 10 years in the air transport industry through the mail service."[17]

The Federal Aviation Commission, which was established by the Black-McKellar Act of 1934, submitted 102 recommendations in its report to Congress of January 30, 1935. It contended that the orderly development of air transportation required two fundamental ingredients. First, in the interest of safety, certain minimum standards of equipment, operating methods, and personnel qualifications should be maintained. Second, "there should be a check in development of any irresponsible, unfair, or excessive competition such as has sometimes hampered the progress of other forms of transport."[18]

Congressman Randolph contended that "unbridled and unregulated competition is a public menace," stating that the air transportation industry would be subjected to such unfortunate economic conditions as "rate war[s], cutthroat devices, and destructive and wasteful practices."[19] Other members of congress emphasized that the legislation was intended to inhibit or prohibit monopolization in the industry.[20]

COMPETITION VS. MONOPOLIZATION

The legislative history of the Civil Aeronautics Act of 1938 indicates that although Congress was generally concerned with "cutthroat," "wasteful," "destructive," and "unrestrained" competition, it was nevertheless opposed to the monopolization of transportation services and sought to insure that such services would be provided in a competitive economic environment. Recommendation 9 of the Federal Aviation Commission focused on the element of competition in entry regulation: "It should be the general policy to preserve competition in the interest of improved service and technological development, while avoiding uneconomic paralleling of routes or duplication of facilities.[21]

THE REGULATORY REGIME CONTEMPLATED UNDER THE PENDING LEGISLATION

It was contemplated that additional carriers would be certificated to perform air transport services in competition with existing carriers where the applicant could demonstrate that "the public interest" would be served. During the hearings, Senator (later President) Harry S. Truman insisted that the proposed legislation was not designed to "throttle" competition in the airline industry.[22] It was predicted that the inauguration of services by a new entrant would not be prohibited, except if the applicant was incapable of performing its operations consistent with the public interest or if the proposed operations would not satisfy a public need and would cause injury to both the existing carriers and the applicant. No carrier would have the right to enjoy exclusivity in the markets it served.[23]

With respect to parallel route authorizations, it was felt that although the practice should be avoided, the governing agency should exercise its discretion in the determination of whether duplicative operating authority should be granted.[24] Moreover, it was anticipated that new authorizations that would "meet an unsatisfied public need" or "materially improve upon the service previously available" would be issued.[25]

It is significant that, in drafting this legislation, Congress explicitly required that competition be considered by the CAB as an element of the public interest.[26] This statutory provision was subsequently interpreted to require that the CAB foster competition as a means of enhancing the development and improvement of air transport services on those routes generating a sufficient volume of traffic to support competing carriers.[27]

PROVIDING ORDER AND STABILITY FOR THE GROWTH OF AN INFANT INDUSTRY

Among the essential purposes of the Civil Aeronautics Act was to shield the air transport industry from the hostile economic forces prevalent in an unregulated economic environment so that the industry could enjoy the stability required for the acquisition of capital and long-term growth.[28] A concern was also expressed that service provided to small communities should be guaranteed.[29]

Additionally, the government sought to protect the operations of small carriers from the dangerous effects of predatory competition. Congressman Randolph contended, "Economic power and reckless management should not be permitted to injure the smaller lines, the employees of the companies, and the public."[30] It was also argued that the position of smaller carriers vis-à-vis larger carriers should be protected. Colonel Gorrell stated: "In spreading out into the regions of light-density traffic and developing smaller communities, the small lines have performed an incalculable service to the country. It must be assured, through certificates, that they may continue to perform such a service, and they must be given an opportunity to protect themselves against even the possibility of oppressive competition."[31]

NOTES

1. *Regulation of Interstate Transportation of Passengers, Mail, and Property by Aircraft, Hearings in S. 3187 by the Senate Comm. on Commerce*, 73rd Cong., 2nd Sess. 1 (1934). *See* Dempsey, *The International Rate & Route Revolution in North Atlantic Passenger Transportation*, 17 COLUM. J. TRANSNAT'L L. 393, 413 (1978).

2. Among the primary proponents of air transport regulation, and the author of the original bills, was Senator Patrick McCarran, who emphasized the significance of the pending legislation by stating,

There was never anything before this country more vital from the standpoint of national development, particularly at this hour of the world's history, and at this hour in our national history, than the legislation which is now pending before this subcommittee, because we are dealing with an infant industry, and we are dealing with it from the standpoint of what it can do for this country commercially, industrially, and as an arm of national defense.

Civil Aviation and Air Transport: Hearings in S. 3659 before a Subcomm. on Interstate Commerce, 75th Cong., 2nd Sess. 7 (1938) [hereinafter cited as *Senate Hearings on S. 3659*].
The significance of the air transport industry is also reflected in remarks by Colonel Edgar S. Gorrell, the president of the Air Transport Association and perhaps the most effective proponent of the pending legislation.
Aviation: Hearings on H.R. 5234 and H.R. 4652 before the House Comm. on Interstate and Foreign Commerce, 75th Cong., 1st Sess. 53 (1937) [hereinafter cited as *House Hearings on H.R. 5234 & H.R. 4652*]. *See* Westwood & Bennett, *A Footnote to the Legislative History of the Civil Aeronautics Act of 1938* and *Afterword*, 42 NOTRE DAME LAW 309, 320 (1967).

3. *Senate Comm. on Interstate Commerce, Air Transport Act, 1937*, S. Rep. No. 686, 75th Cong., 1st Sess. [hereinafter cited as *Senate Committee Report on ATA*]; *Senate Comm. on Interstate Commerce, Air Safety Act, 1937*, S. Rep. No. 687, 75th Cong., 1st Sess. 2 (1937).

4. SENATE SUBCOMM. ON ADMINISTRATIVE PRACTICE AND PROCEDURE OF THE JUDICIARY COMMITTEE, 94th Cong., 1st Sess., CIVIL AERONAUTICS BOARD PRACTICES AND PROCEDURES, 207–8 (Comm. Print 1976) [hereinafter cited as KENNEDY REPORT].

5. *Senate Hearings on S. 3659, supra* note 2, at 30–31. *See Civil Aeronautics Authority on S. 3760 before the Senate Commerce Comm.*, 75th Cong., 2nd Sess. 338–39 (1938); and HOUSE COMM. ON INTERSTATE AND FOREIGN COMMERCE, CIVIL AERONAUTICS BILL, H.R. DOC. 2254, 75th Cong., 2nd Sess. 2 (1938). Between the two alternatives posed by Colonel Gorrell—subsidization or nationalization of the airlines on the one hand and economic regulation on the other—the colonel favored and the Congress adopted the latter.

6. *See* KENNEDY REPORT, *supra* note 4, at 195; R. BURKHARDT,.THE CIVIL AERONAUTICS BOARD 4 (1974); S. RICHMOND, REGULATION AND COMPETITION IN AIR TRANSPORTATION 4 (1961); H. KNOWLTON, AIR TRANSPORTATION IN THE UNITED STATES 4 (1941); C. PUFFER, AIR TRANSPORTATION 2–3 (1941); and L. KEYES, FEDERAL CONTROL OF ENTRY INTO AIR TRANSPORTATION 65–66 (1951).

7. *See* PUFFER, *supra* note 6, at 3; H. KNOWLTON, *supra* note 6, at 6–7; L. KEYES, *supra* note 6, at 65; and KENNEDY REPORT, *supra* note 4, at 199–200.

8. KENNEDY REPORT, *supra* note 4, at 203–4.

9. F. THAYER, JR., AIR TRANSPORT POLICY AND NATIONAL SECURITY 10 (1965).

10. H. KNOWLTON, *supra* note 6, at 9; KENNEDY REPORT, *supra* note 4, at 204. The multitude of problems that arose under this regulatory regime has been adequately explored by a number of commentators and need not be repeated in any depth here. *See* e.g., REPORT OF THE CAB SPECIAL STAFF ON REGULATORY REFORM 31–32 (1975) [hereinafter cited as CAB STAFF REPORT]. *See* S. RICHMOND, *supra* note 6, at 6; R. BURKHARDT, *supra* note 6, at 5–6; H. KNOWLTON, *supra* note 6, at 9; L. KEYES, *supra* note 6, at 63–65; KENNEDY REPORT, *supra* note 4, at 203–9.

11. The Motor Carrier Act was subsequently scattered throughout the provisions of the Interstate Commerce Act, 49 U.S.C. §§ 10101–11916. *See* Dempsey, *Entry Control under the Interstate Commerce Act: A Comparative Analysis of the Statutory Criteria Governing Entry in Transportation*, 13 WAKE FOREST L. REV. 729, 735 (1977). The ICC today regulates freight forwarders and domestic water carriers as well. *See* Dempsey, *The Contemporary Evolution of Intermodal and International Transport Regulation under the Interstate Commerce Act: Land, Sea, and Air Coordination of Foreign Commerce Movements*, 10 VAND. J. TRANSNAT'L L. 505 (1977); 46 ICC PRAC. J. 360 (1979).

12. *Regulation of Transportation of Passengers and Property by Aircraft, Hearings on S. 2 and S. 17 before a Subcomm. of the Senate Comm. on Interstate Commerce*, 75th Cong., 1st Sess. 67 (1937) [hereinafter cited as *Senate Hearings on S. 2 and S. 1760*].

13. *House Hearings on H.R. 5234 & H.R. 4652, supra* note 2, at 163.

14. *Senate Hearings on S. 2 & S. 1760, supra* note 12, at 513.

15. *Id.* at 376–77.

16. *Senate Committee on Interstate Commerce*, 75th Cong., 1st Sess., *Report on Air Transport Act of 1937*, 2 (1937) [emphasis supplied]. *See also Senate Comm. on Commerce*, 75th Cong., 3rd Sess., *Report on Civil Aeronautics Act of 1938* (1938), which emphasized, as among the central objectives of the proposed legislation, the prevention of "bad practices and of destructive and wasteful tactics resulting from the intensive competition now existing within the air carrier industry." *Id.* at 2. *See* S. RICHMOND, *supra* note 6, at 7; Dupre, *A Thinking Person's Guide to Entry/Exit Deregulation in the Airline Industry*, 9 TRANSP. L.J. 273, 280–81 (1977).

17. Quoted in S. RICHMOND, *supra* note 6, at 8.

18. *Senate Comm. on Interstate Commerce, Federal Aviation Commission*, S. Doc. No. 15, 75th Cong., 1st Sess. (1935) [hereinafter cited as *Federal Aviation Commission*].

19. *See* 83 CONG. REC. 6852 (1938) (remarks of Senator McCarran [the author of the original Senate bill]).

20. *See* 83 Cong. Rec. 6370 (1938) (remarks of Senator Borah and Senator Copeland).

21. *Federal Aviation Commission, supra* note 18, at 61. *See Senate Hearings on S. 2 & S. 1760, supra* note 12, at 704. *See* S. Richmond, *supra* note 6; and CAB Staff Report, *supra* note 10.

22. The initial bills contemplated that jurisdiction over entry and rates in the airline industry would be vested in the Interstate Commerce Commission, which already held authority over rail and motor carriers. *Regulation of Transportation of Passengers and Property by Aircraft: Hearings on S. 3027 before a Subcomm. of the Senate Comm. on Interstate Commerce,* 74th Cong., 1st Sess., 100 (1935).

23. *Senate Hearings on S. 2 & S. 1760, supra* note 12, at 503–4.

24. *Federal Aviation Commission, supra* note 18, at 54–55.

25. *Id.* at 52; *see House Comm. on Interstate Commerce, Regulation of Transportation of Property and Passengers by Air Carriers,* H.R. Doc. No. 911, 75th Cong., 1st Sess. (1937).

26. *See* 83 Cong. Rec. 6726 (1938). 49 U.S.C. § 1302(a) (1977).

27. Continental Air Lines, Inc. v. CAB, 519 F.2d 944, 946 (D.C. Cir. 1975).

28. One airline executive stated, "As we see it, the Civil Aviation Authority, when created, will have two broad policy functions: first, to reestablish, develop, and expand American aviation, both domestic and foreign, on a sound basis; and, second, to maintain a healthy condition and growth through proper regulation and control." *Senate Hearings on S. 3659, supra* note 2, at 149. Existing carriers supporting the proposed legislation sought regulation as a means of enhancing stability in the air transport industry. A representative of United Air Lines stated, "If some degree of permanency can be assured and the investments of already-existing carriers be safeguarded, it will allow those air carriers to intelligently and effectively plan for the future." *Id.*

29. *Senate Hearings on S. 3659, supra* note 2, at 170.

30. 83 Cong. Rec. 6501–15 (1938).

31. *House Hearings on H.R. 5234 & H.R. 4652, supra* note 2, at 90.

14

THE TRADITIONAL REGULATORY CRITERIA

In 1938, President Roosevelt signed into law the predecessor of the Federal Aviation Act,[1] which established the Civil Aeronautics Board[2] as an independent regulatory agency designed to provide classic "public utility" type regulation over the air transportation industry, then deemed to be in its infancy. Essentially, the agency was given authority to regulate three broad areas of economic activity:

1. Entry—prescribing which routes shall be flown and which communities receive air service and designating the specific carrier(s) which will be permitted to serve such markets; the authority to grant or deny certificates of "public convenience or necessity"[3]
2. Rates—the authority to suspend or establish air fares and determine whether proposed rates are "just and reasonable"[4]
3. Antitrust—the authority to approve or disapprove a host of intercarrier transactions, some of which are anticompetitive;[5] approval has traditionally conferred immunity from the effects of the Sherman and Clayton acts[6]

As originally promulgated, the Federal Aviation Act authorized the issuance of operating authority to any applicant who was "fit, willing, and able," under circumstances where the transportation in question was "required by the public convenience and necessity."[7] In performing such responsibilities, the CAB was obligated by the act to "foster sound economic conditions" in transportation, to promote "adequate, economical, and efficient service by air carriers at reasonable charges,"[8] to promote "competition to the extent necessary to assure the sound development of an air transportation system,"[9] and to avoid "destructive competitive practices."[10]

In interpreting the statutory concept of public convenience and necessity, the CAB traditionally weighed and balanced a number of criteria[11] (including the relative service benefits of proposed operations and such considerations as historic participation in the involved market, the ease with which the involved segment would integrate with the applicant's existing route structure, and the carrier's needs for subsidy reduction and/or route strengthening). No single criterion was deemed to be controlling.[12] Most such proceedings involved essentially a two-step process: (1) determining the number of carriers the market in question could reasonably and profitably support, and (2) selecting from among the various applicants which carrier(s) should be designated to receive certificate(s) of public convenience and necessity.

In the years immediately after the act was passed, the CAB was principally concerned with the issuance of "Grandfather" certificates under section 401(e)(1) of the act,[13] which required that a certificate of public convenience and necessity be issued to any applicant on proof that during the grandfather period (May 14, 1938–August 22, 1938), the applicant was an air carrier continuously operating over the segment for which operating authority was sought (unless the service provided during such period was inadequate and inefficient).

In a 1941 case, the CAB stated that four questions were to be considered in any application for new service:

1. Will the new service serve a useful public service, responsive to a public need?
2. Can and will this service be served adequately by existing routes or carriers?
3. Can the new service be served by the applicant without impairing the operations of existing carriers contrary to the public interest?
4. Will any cost of the proposed service to the government be outweighed by the benefit that will accrue to the public from the new service?[14]

In later decisions, the CAB emphasized that no criterion was controlling.[15] As the CAB stated in the *Service to Tri City Case:*

The Board has never established a hierarchy among the various carrier selection criteria, but has rather examined all the relevant criteria in reaching a determination in a given case. The Board's policy has been that the weight afforded a particular decisional factor must be determined in the context of the specific needs of the markets and, on occasion, in light of broader public interest considerations.[16]

In the *Sacramento-Denver Nonstop Case,*[17] the Board elaborated, saying, "Only rarely is there but a single reasonable candidate, and quite often the selection of a particular carrier reflects either an applicant's incremental advantage or its ability to combine several factors."[18]

In the *Miami–Los Angeles Competitive Nonstop Case,*[19] the CAB enumerated ten factors it had weighed in determining which, among multiple applicants, should or should not receive certificated authority to serve a particular market:

1. Route integration as evidenced by the ability to convenience beyond-segment traffic

2. Frequencies to be operated over the involved segment

3. The type of equipment to be employed

4. The fares to be charged

5. The identity of the involved points

6. The historic participation in the involved traffic

7. Efforts to promote and develop the involved market

8. The need of the applicant for route strengthening

9. The profitability of the route for the applicants and the existing carriers

10. The potential of diversion of traffic from existing carriers[20]

Traditionally, the CAB sought to improve the competitive posture of smaller carriers vis-à-vis the larger incumbents, and reduce industry concentration, by favoring the issuance of operating authority to the small carriers.[21] This polity of route strengthening "was intended to combat excessive concentration and to maintain a balance of competitive opportunities within the industry by strengthening the smaller carriers who were at a disadvantage because of their route system."[22]

Enhancing the competitive posture of smaller carriers by issuing segments of potentially lucrative route authority was viewed as being "of great importance in perfecting the route structure of the nation."[23] Thus, the CAB frequently scrutinized the competitive positions of various applicants and their relative requirements for route strengthening.[24] By strengthening the smaller carriers, the CAB could bring about a concomitant reduction in their subsidy requirements, which, in itself, became another important criterion of carrier selection.[25]

In other cases, the CAB concluded that the potential service benefits offered by larger carriers outweighed the need of smaller carriers for subsidy reduction. For example, in the *Fort Myers–Atlanta Case*,[26] the CAB recognized that a carrier "having access to the largest volume of support traffic" would be "in the best position to provide the greatest frequency and capacity and flow the maximum number of passengers over the . . . segment [and thereby] convenience the largest number of beyond-segment passengers."[27] Similarly, the ease with which a proposed segment integrated with a carrier's existing route structure was frequently perceived as a factor weighing in favor of the carrier, for such a coherent structure might enable it to convenience a larger segment of the traveling public.

The relative beyond-segment capabilities of the applicants was frequently perceived as an important criterion of carrier selection.[28] For example, the CAB in the *Memphis–Twin Cities/Milwaukee Case*[29] noted:

Flow traffic is important in developing a thin market, because it helps support the frequencies necessary to permit service levels sufficient to attract and hold new customers. In addition, it generates systemwide profits and benefits passengers by increasing single-lane and single-carrier alternatives. Because beyond-traffic contributes significant benefits to carriers and passengers, it has become an important element in selecting a carrier to serve this sort of market.[30]

From time to time, the CAB also made efforts to strengthen the financial posture of even large carriers facing financial difficulties, through the issuance of lucrative segments of operating authority.[31] Concern with the financial health of carriers subject to its jurisdiction frequently led the CAB to view the potential diversion of traffic from incumbent carriers and consequential revenue loss as a factor militating against the issuance of operating authority to a new entrant.[32] Similarly, the CAB scrutinized route proposals to determine whether the inauguration of service might cause financial injury to the applicant or, as the CAB stated, "whether the proposed operation would result in a revenue deficiency that would weaken the carrier and impair its ability to properly serve its route."[33]

NOTES

1. 49 U.S.C. §§ 1301–551 (1979). Actually, as originally promulgated in 1938, it was named the "Civil Aeronautics Act." It was not until 1958, when Congress established the Federal Aviation Administration, that the legislation was given the title "Federal Aviation Act." 72 Stat. 731. The 1958 act was not a substantive recodification, however, insofar as it affected economic regulation by the Civil Aeronautics Board. The Airline Deregulation Act of 1978, 92 Stat. 1705, which is discussed extensively in this book, does not constitute a separate body of legislation but instead amends several pieces of existing legislation, of which the Federal Aviation Act is the most significant.

2. As originally created in 1938, the agency was titled the "Civil Aeronautics Authority." The name was changed the following year.

3. *See* 49 U.S.C. § 1371.

4. *See* 49 U.S.C. §§ 1373, 1482. *See* Gillick, *Recent Development in Airline Tariff Regulation: Procedural Due Process and Regulatory Reform,* 9 TRANSP. L.J. 67, 67–72 (1977).

5. *See* 49 U.S.C. §§ 1378, 1379, 1382.

6. *See* 49 U.S.C. § 1384 (1979). These constitute the three primary areas of economic regulation. Other important responsibilities include entry of foreign air carriers, 49 U.S.C. § 1372, regulation of mail carriage, 49 U.S.C. §§ 1375, 1376, and (since 1978) assuring service to small communities.

7. Former Federal Aviation Act § 401(d)(1), (d)(2), and (d)(3); 49 U.S.C. § 1371 (a)(1), (d)(2), and (d)(3).

8. Former Federal Aviation Act § 102; 49 U.S.C. § 1302.

9. Former Federal Aviation Act § 102(a)(4); 49 U.S.C. § 1302(a)(4) (1977).

10. Former Federal Aviation Act § 102(a)(3); 49 U.S.C. § 1302(a)(3) (1977).

11. Memphis-Tampa/St. Petersburg/Clearwater Subpart N Proceeding, CAB Order 77-5-66 (1977), at 2–3.

12. Fort Myers-Atlanta Case, CAB Order 76-1-81 (1976), at 11, and cases cited therein.

13. Former 49 U.S.C. § 1371(e)(1) (1977).

14. Delta Air Corporation, Service to Atlanta and Birmingham, 2 C.A.B. 250, 251–52 (1940). These criteria are strikingly similar to those developed by the Interstate Commerce Commission in the regulation of common carrier entry of motor carriers in Pan American Bus Lines Operations, 1 M.C.C. 190 (1936). *See* Dempsey, *Entry Control under the Interstate Commerce Act: A Comparative Analysis of the Statutory Criteria Governing Entry in Transportation,* 13 WAKE FOREST L. REV. 729, 735–53 (1977).

15. CAB Order 77-3-132 (1977).

16. *Id.* at 3.

17. CAB Order 77-6-27 (1977).

18. *Id.* at 3. Memphis-Twin Cities/Milwaukee Case, CAB Order 78-6-20 (1978), at 2.

19. CAB Order 76-3-93 (1976).

20. *See id.* at 31.

21. Oklahoma-Denver-Southeast Points Investigation, CAB Order 77-4-146 (1977), at 7; Transpacific Route Investigation, 51 C.A.B. 161, 287 (1968).

22. Oklahoma-Denver-Southeast Points Investigation, CAB Order 77-4-146 (1977), at 15.

23. New York-Chicago Service Case, 22 C.A.B. 52 (1955).

24. Southwest-Northwest Service Case, 22 C.A.B. 52 (1955).

25. Southern Airways Route Realignment Investigation, CAB Order 73-2-90 (1973), at 9-16; *aff'd* Southern Airways *et al* v. C.A.B. 498 F.2d 66 (D.C. Cir. 1974).

26. CAB Order 75-10-119 (1975).

27. *Id.*

28. *See* Sacramento-Denver Nonstop Case, CAB Order 77-6-27 (1977), at 4–5.

29. CAB Order 78-3-35 (1975).

30. *Id.* at 2. *See* Memphis-Tampa/St. Petersburg/Clearwater Subpart N Proceeding, CAB Order 77-5-66 (1977).

31. *See* Oklahoma-Denver-Southeast Points Investigation, CAB Order 77-4-146 (1977), at 16.

32. *See* Memphis-Tampa/St. Petersburg/Clearwater Subpart N Proceeding, CAB Order 77-5-66 (1977), at 4-5; American Airlines, Palm Springs Restriction, 50 C.A.B. 359, 360 (1969).

33. *See* Comment, *An Examination of Traditional Arguments on Regulation of Domestic Air Transport,* 42 J. AIR L. 7 COM. 187, 203–4.

15

CAB REGULATION, 1938–1975: THE CONGRESSIONAL PERSPECTIVE

Congressional scrutiny of arguments for deregulating the airline industry began with a series of hearings in 1975 under a Senate subcommittee chaired by Edward Kennedy. Such congressional analysis was subsequently expanded by Senator Howard Cannon, chairman of the Senate Commerce Committee, who held a parallel series of hearings.

Whether their conclusions were accurate or inaccurate is an issue open to debate. The significance of the instant summarization lies not in the accuracy of these allegations; rightly or wrongly, it is the perspective from which Congress viewed the airline industry and CAB regulation and the foundation on which Congress acted to dismantle the regulatory umbrella.

ENTRY

The legislative history of the 1938 act reveals that Congress intended that the board implement a cautious yet moderately liberal approach to entry, permitting new enterprises to compete as the air transportation market expanded. But entry into the industry was effectively prohibited by the restrictive regulatory policies of the CAB. Between 1950 and 1974, the CAB received 79 applications from firms seeking to obtain operating authority to provide scheduled domestic service. None were granted.[1] Moreover, between 1969 and 1974, the CAB imposed a "route moratorium," a general policy of refusing to grant or even hear any applications to serve new routes.[2] As a result of these policies, the "big four" in 1938—United, American, Eastern, and TWA—were the "big four" of the mid-1970s. In 1938, United controlled 22.9 percent of the market; in 1975, it accounted for 22.0 percent.[3]

Not a single new domestic trunk-line carrier had been authorized. Although there were 16 such carriers "grandfathered" in 1938, there were only 10 such carriers some four decades later. The CAB had not permitted a single bankruptcy. These 16 domestic trunkline carriers of 1938 had merged into the 10 that existed in the mid-1970s; the 19 local-service carriers licensed shortly after WWII had merged into the 9 that existed in 1975.[4]

Congress was misled on this point. Although only 16 airlines were certificated in 1939, 86 more received authority between 1938 and 1975, and hundreds more were granted exemptions.

RATES

Traditionally, the board applied classic rate-making methodology. Classic rate making is ordinarily employed to set the rates of a regulated monopolist, such as a public utility, and utilizes the following formula: costs + reasonable return on investment = revenue requirement.[5]

Before 1978, the CAB followed this approach, with significant modifications developed in its *Domestic Passenger Fare Investigation (DPFI)*.[6] First, it examined the cost and revenue figures not of the individual airlines, but of the industry as a whole. Second, it adjusted these figures to determine what industry costs and revenues would have been had load factors of 55 percent been achieved (i.e., it assumed that planes were flying 55 percent full). To costs determined on this basis was added a 12 percent return on investment. Finally, fares were set at a level adequate to generate this "revenue requirement."[7]

Every three months, the CAB published a compilation of industry cost and revenue figures with these adjustments, simplifying the task of determining what fare level the industry was entitled to set. If a carrier proposed a tariff embracing that fare level, it could be reasonably certain that the CAB would approve the tariff as "just and reasonable."

Congress concluded that this system, although administratively efficient, tended to keep air fares at an unreasonably high level. The load factor level of 55 percent was deemed to be too low. In the California and Texas intrastate markets, carriers regularly achieved 60 to 70 percent load factors, which revealed that passengers would accept moderately more crowded aircraft if they could enjoy correspondingly lower fares.[8] The 55 percent assumption was based on an industry average and did not take into account the ability of individual carriers to exceed this standard or their inability to achieve it, generally or on particular routes.[9]

The board traditionally prohibited selective price reductions by requiring that carriers charge equal fares for equal distances. Thus, it became difficult for carriers to lower fares in less densely traveled markets in order to stimulate demand. The CAB also inhibited across-the-board price cuts by generally refusing to approve such reductions unless, assuming all competitors participated in the reduction, each would achieve the target 12 percent return on investment.[10]

Congress concluded that, by preventing selective price reductions and inhibiting general price cuts, the board had encouraged carrier inefficiency, for it became difficult for the more efficient firms, by lowering their prices, to take business away from the less efficient.[11] Congress also concluded that the board's policies had little effect in stimulating increased industry profits. The level of profitability had fallen well below the board's 12 percent target.[12]

The absence of carrier profits was probably attributable to the fact that the airline industry is structurally competitive. The inability to engage in route and rate competition led various firms to engage in service competition. By purchasing larger aircraft in greater numbers and by increasing frequencies, they tended to lower their load factors. By offering "lavish" in-flight amenities and increasing their advertising budgets and operational expenditures, they tended to diminish their profits.[13] As their costs increased and profits diminished, they tended to seek fare increases, causing prices to spiral upward. (Congress was misled here too; real prices fell steadily under regulation.)

The fundamental deficiency of the board's rate policies during this period was its failure to recognize the elasticity of demand inherent in passenger transportation—that by lowering fares, air carriers might well stimulate new traffic and thereby fill empty seats.[14] The discretionary traveler, one who might take a vacation or visit relatives only if the price was right, was a largely unexploited source of potential revenue.[15]

ANTITRUST

In the late 1960s, excessively optimistic CAB and industry demand projections led the industry to invest in large numbers of widebodied aircraft. But passenger demand failed to live up to these expectations.[16] The diminution of disposable income engendered by the recession of the early 1970s, coupled with the tendency of air carriers to raise their prices, led load factors to drop and carrier profits to turn downward.

In response, a number of the major carriers (e.g., United, TWA, and American) agreed to a collective reduction of service provided on several of the major domestic routes.[17] The board continually approved these agreements between 1971 and 1975, first as an emergency response to overinvestment and excessive capacity and, after 1973, as a necessary response to the fuel shortages and escalating fuel costs that existed after the Arab oil embargo.[18]

Although capacity limitation agreements can theoretically bring about lower carrier costs and correspondingly lower fares, the latter did not materialize. According to Congress, the airline industry was, in fact, the only major industry that raised its prices during the recession of the early 1970s.[19] (Congress was misled here too. With the Arab oil embargo of 1973, aviation fuel costs rose sharply. Fuel accounts for more than 25 percent of industry costs and had to be passed through to consumers.) The report of the Kennedy subcommittee concluded: "The classic regulatory response to defects in regulation is to create more regulation: the Board's response to the problem of excess capacity was to introduce capacity restricting agreements. Yet, to do so in this highly competitive, complex industry brought the consumer the worst of both worlds—high prices and poor service."[20]

Congress found that consumers desire lower-fare service and that increased route and rate competition was likely to induce carriers to offer such lower fares.[21] It recognized the inherent difficulty in applying classical rate and entry regulation to a competitive, economically volatile industry.[22]

It was generally concluded that the traditional system of airline regulation (1) caused airfares to be considerably higher than they otherwise would be, (2) resulted in a serious misallocation of resources, (3) encouraged carrier inefficiency, (4) denied consumers the range of price and service options they would prefer, and (5) created a chronic tendency toward excess capacity in the industry.

The Kennedy subcommittee concluded:

The airline industry is potentially highly competitive, but the Board's system of regulation discourages the airlines from competing in price and virtually forecloses new firms from entering the industry. The result does not mean high profits. Instead, the airlines—prevented from competing in price—simply channeled their competitive energies toward costlier service: more flights, more planes, more frills. . . .

The remedy is for the Board to allow both new and existing firms greater freedom to lower fares and . . . to obtain new routes. This freedom should lead the airlines to offer service in fuller planes at substantially lower prices, a form of service that most consumers desire.[23]

This, in fact, was precisely the policy adopted by the Civil Aeronautics Board under the chairmanship of Alfred E. Kahn.

NOTES

1. Senate Subcomm. on Administrative Practice and Procedure of the Judiciary Committee, 94th Cong., 1st Sess., Civil Aeronautics Board Practices and Procedures, 207–8 (Comm. Print 1976) [hereinafter cited as Kennedy Report].

2. Id. at 6. During the late 1960s, Chairman Secor Brown led the CAB to implement the moratorium on the grounds that there was excessive capacity in the

industry. As a result, no entry applications were even set for hearing for several years. *Id.* at 7.

3. *Id.* at 79–80.

4. *Id.* at 6.

5. *Id.* at 10, 109.

6. *See, e.g.,* CAB Order 74-3-82 (1974).

7. KENNEDY REPORT, *supra* note 1, at 10.

8. *Id.* at 113–15.

9. *Id.* at 10.

10. *Id.* at 10–11, 124–25.

11. Thus, the subcommittee concluded that the board's policies had caused fares to be higher than they would be in a competitive market and had inhibited industry efficiency. *Id.* at 113.

12. *Id.* at 11.

13. *Id.* at 25, 39.

14. *See id.* at 123–24, 128.

15. Note that this argument is almost wholly inapplicable to the transportation of freight, which has relatively little demand elasticity. *See* Waring, *Rate Adjustments on Specific Movements,* TRANSPORTATION LAW INSTITUTE, RATE REGULATION 7 REFORM (1979).

16. KENNEDY REPORT, *supra* note 1, at 35.

17. *Id.* at 143–44.

18. *Id.* at 12–13.

19. *See id.* at 23.

20. *Id.* at 19.

21. *Id.* at 39.

22. *Id.* at 3.

23. *Id.*

THE CAB UNDER ALFRED KAHN: THE ORIGINS OF DE FACTO DEREGULATION

This chapter examines the principal efforts of the Civil Aeronautics Board in the late 1970s to deregulate the domestic aviation industry. As we shall see, the CAB discounted the industry's misgivings and proceeded steadfastly on a course beyond regulatory reform to deregulation.

President Gerald Ford became firmly convinced that the air transportation industry should be substantially deregulated. In 1975, he submitted a deregulation bill to Congress and appointed John Robson as chairman of the CAB. As CAB chairman, Robson reversed many of the anticompetitive regulatory features for which the CAB had been soundly criticized. The route moratorium and the capacity-limitation agreements were terminated. Yet, as a lawyer, he found himself constrained by the provisions of the Federal Aviation Act from advancing too radically in the direction of liberalizing pricing and entry.

His successor, Alfred Kahn, a Cornell University economist who was appointed CAB chairman by President Jimmy Carter, was not so inhibited. By 1978, the CAB had turned sharply. It began to grant operating authority by the bushel-basketful, at first to any carrier that proffered a low-fare proposal and, subsequently, to virtually any "qualified" applicant under an "experimental" policy labeled "multiple permissive entry." The CAB in 1978 amended its rate policies in the *DPFI* by essentially providing downward pricing flexibility, under certain circumstances of up to 70 percent, and upward flexibility of 10 percent. These efforts encouraged carriers to offer the lowest fares in history. The lower fares and the general economic recovery of the mid-1970s stimulated demand, which increased load factors and enabled carriers to realize the highest profits in the history of commercial aviation—at least until 1979, when profits began to plummet, a trend exacerbated by economic recession.

COMPETITION EMBRACED AS THE OVERRIDING POLICY OBJECTIVE

The CAB under Alfred Kahn enthusiastically embraced the observations of the Kennedy subcommittee and those academicians sharing its conclusions.[1] The CAB admitted that the traditional regulatory structure had created significant incentives for service and quality competition but had largely ignored the potential for innovative pricing proposals and rate competition.[2] It acknowledged the public benefits of "increased service frequencies, better connecting possibilities, more extensive single-plane service" and the other quality improvements engendered under the traditional regulatory regime.[3]

Nevertheless, it was felt that this system had left fares to be set at a higher level than they might have been in a freely competitive market and had thereby deprived travelers of low-fare alternatives.[4] To strike a proper balance between service and price, the CAB felt compelled to establish a regulatory environment in which both price and service competition were encouraged.[5] The board believed that lower fares would attract the discretionary traveler[6] and thereby enable carriers to make more efficient use of their equipment.[7] Thus, the following entry policy was adopted:

In determining whether it would be to the public's benefit to authorize competitive service in a market, we must consider the benefits to be derived from fare competition and fare/service variety as well as traditional factors and that in choosing among various applicants for competitive rate authority, carrier proposals to offer significantly lower fares, or a greater variety of price/service combinations deserve far greater weight than they have been accorded in the past.[8]

To create an atmosphere conducive to these objectives, the CAB began to certificate a larger number of carriers than it would have under its traditional criteria, stressing the value of liberal entry as a means of sustaining price competition.[9] And, to accomplish its objective of increasing rate competition between carriers, the CAB began a novel approach to evaluating the low-fare proposals of particular applicants as an entry criterion of significant, even determinative, weight.[10] If none of the applicants had submitted a low-fare proposal, the CAB returned to its traditional carrier-selection criteria.[11]

The board began to emphasize its belief that "competition is the best guaranty that the traveling public will receive service responsive to its needs,"[12] and it adopted a policy that "competition on the basis of fares as well as service is not only permissible, but compelled."[13] The board believed that because "the freedom to enter markets provides the best assurance of price and service competition," it was "actively expanding the opportunities for airlines to serve new routes."[14]

The issuance of operating authority to a number of carriers on a permissive (rather than mandatory) basis also was perceived as a means of stimulating increased price competition.[15] In granting permissive authority, the board left to the business judgment of carrier management the extent to which competitive service would be offered.[16] Mandatory authority was not perceived as an effective means of insuring that a responsive level of service would be provided, for the board had traditionally been rather lax about enforcing such "common carrier" certificate obligations.[17] Moreover, the board did not want to place itself in a position where it would be forced to "compel an airline to provide unsubsidized service" that turned out to be "uneconomic."[18]

Latent permissive authority (i.e., operating authority that had been issued to a carrier but that the carrier was not actively using) was also viewed as posing a beneficial competitive stimulus to incumbents. "It represents a threat of entry and therefore provides a competitive spur to incumbents, if they fail to meet the public's service needs or if the market grows to the point that it can support another airline, the dormant carrier is free to enter at once without the need for a costly and time-consuming certification proceeding."[19] Furthermore, each operating authority had significant value in forcing carriers to adhere to the notion of threshold pricing (i.e., the threat of potential competition would encourage carriers to maintain prices at a level sufficiently low to forestall entry by new competitors; the carrier would, in a market it dominates, set a threshold price—a price above cost but low enough to make the market unattractive to potential competitors).[20] Thus, new entry, or the threat of entry, would be a competitive catalyst to pricing and service competition. This was subsequently to be christened the "theory of contestable markets."

The CAB was convinced that multiple awards, combined with downward pricing flexibility, would insure that the traveling public enjoyed the benefits of carrier innovation and rate competition.[21] Carrier management would have increased freedom to manage its affairs in response to consumer demand. Market forces would enable consumers to enjoy service by those carriers best suited to participate in the traffic. Indeed, the market was viewed as a superior mechanism (vis-à-vis governmental regulation) for selecting both the most efficient and economical participants and the most desirable combination of price and service options.[22] Consumer choice was also perceived as the best means of ascertaining the appropriate number and identity of carriers that should serve any particular market.

CALLS FOR CAUTION AND MODERATION DISMISSED

Although determined to inject more competition into air transportation, the CAB proceeded with caution at first, refusing to overload markets with too many carriers[23] (so as to protect the carriers serving the markets from

the harmful effects of excessive competition) and issuing operating authority to only that number of carriers it believed each market could adequately support.[24] Subsequently, the board became less concerned with
certificating only that number of carriers that the market could profitably
support. In fact, it began (at first implicitly and later explicitly) to authorize a number potentially too large to reasonably maintain profitable operations, saying:

It may happen that one (or more) of the carriers will find it unprofitable to continue operating in this market and will withdraw. Should that occur, we would
interpret that as a sign that the type of service provided by it is not desired by the
public. The choice is more efficiently made by the marketplace than by the Board.[25]

The board quickly began to consider the issuance of permissive authority to all "qualified" applicants,[26] convinced that "market forces would
more likely result in optimum service at optimum fares, for the market
selection process operates continuously and efficiently."[27] Reliance on market
forces is the rule, rather than the exception, in other sectors of the economy. The U.S. Department of Transportation argued that increased competition would lead to lower prices and improved service without subjecting the industry to destructive competition or excessive concentration and
without subjecting passengers to the dangers of unsafe operations.[28]

Several parties argued that the traditional regulatory structure should be
maintained. They contended that the objectives of increased rate and route
competition could be adequately accomplished without the indiscriminate
issuance of permissive authority to all applicants.[29] Automatic route awards
would eliminate the strongest incentive for pricing competition—the existing emphasis on low-fare proposals as a carrier selection criterion.[30] Indeed, a preferable approach to the adoption of a policy of multiple permissive entry might have been to retain carrier selection by stressing policies
of fostering new entrants, rewarding low-fare innovations, and encouraging industry competitive balance by strengthening smaller carriers.[31]

But the traditional system of carrier selection was perceived by the board
as having fostered a less efficient system than a policy of multiple permissive entry, a policy that would permit the marketplace to make ultimate
determinations with respect to price and service. The board asserted that
establishing opportunities for dormant authority would keep the potential
of new entry alive and thereby "keep incumbent carriers on their toes."[32]

Certain parties urged the board not to apply a policy of multiple permissive entry on an indiscriminate, universal basis. They generally emphasized the drastic differences between markets and contended that rational
regulation must be tailored to serve the spectrum of interests existing within
the markets. The needs of individual communities, it was contended, would
continue to vary widely regardless of the regulatory policies ultimately

adopted by the board. A flexible formula adaptable to the facts and circumstances of each case would be a far more rational means of regulation than would adoption of an inflexible general rule that could not be molded to satisfy the peculiar needs of individual markets.[33]

Other parties argued that ad hoc entry deregulation would create a destructive equilibrium through a process of route-by-route freedom of entry while the bulk of the regulatory structure would remain structurally unchanged. A policy of multiple permissive entry would, it was argued, create an irrational economic structure consisting of small enclaves of "free" entry within a comparatively closed and restricted environment. To apply such a policy would create a gerrymandered national route structure in which certain markets would be open to multiple entrants on a permissive basis while other markets would be served by certificated carriers holding mandatory authority to provide service.[34]

Still other groups urged the board to proceed with caution during a gradual transitional period from direct, pervasive regulation to greater reliance on free-market forces.[35] For example, Allegheny Airlines (today, USAir) insisted that after maintaining a "hot-house of protectionism" for 40 years, the board should not move too rapidly to throw the industry to the wolves of the marketplace, for such hasty action could be highly disruptive for consumers and the industry without any compensatory public benefits.[36]

The local-service airlines argued that the entry policies of 40 years of regulation placed large trunk-line carriers in an inherently superior position in terms of route system capabilities and equipment. They urged the CAB to phase in an open-entry policy gradually in a manner that would offer them compensatory route segments and preferential treatment to offset the clearly one-sided economic posture that regulation had established.[37] In response, the CAB stated:

[A] general policy of multiple entry . . . should not be limited to a few routes or areas; . . . it should be extended to the very core of the system and be broad enough (and carried out rapidly enough) to create substantial new competitive opportunities for all segments of the industry, including small trunklines and local service carriers.[38]

A number of small communities expressed the fear that unlimited entry might disrupt, inhibit, or effectively impede continuous or nonstop service and airport construction or expansion. They were also concerned that the "permissive" nature of new authority deprived them of any assurance that service, once inaugurated, would be maintained.[39] Further, there was no assurance that a carrier receiving a permissive authorization would even begin the new service, despite the board's finding that the public convenience and necessity *required* new service.

The CAB was implicitly unconcerned with the fate of those communities whose market demand was insufficient to attract or retain new service. It felt that the market would best distribute carriers and their aircraft according to the laws of supply and demand and that markets unable to generate sufficient traffic to support trunk-line carriers or nonstop service might be able to attract local carriers or multiple-stop service. If not, it was in the best interests of nationwide industry economies and efficiencies that they not be served.[40]

As to the question of airport financing, certain carriers and civic parties argued that distortions in carrier behavior and systemwide market perversities arising as a result of the artificial hybrid of heightened competition in some markets and a close regulatory system in others would impede future efforts to finance and construct the airport facilities necessary to accommodate the type of traffic growth the CAB was seeking to encourage.[41] The board was convinced, saying only that the issuance of permissive authority would not relieve carriers of their contractual obligations at those airports where space was leased.[42] Similarly, although it was pointed out to the board that an inherent barrier to new entry might be the absence of landing slots at major airports (e.g., Washington National and Chicago O'Hare), the board wholly refused to take into account the scarcity of such slots in its certification policies.[43]

In the *Oakland Service Case*[44] and the *Chicago-Midway Low-Fare Route Proceeding*,[45] the board abandoned its traditional approach in entry proceedings in favor of a revolutionary policy of granting permissive operating authority to any qualified carrier that applied for it.[46]

PROFITABILITY OF PROPOSED OR EXISTING OPERATIONS DEEMED IRRELEVANT

The CAB under Alfred Kahn abandoned any effort to scrutinize whether a proposal would be profitable within a reasonable period after its inauguration, leaving the investment decision solely to the discretion of business management.[47] The agency began to permit carriers to experiment freely with their transportation proposals in the marketplace,[48] giving little consideration to the economic injury suffered by incumbent carriers (which might, in fact, have been providing an exemplary level of service at reasonable rates). Potential profitability of proposed operations became an increasingly less important factor as the board issued permissive authority, leaving "the responsibility for providing good service to the public in a way that is profitable to the carrier . . . with the latter's management."[49] By awarding permissive, rather than mandatory, authority to the new entrant, the board would allow the carrier to withdraw from the market should its experimental service prove unprofitable.[50]

The board also became explicitly less and less concerned that existing carriers might suffer economic injury as a result of the implementation of its novel policies.[51] If existing carriers began to suffer financial injury as a result of new competition, the board was confident that they would take steps to reduce their losses or withdraw from the market.[52] Although new entry might well divert traffic and revenue from the incumbent and reduce its profits, so long as increased competition did not impair the carrier's ability to fulfill its certificated obligations, the board was content to permit market forces to run their course.[53] However, even under circumstances where it could be convincingly demonstrated that diversion of traffic and revenue might so jeopardize the economic viability of the incumbent's operation as to cause it to withdraw from the market (in which the CAB was inclined to inject a new entrant) or where its financial condition might be so impaired that it would be forced to terminate all of its operations, there was some doubt whether the board would exhibit some restraint in the application of its new entry approach.[54] In fact, under the former alternative, the CAB began to view incumbent withdrawal from a market as *"prima facie* evidence that a more efficient carrier had replaced a less efficient one, to the long-run benefit of the traveling public."[55] And as to the latter, the CAB did not interpret the Federal Aviation Act as requiring that it "try to guarantee the continued existence of any particular firm in the industry."[56]

Ultimately, the CAB announced that it no longer felt particularly inclined to protect the financial health of the carriers subject to its jurisdiction by moderating its entry policies, saying: "In healthy competition, producers who are inefficient or made bad decisions may fail, but efficient and well-managed producers can operate profitably. . . . The occasional failure can serve a useful purpose, not only by eliminating the inefficient or imprudent operator, but also by flashing a yellow light to others."[57] Diversion of traffic from existing carriers was perceived as having a useful purpose in "signaling consumer preferences to the industry and thereby serving as both an inducement and a prod to innovative and efficient operations."[58] The CAB no longer felt any responsibility to protect the revenues or market shares of any particular carrier subject to its jurisdiction no matter how efficient or economical its historic performance.[59]

POTENTIAL FOR DESTRUCTIVE COMPETITION DISMISSED

Several carriers[60] prophetically contended that the absence of entry controls would lead to a destructive situation: "More carriers will enter markets than the market can sustain, capacity will be offered for which there is no demand at a price which covers the cost of offering it, and all competitors will suffer losses in these markets."[61] This, in turn, would depress industry profits, discourage investment and the introduction of more technologically sophisticated aircraft, and lead to a deterioration in service. The long-term result would be a general oligopolization in the market.[62]

"Destructive" or "cutthroat" competition was defined by the CAB as a competitive situation in which (1) a powerful competitor seeks to drive rivals out of a market through the utilization of predatory tactics, with the hope of securing monopoly profits after they exit, or (2) all competitors operate at a price that consistently fails to meet the costs of even the most efficient.[63] As to the former, the board was convinced that multiple awards would not result in the type of destructive competition that the agency was compelled, by its governing statute, to prevent.[64] The board felt that it had ample alternative means to deal with the problem, means such as its authority over rates and its powers to deal with unfair competition.[65]

The latter type of destructive competition was perceived to exist only under those circumstances where capital was "long-lived and immobile" and where "through miscalculation competitors irretrievably" committed "too much capital to a particular market," a situation thought not to exist in the airline industry.[66] The airline industry was believed to have relatively insignificant economies of scale, low barriers to entry, reasonably elastic demand, and highly mobile resources.

Neither did the CAB believe that the contemporary economic environment was such that the destructive competitive wars that Congress sought to preclude by promulgating the act would occur as a result of unlimited entry.[67] The industry was perceived to be prosperous and stable, with fleet size in approximate equilibrium with demand,[68] thereby depriving any predatory-minded carrier from an opportunity to dump excess aircraft into markets already adequately served by existing carriers.[69] Additionally, it was believed that the capital markets had been disciplined by the traffic recession of the early 1970s and by increased fuel costs and therefore would probably not support irrational or uneconomic service.[70]

The board alleged that its implementation of a policy of multiple entry would not cause significantly more carriers to serve a market than would have served it had the Board employed traditional entry criteria and engaged in carrier selection.[71] This prediction was essential to support two of its other fundamental assertions: (1) that a multiple entry policy would not result in destructive or wasteful competition; and (2) that multiple entry would not result in profligate fuel consumption and concomitantly increased noise and air pollution.

POTENTIAL FOR OLIGOPOLISTIC MARKET DISMISSED

Several carriers were concerned that unlimited entry would lead to a long-term oligopolization of the airline industry. The "big five" (United, American, TWA, Eastern, and Delta) already enjoyed 75 percent of domestic trunk-line operating revenue.[72] Thus, the structural problem of the industry was not that it lacked potential competition but that it was dominated by a relatively small number of large firms, with oligopolistic tendencies already apparent.[73] Such oligopoly power would probably increase as a result of multiple permissive authorizations because large carriers could selectively use their resources to preempt those relatively few markets open for entry and capable of sustaining multiple carrier competition. Concurrently, the potential value of those few routes open for entry would be seriously diluted for newcomers as a result of excessive authorizations.

These parties argued that given the capital requirements of air transportation and the interrelationship of traffic flows that place a premium on the ability of a carrier to marshal traffic support from as many sources as possible, any type of open market structure affords the dominant carriers inherent advantages, which are exceptionally difficult to overcome. The most effective means for an incumbent, particularly a large and financially stable incumbent, to deter new entry would be to demonstrate that it would respond sharply and swiftly to the inauguration of new service. Because potential entry could be deterred by potential response, the elimination of competition through the employment of predatory tactics would be economically rational regardless of the number of entrants certificated by the board.[74] Although the traditional regulatory scheme permitted competition at reasonable costs, avoided destructive route and rate wars, and created meaningful opportunities for smaller carriers, the inevitable result of an implementation of a policy of multiple permissive entry, the parties argued, would be an increase in systems costs, short-term rate wars materially injurious to both the carriers and the public, and a practical limitation on entry opportunities to large, powerful carriers.[75]

The CAB insisted that unlimited entry would not lead to excessive concentration in the industry. It felt that the industry had relatively few economies of scale, beyond those of a relatively low initial threshold.[76] Equipment was viewed as being exceptionally mobile; it could be shifted from market to market or sold. "Therefore, even if more carriers initially move into a market than it can support, there is little irretrievable commitment of resources to prevent one or more . . . from withdrawing or reducing service and turning their attention to other markets where their capital assets can be better used and the public demands better served."[77] The board asserted that there are numerous markets in which smaller carriers compete successfully with larger ones. It added, "We fully expect that the industry will continue to have many healthy members, nor do we fear for a disappearance of profitable expansion opportunities for small and medium-sized carriers."[78]

NOTES

1. The board noted: "Commentators, after studying the cumulative effects of both the Board's and the industry's orientation towards service improvements rather than fare competition, are virtually unanimous in concluding that today's public

would benefit from a system in which there was more fare competition and greater price/service variety. This is our judgement as well." Las Vegas-Dallas/Fort Worth Nonstop Service Investigation, CAB Order 78-3-121 (1978), at 2.

2. *Id.* at 1–2.

3. *Id.* at 2.

4. *Id.* at 3.

5. *Id.* at 3.

6. *Id.*

7. *Id.*

8. *Id.* at 4.

9. *Id.* at 9; CAB Order 78-7-116 (1978), at 1.

10. *See* Miami-Los Angeles Low-Fare Case, CAB Order 78-1-35 (1978).

11. Midwest-Atlanta Competitive Service Case, CAB Order 78-4-13 (1978), at 7; Ohio/Indiana Points Nonstop Service Investigation, CAB Order 78-2-71 (1978), at 27.

12. Midwest-Atlanta Competitive Service Case, CAB Order 78-4-113 (1978), at 2.

13. *Id.*

14. *Id.*

15. Improved Authority to Wichita Case, CAB Order 78-3-78 (1978), at 4. The purported benefits of permissive vis-à-vis mandatory operating authority are discussed in Improved Authority to Wichita Case, CAB Order 78-3-78 (1978), at 5; U.S.-Latin America All-Cargo Service Investigation, CAB Order 78-4-44 (1978); and Baltimore-Detroit Nonstop Proceeding, CAB Order 78-5-112 (1978).

16. Eastern Air Lines-Piedmont Aviation Route Exchange, CAB Order 77-12-76 (1977), at 1.

17. This traditional policy may have been inconsistent with the existing statutory provision of Section 401(j) of the Federal Aviation Act, 49 U.S.C. § 1371(j), which required board approval as a condition precedent to the abandonment of a route.

18. Eastern Air Lines-Piedmont Aviation Route Exchange, CAB Order 77-12-76 (1977), at 2.

19. Improved Authority to Wichita Case, CAB Order 78-3-78 (1978); Atlanta-Charleston Competitive Nonstop Case, CAB Order 78-2-114 (1978), at 3.

20. The level of the threshold price depends on how easy it is for other firms to enter. In contrast, when a monopoly is guaranteed by the inherent market characteristics of the entry, the monopoly firm is free to set an excessively high price and reap monopoly profits, without fear of competition.

21. *See* Las Vegas-Dallas/Fort Worth Nonstop Service Investigation, CAB Order 77-7-116 (1978), at 11.

22. Oakland Service Case, CAB Order 78-4-121 (1978), at 34, 38.

23. *See* Memphis-Twin Cities/Milwaukee Case, CAB Order 78-6-20 (1978), at 2.

24. *See, e.g.,* Cincinnati-Washington Subpart M Proceeding, CAB Order 77-10-4 (1977), at 2; Improved Authority to Wichita Case, CAB Order 78-3-78 (1978), at 3.

25. Ohio/Indiana Points Nonstop Service Investigation, CAB Order 78-2-71

(1978), at 29; *see* Improved Authority to Wichita Case, CAB Order 78-3-78 (1978), at 4; Phoenix-Des Moines/Milwaukee Route Proceeding, et al., CAB Order 78-1-116 (1978), at 29.

26. By "qualified," the Board meant "fit, willing, and able." 49 U.S.C. § 1371(d)(1) (1979).

27. Florida Service Case, CAB Order 78-7-128 (1978), at 4; Air Wisconsin Certification Proceeding, CAB Order 78-8-196 (1978).

28. Brief of the Department of Transportation in Improved Authority to Wichita Case, et al. (on file with the DOT in Docket 28848, April 27, 1978).

29. Brief of North Central Airlines, Inc., in Memphis-Twin Cities/Milwaukee Case, et al. (on file with the DOT in Docket 29186, April 21, 1978); Brief of National Air Lines in Improved Authority to Wichita Case, et al. (on file with the DOT in Docket 28848, April 27, 1978).

30. *See* Brief of Pacific Southwest Airlines, Inc., in Improved Authority to Wichita Case, et al. (on file with the DOT in Docket 28840, April 27, 1978).

31. *Id.*

32. Oakland Service Case, CAB Order 78-4-121 (1978), at 42.

33. *See* Brief of Southern Airways, Inc. in Improved Authority to Wichita Case, et al. (on file with the DOT in Docket 28848, April 27, 1979); Brief of North Central Airlines, Inc., in Memphis-Twin Cities/Milwaukee Case, et al. (on file with the DOT in Docket 29186, April 27, 1978); Comment of the Indianapolis Airport Authority in Las Vegas-Dallas/Fort Worth Nonstop Service Investigation, et al. (on file with the DOT in Docket 29445, April 27, 1978); and Brief of Hughes Airwest, Inc., in Improved Authority to Wichita Case, et al. (on file with the DOT in Docket 28848, April 27, 1978).

34. Brief of Continental Air Lines and Comment of the Houston Pacific in Improved Authority to Wichita Case, et al. (on file with the DOT in Docket 28848, April 27, 1978).

35. *See* Comment of the Indianapolis Airport Authority in Las Vegas-Dallas/Fort Worth Nonstop Service Investigation, et al. (on file with the DOT in Docket 29445, April 27, 1978); and Comment of the Federal Trade Commission in Improved Authority to Wichita Case, et al. (on file with the DOT in Docket 21162, April 27, 1978).

36. *See* Brief of Allegheny Airlines in Ohio/Indiana Points Nonstop Service Investigation, et al. (on file with the DOT in Docket 21162, April 27, 1978).

37. Las Vegas/Dallas/Fort Worth Nonstop Service Investigation, CAB Order 78-7-116 (1978), at 4.

38. *Id.* at 5.

39. *See id.* at 4. Thus, certain local parties believed that should multiple carrier competition prove so uneconomic that all carriers holding permissive authorization withdrew, no carrier would be the first to reenter for fear of suffering the same fate. If a number of carriers held dormant authority to serve the market, there would be some reluctance by any of them to expand the requisite start-up costs and enter the market. The enhanced risk of inaugurating such service would lead carriers to emphasize other markets in their service offerings and neglect newly granted authority. *See* Oakland Service Case, CAB Order 78-9-96 (1978), 37–41, 47–50. Comments of the Louisville and Kansas Parties in Improved Authority to Wichita Case, et al. (on file with the DOT in Docket 28848, April 27, 1978); Brief

of the State of Minnesota in Memphis-Twin Cities/Milwaukee Case (on file with the DOT in Docket 29816, April 27, 1978).

40. *See* Oakland Service Case, CAB Order 78-9-96 (1978), at 34–35.

41. *See* the position of Delta Air Lines quoted in *id.* at 34, and Comment of Houston Parties in Improved Authority to Wichita Case, et al. (on file with the DOT in Docket 28848, April 27, 1978).

42. Oakland Service Case, CAB Order 78-9-96 (1978), at 35.

43. Applications of Colonial Airlines, Inc., CAB Order 78-6-183 (1978), at 4–6; Air Wisconsin Certification Proceeding, CAB Order 78-8-196 (1978), at 4.

44. CAB Order 78-4-121 (1978).

45. CAB Order 78-7-40 (1978).

46. Oakland Service Case, CAB Order 78-4-121 (1978), at 19.

47. Memphis-Twin Cities/Milwaukee Case, CAB Order 78-3-35 (1978), at 1, n.3, and 4.

48. *See* Applications of Colonial Airlines, Inc., CAB Order 78-6-183 (1978), at 14.

49. Cincinnati-Washington Subpart M Proceeding, CAB Order 77-10-4 (1977), at 2.

50. Application of Piedmont Aviation, Inc., et al., CAB Order 78-8-97 (1978), at 15.

51. Service to Richmond Case, CAB Order 77-5-69 (1977), at 5; Greenville/Spartanburg-Washington/New York Subpart M Case, CAB Order 77-10-1 (1977), at 7.

52. Service to Richmond Case, CAB Order 77-45-69 (1977), at 4, n.3; *see* Ohio/Indiana Points Nonstop Service Investigation, CAB Order 78-2-71 (1978), at 8–9.

53. Greenville/Spartanburg-Washington/New York Subpart M Case, CAB Order 77-10-1 (1977), at 8.

54. Ohio/Indiana Points Nonstop Service Investigation, CAB Order 78-2-71 (1978), at 8, n. 8.

55. Greenville/Spartanburg-Washington/New York Case, CAB Order 77-10-1 (1977), at 9.

56. Application of Piedmont Aviation, Inc., et al., CAB Order 78-8-97 (1978), at 9, n. 18.

57. Oakland Service Case, CAB Order 78-4-121 (1978), at 25.

58. Chicago-Midway Low-Fare Route Proceeding, CAB Order 78-7-40 (1978), at 25.

59. *See id.*

60. The carriers that made this argument were, predominantly, smaller carriers.

61. Oakland Service Case, CAB Order 78-4-121 (1978), at 25.

62. *Id.*

63. *Id.*

64. *Id.* at 24–32, 41.

65. Improved Authority to Wichita Case, CAB Order 78-3-78 (1978), at 4.

66. Oakland Service Case, CAB Order 78-4-121 (1978), at 26.

67. *Id.* at 32.

68. *Id.*; CAB Order 78-9-96 (1978), at 48–49; Las Vegas-Dallas/Fort Worth Nonstop Service Investigation, CAB Order 78-7-116 (1978), at 4–5.

69. Las Vegas-Dallas/Fort Worth Nonstop Service Investigation, CAB Order 78-7-116 (1978), at 4.

70. Oakland Service Case, CAB Order 78-4-121 (1978), at 32.

71. *Id.*

72. Brief of Pacific Southwest Airlines in Improved Authority to Wichita Case, et al. (on file with the DOT in Docket 28848, April 27, 1978).

73. Brief of North Central Airlines, Inc., in Memphis-Twin Cities/Milwaukee Case, et al. (on file with the DOT in Docket 29186, April 27, 1978).

74. Comment of National Airlines, Inc., in Improved Authority to Wichita Case, et al. (on file with DOT in Docket 28848, April 27, 1978).

75. *See* Comment of Houston Parties, *supra* note 41.

76. Oakland Service Case, CAB Order 78-4-121 (1978) at 37, citing "R. CAVES, AIR TRANSPORT AND ITS REGULATORS (1962); Crane, *The Economics of Air Transportation,* 22 HARV. BUS. REV. 495 (1945); Loontz, *Domestic Air Line Self-Sufficiency: A Problem of Route Structure,* 42 AM. ECON. REV. 103 (1952); REPORT OF THE CAB SPECIAL STAFF STUDY ON REGULATORY REFORM 102–7 (1975)." *Id.,* n. 46.

77. Las Vegas-Dallas/Fort Worth Nonstop Service Investigation, CAB Order 78-7-116 (1978), at 3.

78. Oakland Service Case, CAB Order 78-4-121 (1978), at 37.

17

THE AIRLINE DEREGULATION ACT
OF 1978

In 1978, Congress passed and President Carter signed into law the Airline Deregulation Act, which (1) dismantled the regulatory umbrella that had traditionally shielded the industry from destructive competition, and (2) abolished the Civil Aeronautics Board (as of 1985).[1]

The legislation substituted increased reliance on competition for classical price, profit, and entry regulation.[2] It reflected the economic view that increased competition in the airline industry would force prices down and eliminate excess capacity; if firms were free to set prices and enter markets without regulatory constraints, they would experiment in offering different combinations of price and service. Thus, the underlying theory of this legislation was that liberalized entry and pricing would force carriers to adhere to the competitive pressures of the marketplace to provide the range of price and service options desired by the public.

THE NEW POLICY DECLARATION

The statutory criteria governing all modes of transportation have traditionally been couched in inherently vague, if not vacuous, terminology. Congress recognized that it had neither the expertise nor the time to fulfill properly its obligations under Article 1, Section 8, of the Constitution: to regulate commerce between and among the several states. Therefore, it created regulatory bodies to develop the requisite expertise and gave them rather wide discretion to regulate the industry as they best perceived the fulfillment of the congressional intent. Furthermore, Congress recognized that the needs of the public would not remain static but that the optimum regulatory structure would evolve to meet the dynamic growth and maturity of our nation's commerce.

Hence, such statutory criteria as "public convenience and necessity," standing alone, have virtually no inherent meaning. Nevertheless, Congress set forth its declaration of policy in Section 102 of the act to indicate more specifically its interest as to precisely how transportation should be regulated, to give the agency some indication of the congressional purpose and the ultimate objectives for which the agency should strive, and to thereby breathe life into what might otherwise be virtually vague statutory phraseology.

The Airline Deregulation Act amended the Federal Aviation Act to establish a new declaration of policy for interstate and overseas transportation.[3] The declaration included 10 subsections that specified the criteria deemed by the board to be consistent with the public interest and the public convenience and necessity.

The first two stressed the importance of safety, emphasizing that this would be a policy objective of the highest priority[4] and that the board would prevent any deterioration in established safety procedures.[5] There can be no doubt that Congress intended that there be no diminution in the board's safety evaluation.

Two of these provisions also dealt with the role accorded to competition as a policy objective. Before the 1978 amendments, this section's only reference to competition was that the CAB should promote "competition to the extent necessary to assure sound development of the air transportation system." Under the Airline Deregulation Act of 1978, competitive market forces (including actual and potential competition) were to be employed "to provide the needed air transportation system, . . . to encourage efficient and well managed carriers to earn adequate profits and to attract capital"[6] and "to provide efficiency, innovation, and low prices, and to determine the variety, quality and price of air transportation services."[7]

Low fares were to be encouraged, as was the adequacy, economy, and efficiency of service, but "without unjust discriminations, undue preferences or advantages, or unfair or deceptive practices."[8] Similarly, the CAB was required to guard against "unfair, deceptive, predatory, or anticompetitive practices" and to avoid "unreasonable industry concentration, excessive market domination" and similar occurrences that might enable "carriers unreasonably to increase prices, reduce services, or exclude competition."[9]

In addition to promoting rate competition as a policy objective, the CAB was also directed to encourage new entry and route expansion by existing air carriers. The board was obligated to strengthen smaller carriers to insure a more competitive and effective industry.[10]

Finally, three other subsections promoted the prompt procedural disposition of regulatory proceedings,[11] encouraged use of satellite airports in urban areas,[12] and attempted to insure that reasonably adequate service be provided to small communities, with federal subsidies where appropriate.[13]

PC & N AND THE BURDEN OF PROOF

Traditionally, entry into air transportation by domestic carriers was governed by two statutory criteria:

1. the proposed service is *required* by the public convenience and necessity (PC & N), and
2. the applicant is fit, willing, and able.[14]

The burden of proof in application proceedings was, under section 556(d) of the Administrative Procedure Act,[15] on the applicant.

The 1978 act left these provisions unchanged for carriers seeking to serve international routes.[16] However, it significantly amended the entry criteria for domestic and overseas transportation (between points located within the territories and possessions of the United States, albeit over international waters) by requiring the CAB to issue a certificate when the proposed service was *consistent with* the PC & N.[17] The fitness standard remained unchanged.[18]

However, the burden of proof was shifted to an opponent (typically an incumbent carrier, which was required to demonstrate that the proposed operations were *not* consistent with the PC & N).[19] In order to deny an application for operating authority, the CAB was required to conclude, "based upon a preponderance of the evidence," that such transportation was "not consistent with the public convenience and necessity."[20] The burden of proof on the fitness issue remained unchanged.[21]

AUTOMATIC MARKET ENTRY

During the first month of 1979, 1980, and 1981, each certified passenger carrier could apply for nonstop route authority between any one pair of points (which had not been protected) by filing a notice.[22] The carrier was not required to demonstrate consistency with the public convenience and necessity. It must, however, have satisfied the fitness test.[23]

Each carrier could also protect from automatic entry one pair of points between which it already held nonstop authority.[24] The Airline Deregulation Act also included an escape clause enabling the board to modify the program if the program caused substantial harm to the national transportation industry or a substantial reduction in service to small and medium-sized communities.[25]

DORMANT AUTHORITY

A certificate authorizing transportation between two points was considered dormant if the certificated carrier had not provided at least five round-trips a week for 13 weeks during the preceding 26-week period.[26] The board was required to award the dormant route within 60 days to the first carrier submitting an application that demonstrated it had satisfied FAA regulations and was able to comply with the CAB's regulations,[27] unless the board concluded that the issuance of such a certificate was not consistent with the public convenience and necessity.[28] However, there was a rebuttable presumption that the authority sought was consistent with the PC & N.[29] If no more than a single carrier served the route, the board was required to suspend the dormant incumbent's authority for a 26-week period, unless it concluded that such suspension was unnecessary to encourage continued service by the newly authorized carrier.[30]

EXPERIMENTAL CERTIFICATES

If the CAB concluded that a test period was required to evaluate proposed new operations, it could issue a certificate for a temporary period.[31] If such a certificate was issued on the basis that the carrier would provide innovative or low-cost transportation, and the carrier failed to provide such service, the board could modify, suspend, or revoke the authority.[32]

OTHER ENTRY PROVISIONS

Carriers were allowed to carry domestic fill-up traffic on flights in foreign transportation. This privilege was limited to one round-trip daily.[33]

Carriers operating aircraft seating fewer than 56 passengers, or with cargo service of 18,000 pounds or less, were exempted from the certificate requirements of section 401.[34] The board's regulations had previously limited the commuter carrier exemption to aircraft seating 30 or fewer passengers.

As of December 31, 1981, the Airline Deregulation Act of 1978 terminated the CAB's licensing function insofar as it determined consistency with the public convenience and necessity.[35] The CAB also lost its jurisdiction over domestic rates in 1983. The board did, however, continue to make fitness determinations until it went out of existence on January 1, 1985, when these responsibilities were transferred to the U.S. Department of Transportation (DOT).

DOT also acquired jurisdiction over the remaining regulatory responsibilities in aviation, including consumer protection, the Essential Air Services small community subsidies program, international entry and rates, and mergers. The last was transferred to the U.S. Department of Justice on January 1, 1989.

NOTES

1. Pub. L. 95-504 (Oct. 24, 1978).

2. The Conference Committee emphasized that the purpose of the act was "to encourage, develop, and attain an air transportation system which relies on competitive market forces to determine the quality, variety, and price of air services." *Conference Report,* Airline Deregulation Act of 1978, Rep. No. 95-1779, 95th Cong., 2nd Sess., 53 (Oct. 12, 1978) [hereinafter cited as *Conference Report*].

3. 49 U.S.C. § 1302(a) (1979).

4. *Id.* § 1302(a)(1).

5. *Id.* § 1302(a)(2). Indeed, the act further provides, "The Congress intends that the implementation of the Airline Deregulation Act of 1978 result in no diminution of the high standards of safety in air transportation attained in the United States at the time of the enactment of this Act." *Id.* § 1301.

6. *Id.* § 1302(a)(4).

7. *Id.* § 1302(a)(9).

8. *Id.* § 1302(a)(3). This provision also encourages coordinated air transport operations, as well as "fair wages and equitable working conditions." *Id.*

9. *Id.* § 1302(a)(7).

10. *Id.* § 1302(a)(10).

11. *Id.* § 1302(a)(5).

12. *Id.* § 1302(a)(6).

13. *Id.* § 1302(a)(8).

14. Former section 401 of the Federal Aviation Act; former 49 U.S.C. § 1371 (1977).

15. 5 U.S.C. § 556(d) (1979).

16. *See* 49 U.S.C. §§ 1302(c), 1371(d)(1)(B) (1979). *See* P. DEMPSEY, LAW & FOREIGN POLICY IN INTERNATIONAL AVIATION (1987).

17. 49 U.S.C. § 1371(d)(1)(A) (1979).

18. *Id.* § 1371(d)(1).

19. *Id.* § 1371(d)(9)(B).

20. *Id.* § 1371(d)(9)(C).

21. *See id.* § 1371(d)(9)(A).

22. *Id.* § 1371(d)(7)(A).

23. *Id.*

24. *Id.* § 1371(d)(7)(C).

25. *Id.* § 1371(d)(7)(D).

26. *Id.* § 1371(d)(5)(A)(O).

27. *Id.* § 1371(d)(5)(A)(D).

28. *Id.* § 1371(d)(5)(F)(i).

29. *Id.* § 1371(d)(5)(F)(ii).

30. *Id.* § 1371(d)(J).

31. *Id.* § 1371(d)(8).

32. *Id.*

33. *Id.* § 1371(d)(6).

34. *Id.* § 1386(b).

35. *Id.* § 1551(a)(1)(A).

18

CAB IMPLEMENTATION OF THE AIRLINE DEREGULATION ACT

INDISCRIMINATE MULTIPLE PERMISSIVE ENTRY EXPLICITLY REJECTED

As the CAB was stripped of much of its regulatory authority, it also lost its chairman, Alfred Kahn, who was designated by President Carter to assault inflation (becoming the nation's "inflation czar"). There he was to preside over the highest inflation in peacetime history. Kahn was replaced as CAB chairman by an Arizona attorney, Marvin Cohen.

The most visible immediate effect of the Airline Deregulation Act was the line of carrier representatives that formed on Connecticut Avenue outside the offices of the CAB. The representatives stood there, exposed to the elements, for the several days between the passage of the act by Congress and the media signing ceremonies of President Carter. With their sleeping bags, folding chairs, and portable radios, scores of airline employees waited patiently in the cold of October for Carter to lay his pen to paper. Like the pioneers of the Oklahoma land rush, the air carriers were poised to storm the CAB to take advantage of the dormant authority provisions of the new act.[1] Within a month, the CAB had awarded operating authority to serve 238 dormant routes.[2] Virtually overnight, carriers such as Braniff had expanded their route systems by as much as one-third (and collapsed as a consequence).

Within two months of the promulgation of the Airline Deregulation Act, the CAB, in the *Improved Authority to Wichita Case,*[3] directly confronted the issue of whether it should adopt a broad policy of issuing multiple permissive authority to *all* "qualified" applicants in markets able to support some service. Its tentative conclusion, rendered before the enactment of the deregulation legislation in *Las Vegas–Dallas/Ft. Worth Nonstop Service Investigation,*[4] had been that the adoption of such a policy would "in most, and possibly in all, instances best meet the transportation goals of the Federal Aviation Act for the present and foreseeable future."[5] Although the CAB had felt confident that the economic and policy issues had been adequately addressed, the legal issues posed serious obstacles to the adoption of de facto deregulation of entry. Thus, the board had been reluctant to go forward with such a radical departure from the traditional regulatory structure, and from the legislative history and the act itself, until it had prepared a comprehensive legal analysis that had at least some possibility of surviving judicial scrutiny. The board's legal staff was actively engaged in the preparation of a legally defensible justification for such a policy when Congress passed the deregulation bill.[6]

Of course, the Airline Deregulation Act laid to rest much of the legal opposition to the adoption of a more liberal entry approach. Under the new act, the CAB would continue to evaluate the PC & N of proposed operating authority applications until 1982. Even during the interim, the burden of proof would be reversed; before an application could be denied, opponents of new entry would be forced to prove, by a preponderance of the evidence, that proposed operating authority was not consistent with the PC & N. This, coupled with the other liberalized entry provisions (e.g., dormant authority, automatic market entry), made it clear that Congress sanctioned the CAB's general policy of moderately liberalized entry. The board interpreted the legislative mandate as confirming and strengthening its "earlier conclusion that a general policy of multiple permissive licensing" was "the approach likely to produce the greatest transportation benefits."[7] It further found that these provisions created a rebuttable presumption in favor of issuing operating authority to any "qualified" carrier that requested it.[8]

In the *Improved Authority to Wichita Case,*[9] the CAB directly confronted the issue of whether it was prepared to abandon its statutory obligation to weight and balance the PC & N in individual operating authority application proceedings. In retrospect, its conclusion appears moderate:

Despite the new Act, however we are not prepared to conclude that a general policy of multiple discretionary entry, if adopted, should be applied universally. There might still be circumstances in which the public interest may be better served by giving only one or less than all qualified applicants immediate authority. (The Act obviously contemplates this possibility by retaining for three years a public convenience and necessity standard for route awards).

For example, it is at least arguable that in some small markets, where no service is feasible without an initial developmental effort or where demand is just on the verge of being able to support service, one airline should be given temporary protection from competition, in the first case, to provide it with the incentive to make the developmental investment and, in the second, to make sure the service is not delayed because potential entrants are scared off by multiple authorizations and the prospect of immediate competition.[10]

INDISCRIMINATE MULTIPLE PERMISSIVE ENTRY IMPLICITLY ADOPTED

The policy purportedly adopted in *Wichita* was not, however, the approach implemented by the board. In every case arising after the promulgation of the Airline Deregulation Act, the CAB rejected carriers' and civic parties' arguments that fewer than all "qualified" applicants should be certificated. The policy implemented was one of indiscriminate, multiple permissive entry, notwithstanding the board's assurances in *Wichita*[11] to the contrary.

A number of smaller carriers argued that Congress intended that the board utilize the interim, three-year period (between the promulgation of the Airline Deregulation Act of 1978 and the elimination of PC & N as an entry criterion on January 1, 1982) as an era of gradual transition during which the board would protect and strengthen the smaller carriers.[12] To this, the board responded that the objective of strengthening was not intended by Congress to be a "justification for noncompetitive awards."[13]

Small carriers maintained that the board was "moving too fast toward deregulation" and that multiple awards would "undermine the goals of strengthening small carriers and avoiding unreasonable industry concentration, excessive market domination, and monopoly power."[14] The board was unconvinced, arguing, "The superior traffic flows available to some large carriers can be largely offset by the advantages unique to the small carriers, such as their ability to develop regional service plans, and, in any event, such advantages as the large carriers may enjoy . . . may well be eroded as the result of free entry in the overall air transportation system."[15]

Smaller carriers argued that certificating more carriers than the market could support would likely encourage large carriers to drive them out, or under. Thus, in the *Southeast Alaska Service Investigation*,[16] Alaska Airlines (ASA) argued that multiple authorizations would endanger its ability to satisfy its certificate obligations, thereby causing "a reduction or loss of service to the smaller southeast Alaskan communities and bush points served by ASA."[17] This argument too was flatly rejected by the CAB, which said that its approach to strengthening small carriers was one of enabling them to take advantage of new route competition rather than shielding them from competition.[18] The Board continued: "We recognize that the greater reliance we now place on competition . . . means that airlines will be increasingly less willing and able to cross-subsidize loss operations with monopoly profits on other routes. . . . We no longer consider this a valid reason for restricting competition." [19]

Even under circumstances where the board recognized that small, remote communities would lose air service by a small carrier as a result of the application of its unrestrained, liberal entry policies, the CAB still refused to modify them, unless it was convincingly established that indiscriminate entry would result in the loss of service that could not be replaced.[20] No party could meet such a standard.

Actually, the board weighed the scales of decision making so heavily in favor of competition that no party was able to convince it that the deleterious consequences of multiple permissive entry outweighed the derived "benefits." For example, small carriers argued (to no avail) that they should be protected in certain markets against entry by large, trunk carriers. Despite the admission in *Iowa/Illinois-Atlanta,* that the market could support only a single carrier, and despite the arguments of the Cedar Rapids, Iowa, parties and Ozark Airlines that only Ozark should be certificated, the CAB proceeded to award operating authority to Northwest as well, saying, "if Ozark were to be driven out of the market by the entry of Northwest . . . we would be inclined to interpret such a result as prima facie evidence that the carrier offering the more attractive combination of benefits had won the competitive battle."[21]

Similarly, in the *Northwest Points–Puerto Rico/Virgin Islands Service Investigation,*[22] in which Eastern argued that the indiscriminate issuance of multiple permissive authority would cause it to suffer diversion of revenue that "could amount to tens of millions of dollars more than the $34 million estimated by applicants," the Board responded:

Diversion from an incumbent is not a significant consideration. . . . Eastern might be driven out of one or more of the markets. . . . Were this to occur . . . we would assume that the carrier offering the most attractive combination of benefits had won the competitive battle, to the ultimate advantage of the traveling public.[23]

In the *Austin/San Antonio–Atlanta Service Investigation,*[24] civic parties argued that multiple licensing would be undesirable because of "the critical shortage of airport terminal space" and that multiple awards "could greatly inconvenience the public by congesting present airport facilities."[25] These arguments too were rejected by the CAB in much the same manner as it had previously rejected a similar argument that the application of its liberal entry policy should be modified to reflect the scarcity of landing slots at certain airports.[26] Here again, the board refused to temper its "liberal certification policy for the sole purpose of trying to avoid possible practical problems that new entrants could pose to airport authorities."[27] Further, the board emphasized, "We are not now inclined to deny entry to any qualified applicant, simply in order to avoid airport congestion."[28]

The decision of the CAB in *Wichita* had been unanimous. But by the time *Austin/San Antonio–Atlanta* had been rendered, CAB Member Richard J. O'Melia had begun to realize that the majority had no intention of deviating from a strict application of a multiple permissive entry policy and had no intention of moderating this policy along the lines suggested in *Wichita*. Member O'Melia vigorously dissented from the majority's decision in *Austin/San Antonio–Atlanta*, saying, "I dissent, because the Board . . . is unnecessarily and . . . woodenly imposing a multiple permissive award policy designed to bring about deregulation today rather than after the transition period prescribed by Congress, and because it appears more concerned with the doctrinal concept of competition than with the real-world demands for air service."[29] Member O'Melia proceeded to cite the policy adopted in *Wichita*, that there "might still be some circumstances in which the public interest may be better served by giving only one or less than all qualified applicants immediate authority." What followed suggested that he felt deceived by the majority's assurances in *Wichita*:

It is because this recognition of an obvious truth was included in *Wichita*, because the policy of multiple permissive awards was not declared to be an inflexible imperative, that I supported and approved the Board's conclusions in that case. And it certainly is not an unpopular proposition; the civic parties uniformly in this case and in [other] cases . . . have begged the Board not to inflict on their respective communities the alleged benefits of multiple permissive awards. Indeed, it should give the Board pause that its multiple permissive policy . . . is being greeted around the country with dismay and outright hostility. If the benefits of multiple permissive authority are that evident, why is there such widespread lack of enthusiasm for it?

The fact is that communities and civic parties recognize that multiple permissive awards provide no assurance of effective and predictable service. . . . [T]he Board is no longer interested in selecting the carrier or carriers that might best serve a market . . . in determining when service will commence . . . [or] with whether service is to be viable, or reliable, or continuous. We are going to turn over those concerns to the marketplace. It is my view that the wholesale abdication of responsibilities during the licensing transition period is not what Congress had in mind and is not consistent with the Deregulation Act.[30]

Without admitting it, the CAB had effectively adopted an indiscriminate policy of multiple permissive entry, for it had systematically rejected every argument that it should moderate its approach. The burdens it placed on opponents of new entry were so onerous that, realistically, they could not be overcome. The board was determined to deregulate, no matter what arguments were made about the deleterious consequences of the blind application of an economic philosophy case in concrete. In the CAB's own words, the board was "determined to extend competition to the very core of the national transportation system."[31]

SHOW-CAUSE PROCEEDINGS: THE FLOODGATES BURST

In the months immediately preceding promulgation of the Airline Deregulation Act, the CAB began to grant new entry opportunities to air carriers through the procedural vehicle of a "show-cause order"—a means of disposing of issues without an oral evidentiary hearing. Thus, it granted hundreds of applications for certificate amendment,[32] route realignment,[33] and even the addition of new segments (when the amendment appeared to be in the nature of restriction removal),[34] through the show-case vehicle. Although the board had begun to issue operating authority more liberally and hastily than ever before, it nevertheless intended to restrict such procedures to only those instances in which it could be claimed that there existed no material facts or complex economic issues.[35] In part, this caution stemmed from the procedural requirement of a "public" hearing set forth in Section 401(c) of the Federal Aviation Act,[36] which the CAB had traditionally interpreted to constitute a requirement for a full "trial-type" evidentiary hearing held before an administrative law judge.

With the promulgation of the Airline Deregulation Act, the obligation for a "public" hearing in routes proceedings was eliminated,[37] and new expeditious procedures were substituted. Within the first month under the new provisions, the board had issued a plethora of "boilerplate" orders setting applications for show-cause disposition, explicitly adopting a policy of multiple permissive entry in these proceedings and thereby creating significant new segments of entry opportunities for air carrier applicants.[38] Soon, the CAB was issuing instituting orders by the bushel-basketful as quickly as its secretaries could type and its copying machines could duplicate. The board was not ashamed to issue orders of incredible redundancy, all employing virtually identical language.[39] The parties and the markets might vary from case to case, as might the name of the proceeding, but the language, the intent, even the ultimate conclusion were essentially the same—authority would be granted to any and all who applied for it.[40]

The haste and carelessness with which the board was issuing massive quantities of certificated operating authority through the show-cause vehicle ultimately led Member O'Melia to register a vigorous dissent in the *Milwaukee Show-Cause Proceeding.*[41] The majority in *Milwaukee* had tentatively decided to grant all applications filed by "qualified" carriers for any conceivable domestic route with which Milwaukee could be linked. Unlike its predecessor orders, in which specific markets at issue were designated, Milwaukee was, virtually, geographically infinite. Member O'Melia was outraged, saying:

This is the first time that the Board has used the show-cause procedure to mount a handout of route awards of undefined geographic magnitude. . . . The door is being thrown open to *any* application for *any* market so long as it involves *Milwaukee* authority.

There are two consequences that particularly concern me. The first is that, rather than the phased and orderly transition to deregulation that Congress mandated, the clear meaning of the Board's action here is instant deregulation. . . .

The second . . . is that we are unnecessarily, improperly, and in a very shameful manner destroying one of the strengths of an administrative agency like the board—its quasi-judicial nature and function. The shameful part is that the destruction is being carried out not with clean direct surgical strokes, but by draining out the reason for being of our judicial process. With no facts to be analyzed, with no law to be interpreted and followed, what is the point of having a judicial process? . . .

Is it worth it to assemble parties, counsel, recorder, and judge in these route cases merely to bear witness to an act of ritualistic genuflection?

Why don't we put an end to this pretense of being a quasi-judicial agency? We are making a mockery of the formal adversary proceeding as the traditional way of determining factual and legal issues in licensing cases. Why don't we discontinue all other pending proceedings on route applications—there must be a couple dozen of them actively being processed—and tell the applicants that we will mail them their route awards after we show-cause them? We don't need a law judge to recite the catechism of multiple permissive authority . . . .

This gutting of our judicial process, this mockery of evidentiary hearings, combined with the telescoping of the transition period, is not, in my opinion, what the Airline Deregulation Act contemplates. . . . And I feel ever so strongly that this is not in the best interest of the consumers, the carriers, and the communities of our country.[42]

Member O'Melia consistently argued, with great fervor, that the majority's approach violated the congressional intent that there be an orderly transition to entry deregulation, that the board was "making a mockery of the administrative process by the meaningless, noncognitive licensing process used to mass produce certificates, but not service; and, most important of all, [that it was] more interested in 'theoretical goals' and 'doctrinal conclusions' than . . . in air service."[43] Although he recognized that many larger markets were receiving additional service as a result of the board's liberalized entry policies, many small and medium-sized communities were not. Moreover, many carriers were complaining that they were not able to continue providing service at smaller communities because the Board's indiscriminate entry policies diluted their strength in larger, lucrative markets.[44] These, and all other arguments made by carriers and civic parties, were consistently rejected by a majority of the board under the conviction that indiscriminate entry was a panacea for any problems that might be created by the application of its multiple permissive entry policy.

Member O'Melia objected to the majority's "blatant disregard of the congressional mandate to administer a transition period and tailor [its] acts to the practical needs of the industry and the public."[45] Further, he saw "only a single-minded push to complete deregulation among [his] colleagues."[46] If the majority was content to proceed in this mindless deregulatory frenzy, he had a novel suggestion for the future regulatory mechanism:

If the recent voting pattern of the Board is to continue, we might just as well reprogram the sausage machine. . . . [T]hen all we have to do is turn the machine to "automatic," and it will turn out multiple permissive awards as efficiently and quickly as we can. That being the case, I have one final suggestion. Assuming, *arguendo*, that the majority is correct, that there is no transition period, then I see no need for any present or future Board members. Our resignations would be a savings to the American taxpayers.[47]

By granting operating authority to any and every carrier that applied for it, despite the possibility of deleterious consequences to carriers and communities, and by refusing even the opportunity to be heard orally, the board had become, in Member O'Melia's estimation, little more than a "sausage machine," grinding out grants of operating authority as fast and thoughtlessly as the wheel would crank.

FITNESS RENDERED IMPOTENT AS AN ENTRY CRITERION

The Traditional Fitness Criteria

Pursuant to Section 401(d) of the Federal Aviation Act,[48] the CAB was directed to issue certificated operating authority where it concluded, inter alia, that the applicant was "fit, willing, and able" to perform the proposed air transportation services[49] and to conform to the provisions of the act and the board's rules, regulations, and requirements.[50] Today, the DOT exercises this authority. Although the act does not define the terms "fit, willing, and able," the board traditionally evaluated three primary factors in its analysis of the applicant's operations: (1) the existence of a proper organizational basis for the conduct of air transportation, (2) the presence of a plan for the conduct of the service made by personnel shown to be competent in such matters, and (3) the availability of adequate financial resources.[51]

The financial posture of an applicant seeking authority to perform air carrier service was of paramount importance in the evaluation of its fitness.[52] Congress intended that only those carriers that could convincingly demonstrate minimum financial strength and sufficient stability to protect the public from abuse or risk should be authorized to perform air transport operations.[53] Although the CAB recognized that the criteria for measuring these requirements could not be determined with mathematical precision, financial posture, experience, operating plans, and compliance disposition have historically proven to be among the most important factors considered.

Traditionally, as important as the evaluation of an applicant's financial posture was the determination of whether an applicant was operationally fit. Both the applicant's experience and its operating proposal were deemed relevant to this issue.[54] Among the multitude of factors evaluated by the board in its determination of whether an applicant was operationally fit[55] were whether (1) the applicant's financial position was relatively secure, and it appeared able to satisfy its obligations as they matured; (2) it possessed a substantial fleet of insured flight equipment; (3) it had established a satisfactory maintenance program; (4) its management held extensive experience in airline operations; and (5) it had satisfactorily demonstrated a willingness and ability to provide the proposed operations with due regard for the protection of the traveling and shipping public (by maintaining sufficient liability and property insurance and by expressing a willingness to adhere to the board's regulations involving reasonable guarantees to the public).[56]

If an applicant failed to submit a reasonably defined plan for its proposed operations, had not demonstrated that those operations would eventually be profitable, and had not proven that its financial condition was sufficient to sustain those services, then, even assuming that the public convenience and necessity required institution, the authority was ordinarily denied.[57]

Fitness in the Post–Airline Deregulation Act Environment: Erosion of the Traditional Standards

As has been indicated, the Airline Deregulation Act did not diminish the fitness issue as a potential barrier to entry in any way. The burden of proving fitness remained with the applicant; and the CAB was obligated to continue its fitness scrutiny of carriers until 1985—long after its PC & N obligations had expired—when such jurisdiction was assumed by DOT. The first two subsections of the new declaration of policy emphasized the overriding importance of safety as a regulatory obligation of the highest priority.

One would have assumed, then, that Congress had intended that the board continue, if not make more stringent, its quasi-judicial interpretation of its fitness responsibilities discussed above. The board did precisely the opposite.

The first important regulatory diminution of the fitness standards came in the *Chicago-Midway Low-Fare Route Proceeding*.[58] In *Chicago-Midway,* the board acknowledged an interrelationship between fitness and safety. Although it argued that operational safety was principally the obligation of the Federal Aviation Administration, it admitted that passengers could reasonably assume that the issuance of operating authority by the CAB represented a determination by the board that the carrier had the requisite personnel, compliance disposition, and financial ability to operate properly.[59] Nevertheless, the board felt compelled to relax the traditional fitness standards so that they would be compatible with the thrust of multiple permissive entry[60] and "would not unnecessarily discourage new entry into the industry in the name of consumer protection."[61]

As a result, the board in *Chicago-Midway* designed a simplified, streamlined test whereby a carrier could easily establish its fitness. The CAB required that an applicant adduce evidence that it

(1) will, before inaugurating its operations, have the managerial skills and technical ability to operate safely;

(2) if not internally financed, has a plan for financing which, if implemented, will generate resources sufficient to commence operations without undue risk to consumers;

(3) has a proposal for operations reasonably satisfactory to meet a part of the demand for service in the city-pair markets embraced in its application; and

(4) will comply with the Act and the rules and regulations promulgated thereunder.[62]

In the *Transcontinental Low-Fare Route Proceeding,*[63] the board further expanded the second criterion of *Chicago-Midway.* Although the board, as recently as 1977, had required a new operator to demonstrate that it possessed "resources commensurate with the nature and scope of the undertaking" sufficient to enable it to operate safely (i.e., that the firm possessed either sufficient capital to operate the proposed service or commitments from investors or lending institutions to provide the requisite capital),[64] the board in *Transcontinental* believed that this requirement "could impose a serious barrier to entry."[65] It therefore eliminated the obligation that a carrier demonstrate its ability to actually obtain the requisite capital to commence reasonably safe operations.[66] An applicant for operating authority need merely proffer a financial plan that, if implemented, would generate sufficient financial resources to commence operations.

The board claimed that relaxation of the fitness criteria would not impair the safe operations of carriers subject to its regulations (and thereby endanger the lives of passengers), saying that if a carrier "cannot operate, the carrier will exist on paper only."[67] True, but would it not be possible for shoestring operators to secure the capital necessary to inaugurate some *de minimis* service for a limited period of time while skimping on equipment, maintenance, and replacement parts? The board had repeatedly emphasized that there are relatively few economic barriers to entry or economies of scale in the airline business. In the highly competitive environment the board was attempting to create, as prices approached marginal costs, would not a real incentive exist for even established incumbents to cut costs and defer maintenance?

CAB Member O'Melia recognized the potentially deleterious consequences that were likely to occur as a result of the deterioration of the traditional fitness criteria. He issued a vigorous dissent on the fitness issue:

My colleagues today have . . . [enshrined] some multiple permissive dogma to control the meaning of "fit, willing, and able" and have gone on to impale the Board on a dangerous notion of what constitutes fitness. The pronouncement on "qualifications is tantamount to a determination that the financial resources of an applicant for route authority have practically no relation to its fitness to provide air transportation. This key determination of what is a critical, statutorily mandated prescription—one which is legislated to endure even after the Board's licensing authority has terminated—has been reached with a sleight-of-hand maneuver that has in terms of its potential impact no parallel in my experience with agency action. . . . From this day forward, an aspiring entrepreneur need only show that in a set of perfect circumstances the proposed operations could be feasible.[68]

For carriers that already provided scheduled certificated operations, the board's fitness scrutiny was perfunctory, at best.[69] Most were not even required to adduce evidence consistent with the *Chicago-Midway* criteria (even as diluted by *Transcontinental*); instead, their fitness was regularly established by "officially noticeable data."[70]

After *Transcontinental,* the board proceeded on a course that further eroded the traditional fitness standards. For example, in the *Florida Service Case,*[71] Administrative Law Judge Dapper, concerned with the poor financial condition of Southeast Airlines, limited its operating authority to a period of one year so that, at the end of this trial period, the board could reexamine the carrier's financial health.[72] This has been the traditional means used by the board to insure that such a poor economic position would not endanger the safety of a carrier's operations. And, traditionally, this has had a prophylactic effect; carriers recognized that if they allowed their financial posture or, more significantly, the safety of their operations to deteriorate further, they would jeopardize renewal of their certificates.

The term-limitation approach was abandoned in *Florida* in favor of the imposition of several conditions intended "to assure that consumers would not suffer unduly if the company's financial condition were to deteriorate further."[73] Such conditions were primarily in the nature of bonding requirements[74] and obligations to file certain documents.[75] Although the board was confident that if Southeast was unable to solidify its poor economic position, it "would terminate the service rather than allow its corporate finances to deteriorate further,"[76] the board made no mention of the possibility that the carrier might instead defer maintenance in order to reduce costs.

NOTES

1. The board had interpreted the new provisions to require the issuance of dormant authority on a "first come, first served" basis. Thus, conceivably, the first individual in line (who, incidentally, represented the nation's largest air carrier, United Air Lines) could have applied for and received *all* of the segments that were dormant. There was no sanction for nonperformance, except perhaps loss of the route to another applicant under the dormant authority provisions once the statutory 26-week period had expired.

2. Improved Authority to Wichita Case, et al., CAB Order 78-12-106 (1978), at 4, n. 1.

3. CAB Order 78-12-106 (1978)

4. CAB Order 78-7-116 (1978).

5. *Id.* at 2.

6. Paul Stephen Dempsey was, in fact, among the attorneys in the board's Office of General Counsel who were delegated the responsibility of preparing the legal justification for application of a general policy of multiple permissive entry.

7. Improved Authority to Wichita Case, et al., CAB Order 78-12-106 (1978), at 6. The board further argued that the new act effectively rendered moot the issue of whether it could issue "permissive" operating authority. The CAB felt that under the new act, Section 401(j), 49 U.S.C. § 1371(j) (1979), had been amended to delete the PC & N requirement of board approval as a condition precedent to route abandonment. The new provision merely required prior notice for termination, suspension, or reduction of service, and no more, or so the board argued. Hence, the board seemed to believe that virtually all operating authority was now to be permissive in nature. *See* Improved Authority to Wichita Case, et al., CAB Order 78-12-106 (1978), at 11–12.

Actually, this conclusion was not compelled by the language of the Airline Deregulation Act. First, Section 401(j)(1)(B), 49 U.S.C. § 1371(j)(B) (1979), provides, "The Board may . . . authorize such temporary suspension of service as may be in the public interest." This language would seem to suggest some requirement of board approval as a condition precedent to a "temporary suspension of service." The new provisions made no mention of what, if anything, was to be done with a proposed termination or reduction in service.

Moreover, the new Act established new procedures under Section 419, 49 U.S.C. § 1389 (1979), to guarantee essential air transportation to small communities. Included among these provisions is one requiring an incumbent carrier (notwithstanding its compliance with the Section 401[j] procedures) to continue to provide "essential air transportation" (*see* Section 419[f], 49 U.S.C. § 1389[f]) to an eligible point (*see* Sections 419[a][1] and 419[b][1], 49 U.S.C. §§ 1389[a][1], 1389[b][1]) for consecutive periods of 30 days, under circumstances where the CAB was unable to find a replacement carrier, even utilizing the inducement of subsidy. 49 U.S.C. § 1389(a)(2)(B)(6) (1979). Under such circumstances, the board was powerless to permit exit; it must require the incumbent to continue its operations. In this sense, operating authority was clearly mandatory, the board's arguments to the contrary notwithstanding.

8. Improved Authority to Wichita Case, et al., CAB Order 78-12-106 (1978), at 10.

9. CAB Order 78-12-106 (1978).

10. *Id.* at 8–9.

11. *Id.*

12. *See* St. Louis-Louisville and San Francisco Bar Area Nonstop Case, CAB Order 79-4-79 (1979), at 4.

13. *Id.* at 8.

14. Norfolk-Atlanta Subpart M Proceeding, CAB Order 79-10-202 (1979), at 1.

15. *Id.* at 3. The board stated, "Multiple awards . . . is the proper remedy for curing high traffic shares." *Id.*

16. CAB Order 78-4-168 (1978).

17. *Id.* at 7.

18. *Id.*

19. *Id.* at 8.

20. *See* Spokane-Montana Points Service Investigation, CAB Order 79-4-80 (1979), at 3.

21. Iowa/Illinois-Atlanta Route Proceeding, CAB Order 78-12-35 (1978), at 5.

22. Order 78-12-105 (1978).

23. *Id.* at 4.

24. CAB Order 79-3-9 (1979).

25. *Id.* at 10 (*quoting* Austin Briefs, pp. 5–6).

26. Applications of Colonial Airlines, Inc., CAB Order 78-6-183 (1978).

27. Austin/San Antonio-Atlanta Service Investigation, CAB Order 79-3-9 (1979), at 11.

28. *Id.* at 12.

29. *Id.,* dissent at 1.

30. *Id.,* dissent at 2.

31. Boise-Denver Nonstop Proceeding, CAB Order 79-5-74 (1979), at 5, n. 21.

32. *See, e.g.,* Application of Hughes Airwest, CAB Order 78-10-120 (1978).

33. *See, e.g.,* In the Matter of United Air Lines, Inc., CAB Order 78-9-59 (1978).

34. *See, e.g.,* Application of Piedmont Aviation, Inc., et al., CAB Order 78-9-148 (1978), at 5, n. 10.

35. *See* Colonial Airlines, Inc., CAB Order 78-6-184 (1978).

36. Former 49 U.S.C. § 1371(c) (1977).

37. 49 U.S.C. § 1371(c) (1977). The board could still set routes application for public hearing, *id.,* Section 1371(c)(1)(A), but it was no longer so compelled.

38. *See, e.g.,* In the Matter of Western Airlines, Inc., CAB Order 78-11-66 (1978).

39. Application of Piedmont Aviation, Inc., et al., CAB Order 79-1-104 (1979), at 5–6.

40. *See, e.g.,* Application of Trans World Airlines, Inc., et al., CAB Order 79-3-48 (1979).

41. CAB Order 79-3-13 (1979).

42. *Id.,* dissent at 1–6 [citation omitted]. *See also* Northern Tier Show-Cause Proceeding, CAB Order 79-3-60 (1979), dissent.

43. Boise-Portland/Seattle/Spokane Show-Cause Proceeding, CAB Order 79-8-160 (1970), dissent at 1.

44. *Id.*

45. Application of American Airlines, Inc., CAB Order 79-10-186 (1979), dissent at 4.

46. *Id.,* dissent at 5. The board, he pointed out, was no longer concerned

about the public's air travel convenience or its necessity for reliable service to all points on the national air transportation system. Rather, we are only interested in a grand experiment in transportation economics—complete deregulation of air transport. If the experiment fails because carriers can no longer afford to serve marginal markets, that fact will merely be a footnote in some professor's textbook.

Id., dissent at 4. This is "some professor's textbook."

47. *Id.,* dissent at 5. If the voting patterns in favor of indiscriminate entry were to continue, Member O'Melia stated, "We have become no more than overpaid rubberstamps." *Id.*

48. 49 U.S.C. Section 1371(d) (1979).

49. United Air Lines v. Civil Aeronautics Board 278 F.2d 446, 449 (D.C. Cir. 1960).

50. This statutory language is almost identical for both scheduled and supplemental carriers (compare 49 U.S.C. Sections 1371 [d][1], [d][2], and [d][3] [1979]).

51. Braniff Airways v. Civil Aeronautics Board, 147 F.2d 152, 153 (D.C. Cir. 1945).

52. Western Air Lines, Inc. v. Civil Aeronautics Board, 495 F.2d 145, 154–55 (D.C. Cir. 1974).

53. Supplemental Air Service Proceeding, 45 C.A.B. 231, 267 (1966).

54. *See, e.g., id.* and Pennsylvania Cent. Air., Youngstown-Erie-Buffalo Op., 1 C.A.A. 811 (1940).

55. 37 C.A.B. 96, 97–99 (1962).

56. *See also* Johnson Flying Service, Inc., Interim Certificate, 37 C.A.B. 120, 121–22 (1962).

57. Airline Transport Carriers, Inc., d/b/a California-Hawaiian Airlines, Supplemental Air Service Case, 39 C.A.B. 200, 303 (1963).

58. CAB Order 78-7-40 (1978).

59. *Id.* at 49–57.

60. *Id.* at 49–50.

61. Transcontinental Low-Fare Route Proceeding, CAB Order 79-1-75 (1979), at 25.

62. Chicago-Midway Low-Fare Route Proceeding, CAB Order 78-7-40 (1978), at 50.

63. CAB Order 79-1-75 (1979).

64. Eugene Horbach, Acquisition of Modern Air Transport, CAB Order 77-3-88 (1977), at 7–9.

65. CAB Order 79-1-75 (1979), at 26.

66. *Id.*

67. *Id.*

68. *Id.,* dissent at 1.

69. *See, e.g.,* Phoenix-Las Vegas-Reno Nonstop Service Investigation, CAB Order 78-12-38 (1978), at 4.

70. *See* Milwaukee Show-Cause Proceeding, CAB Order 79-3-13 (1979), at 6.

71. CAB Order 79-9-177 (1979).

72. *Id.*
73. *Id.*
74. *Id.*, appendix C, at 1.
75. *Id.*, appendix C, at 2.
76. *Id.*, appendix C, at 1.

19

THE DEMISE OF THE CIVIL AERONAUTICS BOARD

The experience of the Civil Aeronautics Board reveals that economic regulation can be like a swinging pendulum. Over a period of time, regulatory philosophy can swing in either direction; and, it can swing too far. In one direction, the pendulum may swing in favor of protecting the industry from the deleterious effects of "excessive," "wasteful," or "destructive" competition. This was the regulatory philosophy that characterized the CAB between 1938 and 1976, when the board gave excessive protection to tightly regulated carriers against route and rate competition. Between 1976 and 1978 was an interim period for the board, when the pendulum began to move away from protectionism in favor of increased reliance on free market forces. After 1978, the pendulum swung fully in the direction of unlimited, unrestrained competition.

Both extremes violated the congressional intent. That intent has recently been described by the U.S. Supreme Court as imposing on transport regulatory agencies the obligation "to strike a fair balance between the needs of the public and the needs of the regulated carriers."[1] This "fair balance" was lost both in the era of excessive protectionism (1938–75) and the era of excessive competition (1979–present).

In enacting the 1938 legislation, "Congress made it clear that, while it was moving to safeguard against the excesses of destructive and unrestrained competition, it was in favor of the competitive principle and opposed to a principle of monopoly."[2] In its determination to protect the industry from excessive competition, the CAB refused to permit the entry of new trunk-line carriers (although more than 80 local service and other airlines had been authorized); it refused to allow the bankruptcy of a single inefficient carrier; it generally discouraged significant airfare competition; and in the early 1970s, it imposed a moratorium on the issuance of new operating authority. As a result, by 1975, there were fewer competitors; the industry suffered from excessive investment, excessive nonprice competition (i.e., "frills"), excessive capacity, and inadequate profits. In general, the board failed to allow the industry to enjoy the beneficial effects of regulated competition—the increased economies and efficiencies of operation that would have arisen as a result of modestly increased pricing and entry competition.

The interim period (1976–78) proved that carriers, encouraged to compete, would offer passengers new, innovative price and service options. This would, in turn, stimulate passenger demand (which was inherently elastic) and thereby enable carriers to fill empty seats on aircraft. Although the board during this period arguably exceeded the congressional intent and the perimeters of the Federal Aviation Act, increased load factors and decreased "frills" led to the highest industry profit levels in history. Carriers were encouraged to improve the economy and efficiency of their operations; passengers were permitted to enjoy airfares set at a more competitive level. It was a win-win situation for both consumers and carriers.

Yet, the board after 1978 violated the congressional will at least as reprehensibly as it did during the worst excesses of the protectionist era. Certainly, Congress intended that air transportation be deregulated, and it established a specific time table for the elimination of regulatory scrutiny of various carrier activities. But clearly too, Congress did not intend for the board to deregulate entry until 1982, or pricing until 1985; it implicitly designated 1978–85 as a transition period during which the board would gradually expose the highly regulated common carrier system to the rigors of the marketplace and allow communities to adjust to the evolving traffic patterns of an air carrier system that was responding to the needs and demands of the market. Congress also encouraged the CAB to employ the transition period as a vehicle to insure continued and viable service at small communities and to enhance the possibility of long-term competition by strengthening small carriers. Had Congress intended that there be no transition period, it would most certainly have opened the floodgates in 1978.

Deregulation should not be viewed as an end in itself; it should instead be perceived as a means to an end, a tool with which to secure a much more important objective—competition. Viewed from this perspective, regulatory means may frequently accomplish the objective of enhanced competition more efficiently than their reckless abandonment. For example, long-term competition could have been most effectively enhanced by strengthening small carriers during the transitional period (strengthening would likely have increased their ability to withstand the aggressive and predatory competition of larger carriers). Although unlimited entry may have increased short-term competition, it unnecessarily impaired the ability of smaller carriers (many of which owe their modest size to the regulatory policies of the CAB) to compete on a long-term basis.

Many small communities also urged the board to shield a carrier or two from unlimited entry in particular markets so that they might be encouraged to provide service. This the board has refused to do, awarding operating authority to every applicant no matter how small the market. Again, the loss of competitive service (or *all* service) in many of these markets had an unfortunate long-term impact on competition.

As we shall see, however unsatisfactory the profits or concentrated the industry became under regulation, they would pale in significance when compared to the results of deregulation. Under deregulation, the industry would lose all the profits it had earned since the Wright brothers' flight at Kitty Hawk.

NOTES

1. Trans Alaska Pipeline Rate Cases, 436 U.S. 631, 653 (1977).
2. Continental Air Lines v. CAB, 519 F.2d 944, 953 (1975).

Part III
THE RESULTS OF DEREGULATION

20

CONCENTRATION

ALLOCATIVE EFFICIENCY, COMPETITION, AND CONTESTABILITY

Deregulation's proponents believed that, freed from the shackles of government, the airline industry would become more competitive, providing the range of price and service options dictated by consumer demand, tapping the elasticities of demand with lower prices, filling capacity, enhancing efficiency, and improving profitability. They also believed that neither safety nor small community access would unduly suffer.[1]

Destructive competition, whose purported existence gave birth to the regulation of the airline industry in the 1930s, was deemed unlikely to occur.[2] But this apparent consensus among economists concealed a basic difference about what a "healthy competitive environment" required. An old joke has the borrower of a jar returning it broken and being asked to explain. He responds that he never borrowed it and, moreover, that it was broken when he got it. There is a similar conflict between the two views on why deregulation would stimulate competition, conflict that appeared among its advocates, sometimes in the same person.

The "traditionalist" view, as it might be called, whose adherents in the 1970s included many free market economists, held that competitive pricing required a sizable number of competitors. Based on some academic studies that failed to find significant economies of scale[3] in air transportation, the adherents argued that a deregulated industry *would* have sufficient competitors to satisfy the traditional notion of workable competition. In the absence of any cost advantages of big firms over small, there would be no motive to merge to achieve such nonexistent economies. This line of argument, then, denied that the air transport industry was a "natural monopoly" (or oligopoly) due to falling unit costs. If costs do not fall over some dimensions, then the view of the industry as being prone to bouts of destructive competition—the view that motivated the early architects of airline regulation—was also called into question. The tendency of prices to approach marginal cost when there is unrestrained competition would not then imply that losses were inevitable with more than a few competitors or the corollary that the prolonged presence of a sizable number of competitors was unlikely.

A second argument for deregulation was based on the notion of "contestability."[4] Some deregulation proponents did not deny that air transport had significant economies of scale, scope, or density and other natural monopoly characteristics, but they insisted that these characteristics need not be a problem because a natural monopolist would be forced to price at cost by the threat of *potential* entry. Thus, markets that were not competitive in the traditional sense of having many competitors might yet be "contestable" under certain conditions, conditions that the airline industry was alleged to fulfill.

There are three key assumptions to this theory. First, the potential entrant has access to the same technology as the incumbent (there are no absolute cost advantages for the incumbent). Second, entry into and exit from a particular market is costless—there are no "sunk" costs involved. Third, consumers respond to a price reduction on the entrants' part more quickly than incumbents can respond with a matching price cut. If these assumptions were satisfied, the mere *threat* of a postentry price matching by the incumbent would not deter entry. Unless prices always remained at cost, there would be an incentive for costless entry to grab some of the monopoly rent, for however short a period, followed by costless exit when the incumbent matched the entrant's lower price. In the airline example, the potential entrant could fly in its "capital on wings" to grab the rent that could be captured by a slight undercutting of the incumbent and then fly out when the incumbent actually matched, thus avoiding a price war and the associated losses altogether. This possibility would then force the natural monopolist to price at cost at all times.

Alfred Kahn's writing provides instances of both these arguments, despite the logical tension between them. Because he was articulate and passionate about deregulation, we turn to him for instances of each. First, traditionalism. A decade ago, Kahn dismissed fears that the industry would become highly concentrated—large airlines, he argued, had no advantages over small ones. Testifying before a House subcommittee in 1977, Kahn was asked the following question by Congressman Roman Hruska:

You are going to invite into the area of new entry the severest competition between airlines who service that particular market and ultimately the big will eat the little, and those who are able to withstand the severe competition and the reduced fares—

even below operating expenses—will prevail. Then the airlines that cannot prevail, of course, will have to go out of business or do something else.

After that transition period then you are going to see the air fares go back up again and the big will control the airline industry.[5]

Kahn dismissed these fears as unfounded:

First, the assumption that you are going to get really intense, severe, cut throat competition just seems to me unrealistic when you are talking about a relatively small number of carriers who meet one another in one market after another. We don't find in American industry generally when you have a few relatively large carriers competing with one another that they engage in bitter and extended price wars.

But number two, the fear that the big will eat the little, that is one that I would really like to nail. . . . [W]hile the average certificated carriers in the United States stock is selling at about two-thirds of book value . . . three of the five biggest carriers' . . . stock is selling at 33 to 37 percent of book value. . . . That means to me the investors do not believe that prediction.[6]

Similarly, in 1977 congressional hearings Kahn said, "I do not honestly believe that the big airlines are going to be able to wipe out the smaller airlines, if only because every study we have ever made seems to show that there are not economies of scale."[7]

True to his traditionalism, Kahn is not unconcerned about the substantial concentration that exists in air transport today, contrary to his expectations. Today Kahn admits that, in advocating deregulation, he misperceived the advantages of the large firms in the airline industry.[8] Now he says, "We underestimated the importance of economies of scale and scope."[9] Elsewhere, Kahn has conceded, "We advocates of deregulation were misled by the apparent lack of evidence of economies of scale."[10] In a 1988 article, he admitted that prices are likely to rise, saying, "I have little doubt that . . . the disappearance of most of the price-cutting new entrants and the marked reconcentration of the industry—will produce higher fares."[11] Similarly, in congressional testimony in 1987, Kahn said, "The industry has become more concentrated at the national level because of mergers and airline failures, and that means in my judgment that price competition may well become less severe in the years ahead."[12]

The trouble is that transportation has simply turned out not to be the ideal model of perfect competition that the traditionalist proponents of deregulation insisted it was. There appear to be significant economies of scale, scope, and density, which will be discussed in detail in the next section.

For an example of the nontraditionalist view (the view that airline trans-

port, though naturally concentrated, nevertheless exhibits "contestability"), we turn to Alfred Kahn again. In the late 1970s, Kahn proclaimed:

Almost all of this industry's markets can support only a single carrier or a few:
their natural structure, therefore, is monopolistic or oligopolistic. This kind of
structure could still be conducive to highly effective competition if only the government would get out of the way; the *ease of potential entry into those individual
markets, and the constant threat of its materializing,* could well suffice to prevent
monopolistic exploitation.[13]

Entry, or more precisely the *threat* of potential entry, would keep monopolists from extracting monopoly profits. This was the essence of contestability theory. In 1977, Kahn testified before a House subcommittee:

A realistic threat of entry by new and existing carriers on the initiation of management alone is the essential element of competition.
    It is only this threat that makes it possible to leave to managements a wider
measure of discretion in pricing. It is the threat of entry that will hold excessive
price increases in check.[14]

In a recent interview, Alfred Kahn noted, "Certainly one of the assumptions behind airline deregulation was that entry would be relatively easy."[15]
    As with the traditionalist prediction of many competitors and few size
economies, the actual deregulation experience has seemed to mock the
nontraditionalist scenario of "contestable" airline markets. Testifying before the Senate Judiciary Committee in 1987, Kahn appeared far less enthusiastic about the potential benefits of contestability:

I attack the easy assumption of the ideologues of *laissez faire* that contestability
takes care of everything; that private parties cannot monopolize airline markets
because the minute they raise their price two bits, there will be a rush of competitors into the market.
    I know of seven studies now of airline pricing since deregulation. They all con-
clude that while, yes, airline markets are relatively easy to enter, the potential entry
of competitors is no substitute for competitors already there. . . .
    Now, the view that contestability of airline markets makes antitrust enforcement
unnecessary is very close to the position that DOT is taking [in the airline merger
cases].
    Contestability is not a sufficient protection, in my opinion, and anybody who
looks at the airline industry certainly knows that the likelihood and opportunity
of entry, particularly by new carriers—low-cost, price-cutting carriers—has greatly
diminished in recent years and is likely to remain much lower than before.[16]

We will see that both the traditionalists and the nontraditionalists were
wrong: after a preliminary bout of classically destructive competition, deregulation has produced a highly concentrated oligopoly. Such concentration followed a rash of mergers and expansions directed at capturing the

scale economies that the traditionalists denied existed. Further, this oligopoly, contrary to the nontraditional view, fails to act like a competitive firm,
pricing at cost, but exploits its market power.

INDUSTRY ECONOMIC ANEMIA

Although destructive competition in the airline industry during the 1930s
was a major rationale for economic regulation in this industry, deregulation's proponents insisted that deregulation would not create destructive
competition. Kahn again can set the scene for us. In a speech before the
New York Society of Security Analysts in 1978, he characterized the opposition to airline deregulation as follows:

The most general fear about [airline deregulation] is that when the CAB withdraws
its protective hand from the doorknob, the door will open to destructive competition—to wasteful entry and cut-throat pricing—that will depress profits, render the
industry unable to raise capital, and so cause a deterioration in the service it provides—on the whole, it must be admitted good service.[17]

Kahn saw the fear as unrealistic. Testifying before a congressional committee, he insisted, "I just do not see any reason to believe that an industry
which is potentially rapidly growing, for which there is an ever-growing
market, cannot prosper and attract capital."[18] Kahn scoffed at deregulation's opponents, who believed that "there is something about airlines that
drives businessmen crazy—that once the CAB removes its body from the
threshold, they will rush into markets pell-mell, en masse, without regard
to the size of each, how many sellers it can sustain, and how many others
may be entering at the same time."[19] But in fact, as a decade of empirical
evidence reveals, deregulation *has* brought about cutthroat pricing, a miserable level of industry profitability, insufficient capital to reequip aging
fleets, and a deterioration of service.

Since deregulation began, the airline industry has suffered the worst economic losses in its history.[20] This period of economic anemia began before
the onset of the economic recession of the early 1980s and ascending fuel
prices and continued steadfastly afterward.[21] The airline industry's average
annual net profit margin over the first 11 years of deregulation was a meager 0.7 percent, compared with 4.5 percent for other U.S. industries.[22] By
1991, the airline industry had lost all the profit it earned since the Wright
brothers' inaugural flight at Kitty Hawk, plus $1.5 billion more.

Ten years after he persuaded Congress to pass the Airline Deregulation
Act, Alfred Kahn wrote, "There is no denying that the profit record of the
industry since 1978 has been dismal, that deregulation bears substantial
responsibility, and that the proponents of deregulation did not anticipate
such financial distress—either so intense or so long-continued."[23]

As noted above, deregulation was largely premised on the assumption that there were no significant economies of scale or barriers to entry in the airline industry. New competitors, it was argued, would spring up to challenge the entrenched incumbents, and the industry would become hotly competitive. In the short run, more than 120 new airlines appeared, although most were small commuter lines.[24] This flood of entry caused prices to spiral downward. A short-term boon for consumers, the price competition that emerged from deregulation was an unmitigated catastrophe for the airline industry and therefore, in the long run, for consumers as well. In the long run, more than 200 airlines have gone bankrupt or been acquired in mergers,[25] and only 74 carriers remain.[26] Only 8 are of significant size, and only about half of those are viable. Among the casualties are such darlings of deregulation as Air Florida, Freddie Laker's Skytrain, Midway, and Donald Burr's People Express. Kahn once pointed to these new upstart airlines as evidence that deregulation was a brilliant success. But they have all since dropped from the skies. As this book goes to press, America West remains, but languishes in bankruptcy.

The price wars, the erosion of profitability, and the industry shakeout that occurred in the aftermath of deregulation provided a textbook illustration of the unique economic characteristics that make transportation inherently vulnerable to price wars and excess capacity. Transportation firms sell what is, in essence, an instantly perishable commodity. Once an aircraft taxis down the runway, any unused capacity is lost forever. Empty seats cannot be warehoused and sold another day, as could, say, canned beans. This inevitably leads to distress-sale pricing during weak demand periods or when excess capacity, created by unlimited entry, abounds.

The short-term marginal cost of adding another passenger to a scheduled flight is virtually nil—printing another ticket, adding another meal and a few drops of fuel, for example. Any ticket sold makes some contribution. Hence, strong incentives exist to sell empty seats for whatever will lure a bottom to fill them.[27] Carriers competing head to head spiral downward in destructive competition. In such circumstances, although carriers cover short-term marginal costs, fixed costs are necessarily ignored.

It was these rather unique and brutal characteristics of air transport that led to distress-sale pricing in the early 1980s, following deregulation. To survive this darkest financial period in the history of domestic aviation, carriers had no choice but to slash wages, trim service and maintenance, and defer new aircraft purchases. The insistence on the part of the deregulators in seeing air transport as just another industry, an almost willful ignorance on their part of the historical experience of destructive competition in transportation—the experience that led to regulation in the first place—has had grave but perfectly predictable consequences.

Airlines needed monopoly opportunities to stem the economic brutality of destructive competition, so they merged and developed hub-and-spoke systems, giving them regional and city-pair market power. It is natural for firms facing extinction to seek out or create monopoly market opportunities to afford them the market power to raise prices. Thus, the large number of industry bankruptcies and mergers and the growth of national and regional (hub) concentration owe their existence to the destructive competition unleashed by deregulation.

CONCENTRATION

National Concentration

The intense destructive competition unleashed by deregulation has reduced the number of major competitors at the national level through waves of bankruptcies and mergers, to the point that the airlines have become, in the words of Alfred Kahn, an "uncomfortably tight oligopoly."[28]

There were 51 airline mergers and acquisitions between 1979 and 1988. More than 20 of those were approved by DOT after 1985, when it assumed jurisdiction over mergers. Fifteen independent airlines operating at the beginning of 1986 had been merged into six megacarriers by the end of 1987. The six largest airlines increased their passenger share from 71.3 percent in 1978 to 80.5 percent in 1990.[29] The eight largest airlines accounted for 81 percent of the domestic market in 1978 and 95 percent in 1991.[30]

The Department of Transportation approved *every* airline merger submitted to it after it assumed the Civil Aeronautics Board's jurisdiction over mergers, acquisitions, and consolidations on December 31, 1984. The Airline Deregulation Act of 1978 insisted that the agency guard against "unfair, deceptive, predatory, or anticompetitive practices" and avoid "unreasonable industry concentration, excessive market domination" and similar occurrences that might enable "carriers unreasonably to increase prices, reduce services, or exclude competition."[31] But these admonitions fell on deaf ears at DOT, which never met a merger it didn't like.

For example, DOT approved Texas Air's (i.e., Continental and New York Air) acquisition of both People Express (which included Frontier) and Eastern Airlines (which included Braniff's Latin American routes),[32] United's acquisition of Pan Am's transpacific routes, American's acquisition of AirCal, Delta's acquisition of Western, Northwest's acquisition of Republic (itself a product of the mergers of North Central, Southern, and Hughes Airwest), TWA's acquisition of Ozark, and USAir's acquisition of PSA and Piedmont.[33]

Nor are these likely to be the last of the mergers. By the end of the century, there may be as few as nine or ten global megacarriers.[34]

We are left with a situation aptly summarized by airline analyst Morton Beyer:

The 11 major airlines have shrunk to eight; the eight former local service carriers are now two and they are trying to merge; the eight original low-cost charter airlines have been reduced to one, through bankruptcy and abandonment; 14 former regional airlines have shrunk to only four; over 100 new upstart airlines were certificated by the CAB and about 32 got off the ground and most of those crashed, leaving only a handful still operating; of the 50 top commuters in existence in 1978, 29 have disappeared. . . .

Today, the top 50 commuter carriers who constitute 90 percent of that industry are captives of the major carriers, in part or in total owned, controlled, and financed by the giant airlines and relegated to serving the big airlines at their hubs.[35]

Hub Concentration

Kahn blames the emergence of what he characterizes as an "uncomfortably tight oligopoly"[36] in domestic air transportation on the Department of Transportation's permissive approach to airline mergers. "They have been *permitted* by a totally, and in my view indefensibly, complaisant Department of Transportation. It is absurd to blame deregulation for this abysmal dereliction."[37] But, as becomes particularly clear on examining the deregulation-induced growth of "fortress" hubs, mergers and acquisitions alone cannot explain the growing concentration of the industry. Even without mergers, the trend was to reconfigure routes in such a way as to constitute a de facto parceling out of airports among ostensible competitors. Lax antitrust policy only aggravated this basic trend.

All but three hub airports are now dominated by a single airline, with more than 60 percent (and sometimes 90 percent) of landings, takeoffs, gates, and passengers. Since deregulation, all major airlines have created hub-and-spoke systems, funneling their arrivals and departures into and out of hub airports, where they dominate the arrivals, departures, and infrastructure.[38] Whereas entry and exit regulation formerly constricted their geographic operations, deregulation has freed airlines to leave competitive and smaller markets and consolidate their strength into regional hub and city-pair market monopolies and oligopolies. The destructive competitive environment of deregulation has led them to seek out monopoly opportunities to stem the hemorrhaging of dollars. Ironically, a lax antitrust policy may have saved the industry from a plethora of bankruptcies. But as the dust settles on the bankruptcies and mergers of deregulation and on the hub consolidation facilitated by unlimited entry and exit, we see a horizon devoid of meaningful competition.

Clearly, the merger of Northwest and Republic resulted in sharply increased levels of concentration at Minneapolis/St. Paul and Detroit; and equally clearly, the same happened at St. Louis when DOT approved the merger of TWA with Ozark Airlines. But as table 20.1 reveals, massive hub concentration has occurred at a large number of cities where no merger had a significant impact.

Table 20.1
Single-Carrier Concentration at Major Airports: Pre- and Post-Deregulation

Airport	1977	1987
Baltimore/Washington	24.5% USAir	60.0% USAir*
Cincinnati	35.0% Delta	67.6% Delta
Detroit Metropolitan	21.2% Delta	64.9% Northwest
Houston Intercontinental	20.4% Continental	71.5% Continental
Memphis	40.2% Delta	86.7% Northwest
Minneapolis/St. Paul	45.9% Northwest	81.6% Northwest
Nashville Metropolitan	28.2% American	60.2% American
Pittsburgh	43.7% USAir	82.8% USAir
St. Louis-Lambert	39.1% TWA	82.3% TWA
Salt Lake City	39.6% Western	74.5% Delta
AVERAGE	33.8%	73.2%

*includes Piedmont

Source: CONSUMER REPORTS, June 1988, at 362–67.

To these figures add the excessive levels of concentration that have also emerged in the monopoly hubs of Cincinnati (now 88 percent Delta), Charlotte (95 percent USAir), Dallas Love (91 percent Southwest), Dayton (64 percent USAir), Detroit (73 percent Northwest), Houston (80 percent Continental), Newark (53 percent Continental), Philadelphia (53 percent USAir), Raleigh (82 percent American), and Washington Dulles (68 percent United) as well as the duopoly hubs of Chicago (85 percent American and United), Dallas (93 percent American and Delta), and Denver (83 percent Continental and United).[39] Even Chicago O'Hare, Dallas and Atlanta Hartsfield are increasingly dominated by a single carrier. In 1977, United had 29 percent of all boardings in Chicago; by 1988, it had 50 percent.[40] American controls 61 percent of Dallas/Ft. Worth. Even before the bankruptcy of Eastern, Delta controlled 62 percent of Atlanta (it now has nearly 90 percent).[41] After Frontier was absorbed, first by People Express and then by Continental (Texas Air), no hub airport enjoyed the three-carrier competition that had existed at Denver.[42]

Indeed, *the explanation for significant levels of hub concentration at all but Detroit, Minneapolis/St. Paul, and St. Louis is not DOT's generous approval of airline mergers but simply the entry and exit opportunities unleashed by deregulation.* Carriers adopting particular cities as hubs have increased frequencies and leased more gates while incumbent airlines have quietly exited in favor of market dominance opportunities of their own in other hub airports.[43] Freedom to enter and exit markets is the very heart of deregulation, and it is responsible for concentration at more hub airports than is the DOT's "dereliction," "abysmal" though this clearly is.[44] The CAB would almost certainly not have approved the widespread entry and abandonments that produced this massive hub concentration.

A study prepared by Dr. Julius Maldutis confirms the high levels of hub concentration resulting from deregulation. Maldutis reviewed concentration levels at 50 of the nation's busiest airports between 1977 and 1987, calculating the Herfindahl-Hirschman Index (HHI) for each. The HHI is the methodology employed by the U.S. Department of Justice for determining acceptable levels of concentration for antitrust review. It provides a measure based on squaring the market shares of individual firms and adding them together. For example, a firm with a 100 percent monopoly would have an HHI of 10,000. Under the Justice Department's analysis, an HHI below 1,000 is presumed unconcentrated; an HHI of between 1,000 and 1,800 is believed moderately concentrated; and an HHI of above 1,800 is deemed highly concentrated. By 1987, 40 of these 50 airports had an HHI above 1,800; in other words, 80 percent of these airports were highly concentrated. Moreover, Maldutis calculated the weighted average of concentration for all 50 airports, finding that it rose from an HHI of 2,215 in 1977 to 3,513 in 1987.[45] This corresponds to a decline in the number of "effective"[46] competitors in the average of the 50 airports from 4.51 in 1977 to 2.85 in 1987.

Hub concentration translates into escalating fares. The *New York Times* has observed, "Passengers who live in a hub city and begin their flight there end up paying higher fares, in some cases 50 percent more than they would had deregulation not occurred."[47] The General Accounting Office found that, after its merger with Ozark, TWA increased fares 13 to 18 percent on formerly competitive routes radiating from St. Louis.[48] A similar study compared fares in markets radiating from Minneapolis-St. Paul, in which Northwest and Republic formerly competed, and found that rates rose between 18 and 40 percent.[49]

In 15 of the 18 hubs in which a single carrier controls more than 50 percent of the market, passengers pay significantly more than the industry norm.[50] A recent study by the U.S. Department of Transportation of 9 hub airports (see table 20.2) found that fares at all but 2 increased faster between 1985 and 1988 than did the airline component of the Consumer Price Index (11.1 percent).

The General Accounting Office compared 1988 fares at 15 concentrated[51] hub airports with fares at 38 unconcentrated airports and found average fares 27 percent higher at the hubs.[52] The higher fares at concentrated

Table 20.2
Airline Hub Market Shares and Price Increases between 1985 and 1988

Hub Airport	Dominant Carrier	Fare Increases
Atlanta	Delta (62%)	5%
Charlotte	Piedmont (89%)	34%
Cincinnati	Delta (81%)	25%
Detroit	Northwest (62%)	27%
Minneapolis	Northwest (77%)	21%
Pittsburgh	USAir (80%)	-6%
Raleigh	American (67%)	35%
St. Louis	TWA (83%)	22%
Salt Lake City	Delta (77%)	26%

Source: WASHINGTON POST, February 5, 1989, at H2, col. 5.

airports do not reflect a premium for nonstop service, since the average number of coupons per traveler at concentrated airports was virtually identical to that at the comparison unconcentrated airports (2.26 vs. 2.28 coupons). And the difference persisted when average trip length was controlled for by excluding from the comparison group of airports those where average trip length was significantly longer than at concentrated airports. Thus neither a higher proportion of nonstops nor a higher proportion of short-haul (and thus more costly) flights can explain the fare premium at concentrated airports, GAO concluded. GAO also found that the increase in fares from 1985 to 1988 was generally greater at concentrated airports and that the increase in fares was especially dramatic when a carrier established dominance during the period (providing further confirmation of the effect of concentration on the fares that had been documented in GAO's earlier study of airfares at St. Louis after the TWA-Ozark merger). Finally, the study found that in 13 of 15 of the concentrated airports, the dominant carrier had higher fares, in some cases very much higher, than other carriers at the same airport.

A recent study by Severin Borenstein[53] found that the relationship between airport dominance and the level of fares stands up to sophisticated econometric analysis that controls for cost and quality effects on fares. His estimates imply that "a 10 percent increase in the average endpoint enplanement share for an itinerary would lead to a 4.3 percent increase in average fare."[54]

City-Pair Concentration

Many defenders of deregulation dismiss critics' concerns about the un-precedented levels of *national* concentration permitted by airline deregu-lation, on the grounds that the relevant markets are not national but are "city-pair" markets—the market for air transport between a particular pair of cities. Thus a Congressional Budget Office (CBO) study of airline deregulation[55] contended:

While there has been a substantial increase in industry concentration since 1983, there has not been a corresponding increase in concentration at the market level. . . . The effective number of carriers serving [city-pair] markets of more than 200 miles with 25 or more passengers per day has grown from 2.4 carriers in 1983 to 2.5 carriers in 1987.[56]

The CBO did not provide data on the earlier period (1978–83) but characterized the evidence as indicating a significant increase in competi-tion over the period as a whole. Since the latter part of the period saw an increase of a scant one-tenth of a competitor, any "significant" increase would have to have come in the earlier period, before the consolidation of the industry after 1983. In a later section of its report, the CBO claimed (without citation) that at the time of passage of the Airline Deregulation Act, "the average city-pair with non-stop flights was served by 1.4 car-riers." Using this figure, we can clearly see that for all practical purposes, new entry had all but ceased by 1983 and that the "significant" increase in competition in question amounts to a change from an effective monop-oly (1.4 competitors), but from a *regulated* monopoly to an *unregulated* duopoly in the average city-pair market. Given the doubts that have arisen on the score of the "contestability" of airline markets, and thus the du-bious role of potential entry in disciplining the actions of incumbent car-riers, it is difficult to take a great deal of solace in the "increased compe-tition" in the average market, for which deregulation is, by this measure, responsible.

Furthermore, there are problems with market definition. The figures pre-sented above, and those used by most of the proponents of deregulation, pertain to the provision of *single-carrier* service between two cities, either nonstop or indirectly through connections over the carrier's hub. This is, from one perspective, too broad a focus and, from another, too narrow.

If it is believed that nonstop service has unique attractions making it a separate market compared with connecting service, the direction of change in concentration in this narrower market is reversed. The effective number of carriers providing nonstop service *fell* in the period from 1983 to 1987 (the period for which the CBO study gave data)[57] for the average city-pair

Table 20.3
Number of City-Pair Markets Receiving Service by One or More Scheduled
Carriers

No. of Carriers	Number of Markets	
	Oct. 1978	July 1988
1	4,093	3,481
2	899	1,054
3	233	413
4	80	192
5	21	83
6	14	45
7	9	22
8	6	14
9	2	4
10+	2	6
TOTAL	5,359	5,314

Source: Department of Transportation, analysis of Official Airline Guide Data, printed in
TRAFFIC WORLD, Dec. 5, 1988, at Supp. B.

market. On the other hand, a broader definition of the market for air
transport between two cities would need to include not just single-carrier
connecting service but also interlining possibilities, which have been drast-
ically reduced in the deregulation period due to the rise of hub and spoking coupled with the tendency toward hub dominance noted above.[58] The
arbitary definition of the market, which *includes* single-carrier connecting
flights but *excludes* interline connections, thus biases the resulting picture
of changes in concentration toward a showing of more competition.

Like the CBO, Alfred Kahn has insisted that the airline industry is more
competitive after deregulation because there are now fewer monopoly *city-
pair* markets, despite the increase in industry concentration.[59] Table 20.3
sustains this claim.

But the same caveats made with regard to the CBO's argument apply
here. It is true that the overall number of monopoly markets has fallen
since deregulation. But remember that, under regulation, a monopolist cannot
extract monopoly rents from buyers because its rates are required by law
to be "just and reasonable." Neither telephone companies nor electric util-

ities can charge monopoly rates despite their monopoly position because their rate and service levels are regulated by governmental agencies. But an unregulated monopoly can charge whatever the market will bear.

In 1978, single firms, which dominated 76 percent of America's city-pair markets, were limited by the Civil Aeronautics Board to charging "just and reasonable" rates and earning no more than a reasonable return on investment. Driven by productivity improvements and by technological growth, real airfares fell significantly during the four decades of regulation. But today, monopoly carriers in nearly two-thirds of America's city-pair markets can charge whatever the market will bear. At the time the Airline Deregulation Act was before Congress, Kahn urged, "No automatic [pricing] freedom should be allowed in markets dominated by a single carrier."[60] Today, nearly two-thirds of our nation's city-pairs are unregulated monopolies.

Nor are duopolies hotbeds of competition. Two firms may implicitly agree to lethargic price and service competition, enjoying in effect a "shared monopoly." In 1978, 93 percent of America's markets were *regulated* monopolies or duopolies; in 1988, 85 percent of America's markets were *unregulated* monopolies or duopolies. Statistically, that suggests an improvement. But again, remember that today, no government agency protects the public against monopoly pricing and the extraction of monopoly profits.

Thus, whether we look at national, airport, or city-pair measures of concentration, the traditionalist argument for deregulation seems to have been refuted by the empirical evidence. Economies of size (scale, scope, and density), the putative absence of which was at the heart of the traditionalist case for deregulation, seem to be pervasive. Former DOT Assistant Secretary Matthew Scocozza confessed, "To be very honest, in 1978 we envisioned that there would be a hundred airlines flying to every major hub."[61] We turn now to the evidence for the nontraditionalist case, which depended on the ease of potential entry to discipline the behavior of even a natural monopoly or duopoly.

CONTESTABILITY MYTHOLOGY DEBUNKED

For several reasons, it is unlikely that a new entrant will emerge to rival the megacarriers. First, the infrastructure of gates, terminal facilities, and— at four of America's busiest airports (Chicago O'Hare, Washington National, and New York's LaGuardia and Kennedy)—landing slots has been consumed. Sixty-eight percent of our airports have no gates to lease to a new entrant.[62] Even if an incumbent would be willing to lease a gate to an upstart airline (and at a carrier's hub, few are so willing), the incumbent could nevertheless exact monopoly rents for the lease. For example, at Detroit, Northwest charges sublessee Southwest Airlines 18 times what Northwest itself pays for the space. The decision of DOT to allow carriers to buy and sell landing slots means that the deeper-pocket carriers can purchase market share and thereby enjoy market power to reap monopoly profits.[63]

Second, United and American, the largest airlines, today own the largest computer reservations systems.[64] Many critics argue that such vertical integration offers the incumbents the potential to enjoy various forms of system bias (including screen bias, connecting point bias, and data-base bias).[65] As the GAO, among others, has concluded, the airline-owned systems are so dominant that they stifle competition in the industry.[66] An airline that owns a CRS has a 13 to 18 percent greater likelihood of selling its tickets through its system.[67] United and American own the dominant computer reservations systems, which together account for 77 percent of passenger bookings.

Moreover, the advantage of being listed in the computer as an on-line connection with one of the major airlines has led 48 of the 50 small carriers to affiliate themselves with the megacarriers, renaming their companies (to, for example, United Express, Continental Express, or American Eagle) and repainting their aircraft in megacarrier colors. Ninety percent of the 31.7 million passengers who flew aboard regional airlines in 1987 were carried aboard code-sharing airlines.[68] The small carriers have become, in effect, franchisees of the behemoths of the industry and are therefore an unlikely source from which new competition will spring. They are also declining in number. The regional airlines, peaking at 246 in 1981, had dwindled to 168 by 1987.[69] Sophisticated computers also give airlines the ability to adjust, on an hourly basis, the number of seats for which discounts are offered, depending on passenger demand.[70]

Third, large airlines have more attractive frequent-flyer programs, which serve to capture business travelers, the most lucrative segment of the market. Once addicted to a carrier's frequent-flyer program and having some investment in accumulated mileage, business travelers often prefer that carrier over its rivals even when the rivals' flights are cheaper, especially since most business travel is not paid for by the individual flying but by his or her firm.

The brand loyalty created by frequent-flyer programs makes it very difficult for a potential rival to find a niche. Even those potential customers without previously accumulated frequent-flyer mileage with the incumbent will be less willing to accumulate future mileage with a new carrier offering travel to decidedly less exotic destinations. Let us say that we could find a major airport with sufficient capacity to allow us to establish a hub. How could, say, an Air Omaha lure passengers away from its rivals' frequent-flyer programs with their free trips to Hawaii when ours could offer only a free weekend in Cedar Rapids?

Not only are the frequent-flyer programs creating passenger loyalty, but commission overrides—bonuses paid to agents who generate some target revenue level for a carrier—are generating travel agent loyalty.[71] Hence, both the passenger and the agent often prefer a more expensive, established airline to a discount carrier.

Fourth, although new entrants enjoyed significantly lower labor costs in the inaugural years of deregulation, the squeeze on carrier profits unleashed by deregulation has forced management to exact serious concessions in terms of labor wages and work rules. Some, such as Continental, Eastern, and TWA, effectively crushed their unions. Others, such as United, American, and Delta, established two-tier pay scales, with B-grade pay for newly hired employees. Thus, the margin of labor cost between a new entrant and an established airline has been significantly narrowed.

Fifth, incumbents have shown that they will not sit idly by while new rivals rob them of market share. When the new entrants offer lower fares, the incumbents almost always match them. This destroys the new rival for a number of reasons. For example, suppose our new carrier, Air Omaha, does some calculations and finds that if it offers a $49 fare between Omaha and Minneapolis, it will fill about 70 percent of its seats because the incumbent, Northwest, offers no fare so low.[72] Because of lower labor costs and the use of leased, relatively old equipment, let us assume Air Omaha's break-even load factor is a modest 55 percent.[73] So, Air Omaha begins operations and rolls in a healthy profit, right?

Wrong. Northwest matches the $49 fare, and Air Omaha's load factors drop to, say, 35 percent, well below its break-even load factor. Not only can Northwest withstand the loss because of its deeper pocket, but the discount fare actually costs it little, because the fare is offered only to passengers traveling between the two points (origin and destination traffic). Remember, Northwest has a major hub in Minneapolis, and most of its passengers are traveling from or to points beyond—in industry jargon, they constitute "beyond-segment feed." They are not offered the bargain fare. Thus, only a portion of Northwest's passengers are given the discount. Moreover, many of the business travelers in the city-pair market will be willing to pay more than $49 because they are addicted to Northwest's frequent-flyer program. Air Omaha must eventually exit the market, for ordinarily only a carrier with a hub at the other end point can successfully challenge a rival at its hub.

Finally, with more than 150 airlines having failed since 1978, many having been pushed into the abyss of bankruptcy by the predatory behavior of their larger rivals, investor confidence in new airline ventures has largely evaporated.[74]

Hence, significant new entry is highly unlikely in the deregulated airline industry.[75] The incumbent carriers' dominance of gates, terminal space, landing and takeoff slots, computer reservations systems, and the most attractive frequent-flyer programs makes it unlikely that new entrants will emerge to challenge the megacarriers. In fact, no major carrier has emerged since 1985.[76] Only two of deregulation's children achieved major carrier status and survived into the 1990s. Of them, by 1991, Midway was dead, and America West was in bankruptcy.

More and more observers are concluding that the airline industry post-deregulation is not "contestable" in the sense required for the theory to apply.[77] Entry barriers are pervasive, especially at hub airports. As one commentator noted:

Entry into the industry by new carriers seems remote, and entry onto new routes is far more difficult than many envisioned it would be with deregulation. Many airline observers thought that the 1978 deregulation of pricing and entry would make airline markets "contestable." That is, airlines could engage in "hit-and-run" entry into each other's markets in response to profit opportunities—simply by shifting a plane from one route to another. Instead the evidence compiled in . . . a large body of solid research by economic and legal scholars . . . demonstrates that incumbent airlines are frequently able to charge higher prices on routes where other carriers face barriers to entry.[78]

Here again, as with traditionalists and scale economies, deregulation's nontraditionalist proponents overestimated the competitive nature of the industry. As Charles Rule, assistant attorney general for antitrust, recently observed: "Most airline markets do not appear to be contestable, if they ever were. . . . [D]ifficulties of entry, particularly on city pairs involving hub cities, mean than hit-and-run entry is a theory that does not comport with current reality."[79] Even Kahn has admitted as much:

Certainly one of the assumptions behind airline deregulation was that entry would be relatively easy. . . . *We believed* that while *entry* should be legally free and *would be relatively easy, we never thought that would provide adequate protection in markets* that are naturally monopolistic or oligopolistic—*that just won't support more than one or two carriers.* But what happened was that the ideologues began simplistically to parrot the word "contestability" as though it were a substitute for looking at the realities, even if the realities were manifestly changing, even if survival of the new entrants was becoming more and more questionable, as more and more of them were going out of business, and even as it became clear that domination of hubs was increasingly unchallengeable by new entrants.[80]

But even if new entry is unlikely, why should we be concerned with the high level of concentration that has emerged in the airline industry under deregulation? After all, Coke and Pepsi dominate the soft drink industry, and don't we still have price competition between them? Although other American industries are dominated by huge firms, transportation is different in the way it influences the economy. As former TWA vice president Melvin Brenner put it:

Other industries, even when comprised of only a few large firms, do not usually end up with a one-supplier monopoly in specific local markets. But this can happen in air transportation.

Moreover, because of the nature of transportation, a local monopoly can do greater harm to a community than could a local monopoly in some other industry. This is because transportation is a basic part of the economic/social/cultural infrastructure, which affects the efficiency of all other business activities in a community and the quality of life of its residents. The ability of a city to retain existing industries, and attract new ones, is uniquely dependent upon the adequacy, convenience, and reasonable pricing of its airline service.[81]

NOTES

1. *See, generally,* Hardaway, *Transportation Deregulation,* 14 TRANSP. L. J. 101 (1985).

2. A 1978 Senate committee report on federal regulation provided a fairly typical summary of those attributes of destructive competition deemed not likely to surface in a deregulated air and motor carrier industry:

A justification sometimes offered for regulation is that in the absence of regulation competition would be "destructive." In other words, without regulation, an industry might operate at a loss for long periods. . . . When there is excess capacity in a competitive industry . . . prices can fall far below average cost. This is because individual producers minimize their losses by continuing to produce so long as their variable (avoidable) costs are covered, since they would incur their fixed (overhead) costs whether they produced or not.

What is "destructive" about large and long-lasting losses? Some economists have suggested that they would result in long periods of inadequate investment and slow technical progress which in turn might lead to poor service and periodic shortages. . . .

Another scenario that has sometimes been suggested is that periods of large losses will result in wholesale bankruptcies and the shakeout of many small producers with the result that the industry in question becomes highly concentrated in a few large firms. . . .

A third and related notion is the possibility that powerful firms might engage in predation. . . .

"Destructive competition" seems . . . unlikely in the cases of airlines and trucks.

Study on Federal Regulation, Report of the Sen. Comm. on Government Affairs, 96th Cong., 1st Sess., 13–15 (1978).

3. Economies of scale are realized when increases in total production simultaneously decrease unit costs. As the scale of production grows, the enterprise becomes more efficient. The cost savings resulting from economies of scale can be attributed to: (1) indivisibilities—a large capital-intensive piece of equipment operates most efficiently at full capacity; and (2) division and specialization of labor—highly specialized labor is more productive labor.

A concept related to economies of scale is economies of scope. The unit cost of producing one more item may be diminished when the scope of activity broadens. For instance, advertising costs per unit of serving a particular city-pair market are lower the more city-pair markets served, due to quantity discounts in media purchasing. *See* J. BAIN, BARRIERS TO NEW COMPETITION (1956); R. HEILBONER & L.

THUROW, ECONOMICS EXPLAINED (1987); and W. SHEPARD, ECONOMICS OF INDUSTRIAL ORGANIZATION (1979).

A related concept is economies of density. By combining passengers and groups of passengers, an airline can carry the aggregation of passengers more cheaply than if it carried those passengers separately. Through careful scheduling of flights, consolidating operations, and routing passengers through its "hub," an airline streamlines its system, making it more dense and thus more efficient. The "hub-and-spoke" scheme employed by all of the major airlines is testimony to this phenomenon. For example, an airline that carries 100 passengers in a single plane to a destination—as opposed to carrying 50 passengers in two aircraft to that same destination—is making use of economies of density. See A. FRIEDLAENDER & R. SPADY, FREIGHT TRANSPORTATION REGULATION: EQUITY, EFFICIENCY, AND COMPETITION IN THE RAIL AND TRUCKING INDUSTRIES (1980); A. LAMOND, COMPETITION IN THE GENERAL-FREIGHT MOTOR CARRIER INDUSTRY (1980).

As Melvin Brenner notes, the failure of researchers before deregulation to find size advantages in air transport was partly a result of regulation itself, which kept large firms from exercising their advantages vis-á-vis small firms. See Brenner, *Airline Deregulation—A Case Study in Public Policy Failure*, 16 TRANSP. L. J. 186–88 (1988).

4. *See, generally*, W. BAUMOL, J. PANZAR, & R. WILLIG, CONTESTABLE MARKETS AND THE THEORY OF INDUSTRY STRUCTURE (1982).

5. *Aviation Regulatory Reform, Hearings before the Subcomm. on Aviation of the Committee on Public Works and Transportation*, 95th Cong., 1st Sess. 178 (1977) [hereinafter cited as *1977 House Hearings*].

6. *Id.* at 178–79 [emphasis supplied].

7. *Id.* at 1137.

8. Testimony of Alfred Kahn before the California Public Utilities Commission 6190 and 6223 (Jan. 31, 1989) [hereinafter cited as Kahn Oral Testimony], at 6247–48.

9. *Id.* at 6201.

10. Kahn, *Surprises from Airline Deregulation*, 78 AEA PAPERS AND PROCEEDINGS 316, 318 (1988) [hereinafter cited as *Surprises from Deregulation*].

11. Kahn, *Airline Deregulation—A Mixed Bag, But a Clear Success Nevertheless*, 16 TRANSP. L.J. 229, 236 (1988) [hereinafter cited as *A Mixed Bag*]; Hamburger, *Fares Rose with NWA's Dominance*, MINNEAPOLIS STAR TRIBUNE, Dec. 1988, at 9A.

12. *Safety and Re-Regulation, Hearings before the Senate Commerce Comm.*, 100th Cong., 1st Sess. 155 (1987).

13. A. Kahn, Talk to the New York Society of Security Analysts 24 (Feb. 2, 1978) [hereinafter cited as Kahn 1978 Speech] [emphasis supplied].

14. *1977 House Hearings, supra* note 5, at 1111. [emphasis supplied].

15. *Interview with Alfred E. Kahn*, ANTITRUST, Fall 1988, at 4, 6 [hereinafter cited as *Kahn Interview*]. According to Kahn, in deregulating the airlines, "we emphasized the contestability of airline markets and thought people would be well protected by the possibility of entry, because airplanes can move." He added, "Well, I think we exaggerated that." *Surprises, But Few Regrets*, REASON, Feb. 1989, at 35, 36.

16. *Airline Deregulation, Hearing before the Subcomm. on Antitrust of the Senate Judiciary Comm.*, 100th Cong., 1st Sess. 64 (1987) [hereinafter cited as *Senate Hearings on Deregulation*].

17. Kahn 1978 Speech, *supra* note 13, at 14.

18. *1977 House Hearings, supra* note 5, at 133.

19. Kahn 1978 Speech, *supra* note 13, at 15.

20. Brenner, *supra* note 3, at 200–201. He says: "The eight years [1979–86] of deregulation comprise the worst financial period in airline history. The cumulative industry operations in those eight years generated a loss of over $7 billion."

21. See Dempsey, *Transportation Deregulation—On a Collision Course*, 13 TRANSP. L.J. 329, 342–52 (1984).

22. Ott, *Industry Officials Praise Deregulation, But Cite Flaws*, AV. WEEK & SPACE TECH., Oct. 31, 1988, at 88. In 1988, the industry's profit margin stood at 1.3 percent, compared with 2 percent a decade earlier. Stockton, *When Eight Carriers Call the Shots*, N.Y. TIMES, Nov. 20, 1988, at 3–1. Alfred Kahn maintains that the airline industry's profit margin "fell to a puny 0.10 in the 1979–87 period." *Surprises from Deregulation, supra* note 10, at 316, n.1.

23. *A Mixed Bag, supra* note 11, at 248. Kahn recently said: "I found it distressing in the middle of this. I hated to be responsible for the industry suffering so. I wanted to be sure that it would always be financially healthy and able to attract capital." Kahn Oral Testimony, *supra* note 8, at 6247–48.

24. *The Frenzied Skies*, BUS. WEEK, Dec. 19, 1988, at 70, 72.

25. Pelline, *Bumpy Ride under Deregulation*, SAN FRANCISCO CHRONICLE, Oct. 28, 1988, at 21. One source estimates that 214 airlines have disappeared from the market. Hamilton, *Is the Airline Industry on the Verge of Going Global?*, WASHINGTON POST, Dec. 11, 1988, at K1.

26. Pelline, *supra* note 25.

27. The difficulty airlines face is in managing yield in a way that lures only those passengers not otherwise likely to fly—hence, Saturday stay-over requirements, which are unappealing to business travelers.

28. Cited in Brenner, *supra* note 3, at 188.

29. *Id.* AVIATION DAILY, Jan. 23, 1991, at 149.

30. *Focus: A Decade of Deregulation*, TRAFFIC WORLD, Dec. 5, 1988, at Supp. D, updated by *Happiness Is a Cheap Seat*, ECONOMIST, Feb. 4, 1989, at 68, and AVIATION DAILY, Jan. 23, 1991, at 149.

31. 49 U.S.C. § 1302(a)(7). *See* Dempsey, *The Rise & Fall of the CAB*, 11 TRANSP. L.J. 91, 135 (1979).

32. The DOT did require that some shuttle routes be sold off in the northeastern corridor, but otherwise the Eastern acquisition by Texas Air passed through unmolested. *See* Dempsey, *Monopoly I$ the Name of the Game*, 21 GA. L. REV. 505, 538 (1987).

33. To these mergers of passenger carriers add the major air cargo acquisition of Seaboard by Flying Tigers and the acquisition of Tigers by Federal Express, as well as the acquisition of Emery and Purolator by CF Air. Moreover, concentration levels in the passenger industry are even more pronounced when one recognizes that before deregulation, America had a healthy charter airline industry that enjoyed significant market share. Under deregulation, it has very nearly vanished. See Brenner, *supra* note 3, at 184. In 1977, nonscheduled airlines had 43,000 domestic

departures, compared with 18,000 in 1986. FEDERAL AVIATION ADMINISTRATION, AIRPORT ACTIVITY STATISTICS OF CERTIFICATED ROUTE CARRIERS 747–49 (1977), FEDERAL AVIATION ADMINISTRATION, AIRPORT ACTIVITY STATISTICS OF CERTIFICATED ROUTE CARRIERS 798–800 (1986). However, pre-deregulation rates of passenger growth were even larger.

34. Stockton, *supra* note 22, at 6.

35. *Senate Hearings on Deregulation, supra* note 16, at 61–62 (testimony of Morton S. Beyer).

36. Kahn, *Despite Waves of Airline Mergers, Deregulation Has Not Been a Failure*, DENVER POST, Aug. 31, 1986, at 3G.

37. *A Mixed Bag, supra* note 11, at 234 [emphasis in original].

38. Dempsey, *Fear of Flying Frequently*, NEWSWEEK, Oct. 5, 1987.

39. Brenner, *supra* note 3, at 190, updated and expanded by Ott, *Congress, Airlines Reassessing Deregulation's Impact*, AV. WEEK & SPACE TECH., Nov. 9, 1987, at 163; Hamilton, *The Hubbing of America: Good or Bad?* WASHINGTON POST, Feb. 5, 1989, at H1, H2 col 5; and Cushman, Jr., *Two Studies Conflict on Airline Fares*, N.Y. TIMES, June 7, 1989, at A14. AVIATION DAILY, Dec. 13, 1991, at 462–64.

40. *The Frenzied Skies, supra* note 24, at 72. AVIATION DAILY, Dec. 13, 1991, at 462-64

41. Hamilton, *supra* note 39, at H1, H2, col. 5. AVIATION DAILY, Dec. 13, 1991 at 462–64.

42. Dempsey, *supra* at note 32, at 592–93.

43. *See* Stockton, *supra* note 22, at 3–6.

44. Dempsey, *Deregulation Has Spawned Abuses in Air Transport*, AV. WEEK & SPACE TECH., Nov. 21, 1988, at 147.

45. *Hearings before the U.S. Senate Comm. on Commerce, Science and Transportation on Airlines* 170–95 (1987) (statement of Julius Maldutis).

46. This is the number of *equal-size* competitors that would produce the same Herfindahl-Hirschman Index as is observed in a market. It is the reciprocal of the HHI in proportionate form.

47. Stockton, *supra* note 22.

48. GENERAL ACCOUNTING OFFICE, AIRLINE COMPETITION 2, 3 (1988).

49. Hamburger, *supra* note 11, at 1A.

50. Hamilton, *supra* note 39, at H2.

51. Concentrated airports were defined as those where one airline handled at least 60 percent of enplanements.

52. GENERAL ACCOUNTING OFFICE, AIR FARES AND SERVICE AT CONCENTRATED AIRPORTS (1989).

53. S. Borenstein, Hubs and High Fares: Airport Dominance and Market Power in the U.S. Airline Industry (University of Michigan: Institute of Public Policy Studies, March 1988).

54. *Id.* at 23.

55. *Policies for the Deregulated Airline Industry*, Congress of the United States, Congressional Budget Office, July, 1988.

56. *Id.* at 15–16.

57. *Id.* at 16.

58. Kahn, in *A Mixed Bag, supra* note 11, at 233, fn. 7, recognizes his figures'

bias toward overstating the increase in competition after 1978 due to their exclusion of interlining.

59. *Surprises from Deregulation, supra* note 10, at 319.

60. *A Mixed Bag, supra* note 11, at 232.

61. *The Frenzied Skies, supra* note 24, at 73.

62. *See* Hardaway, *supra* note 1, at 49.

63. *Id.*

64. United sold half of its Apollo/Covia system for $500 million in 1988 to USAir and five foreign airlines. O'Brian, *Delta, AMR's American Airlines Plan to Merge Computer Reservations Systems,* WALL ST. J., Feb. 6, 1989, at B10, col. 1. TWA had previously sold half of its CRS to Northwest Orient. And Eastern's system had been transferred to a Texas Air subsidiary for a paltry $100 million. Castro, *Eastern Goes Bust,* TIME, Mar. 20, 1989, at 52.

65. *See* Saunders, *The Antitrust Implications of Computer Reservations Systems,* 51 J. AIR L. & COM. 157 (1985).

66. Hamburger, *Fighting Back Begins as Costs Go Up, Up and Away,* MINNE-APOLIS STAR TRIBUNE, Dec. 24, 1988, at 6A.

67. *Id.* These statistics were quoted by Michael Levine.

68. *Dereg's Falling Stars,* OAG FREQUENT FLYER, Aug. 1988, at 28.

69. *Id.*

70. GENERAL ACCOUNTING OFFICE, AIRLINE COMPETITION: IMPACT OF COMPUTERIZED RESERVATIONS SYSTEMS (1986).

71. Rose, *Travel Agents' Games Raise Ethical Issue,* WALL ST. J., Nov. 23, 1988, at B1.

72. Between 1984 and 1988, the industry-average domestic load factor ranged between 59 percent and 62 percent. AMR CORPORATION, ANNUAL REPORT 9 (1989).

73. In contrast, United's break-even load factor between 1986 and 1988 ranged between 62 percent and 64 percent. UAL CORPORATION ANNUAL REPORT 1 (1988).

74. *See* Russell, *Flying among the Merger Clouds,* TIME, Sept. 29, 1986, at 56–57.

75. *See* Dempsey, *supra* note 32.

76. Hamburger, *supra* note 11, at 9A.

77. *See especially* Levine, *Airline Competition in Deregulated Markets: Theory, Firm Strategy, and Public Policy,* 4 YALE L. REG. 393 (1987).

78. Guerin-Calvert, *Hubs Can Hurt on Shorter Flights, at Crowded Airports,* WALL ST. J., Oct. 7, 1987, at 28.

79. C. Rule, Antitrust and Airline Mergers: A New Era, 15, 18 (speech before the International Aviation Club, Washington, D.C., Mar. 7, 1989), republished at 57 TRANSP. PRAC. v. 62 (1989). On January 1, 1989, the Justice Department took over the largely latent airline merger authority of the DOT.

80. *Kahn Interview, supra* note 15, at 7 [emphasis added].

81. Brenner, *supra* note 3, at 189.

21

PRICING

A basic tenet of economics states that in perfectly competitive markets, the laws of supply and demand work to insure that prices are automatically driven down to the level of marginal costs, thus resulting in the lowest equilibrium prices for commodities in that market. A perfectly competitive market, though, presumes large numbers of independent buyers and sellers interested in exchanging a standardized product.[1] In this way, no individual buyers or sellers could influence the market pricing mechanism by their own unilateral actions.

In the debates before deregulation, it was hypothesized that the unregulated airline industry would approximate the behavior of a perfectly competitive market (if one also assumed the validity of the theory of contestable markets), and prices would be driven downward. The former chairman of the CAB, Alfred Kahn, once argued that deregulation would bring about cost-based pricing.

But after a decade of deregulation, pricing seems to reflect only the level of competition in any market, rather than costs. There appears to be a positive correlation between more competition and lower prices and between fewer competitors and higher prices.[2] Indeed, in many hub markets, there appears to be an inverse relationship between prices and costs. As the industry becomes more highly concentrated, prices ascend.[3]

Admittedly, competition has enabled some users (particularly discretionary travelers in major long-haul airline markets) to enjoy lower prices.[4] But these benefits have been unevenly distributed because business travelers, and passengers flying to small towns or from hubs dominated by a single carrier, pay relatively higher prices for poorer service. In addition, as noted above, the unprecedented concentration emerging as a result of massive bankruptcies and mergers threatens to make the low prices enjoyed in large, competitive markets a short-term phenomenon. Deregulation inevitably eradicates some of the important benefits derived from the traditional scheme of economic regulation, including the prohibition against pricing discrimination.[5]

Moreover, as will be shown in what follows, the aggregate benefits from fare reductions may very well have reached zero in 1988. Holding fuel prices constant, the real yield or revenue per passenger mile (a commonly used measure of average fares) being paid in 1988 was exactly what a projection of the pre-deregulation (downward) trend would have given for the same year. This reflects a one-time drop in the years immediately following deregulation, coupled with a slower rate of decline of fuel-adjusted real revenues per passenger mile after deregulation than before. The rate of decline is so much lower that the pre-deregulation downward trend in fuel-adjusted fares "caught up" with the actual levels by 1988—despite the early decline of approximately 13 percent in real terms. The gains from deregulation have proven short-lived indeed; they are already a thing of the past. A preliminary estimate for 1989 indicates that consumers are paying 2.6 percent more per mile than the projection of the pre-deregulation trend.

Growing consumer irritation with the deregulated airline industry is reflected in public opinion polls. In 1984, consumers were asked, "Should airlines be allowed to raise or lower their fares on their own, or should they be required to get government permission?" Only 35 percent responded that airlines should be required to get the government's permission. However, as consumers became more acquainted with deregulation, they became less enamored with it. In 1987, when asked the same question, almost half were willing to opt for more government rate regulation.[6] Even Alfred Kahn has admitted that the time has come to consider price ceilings in markets dominated by a single carrier.[7]

PRICE SAVINGS

Most proponents of deregulation point to what they claim are significant price reductions enjoyed by consumers during the past decade. Kahn, for example, claimed that inflation-adjusted fares had dropped 30 percent through 1988.[8] Kahn's figures run from 1976, before he was appointed chairman of the CAB and two years before promulgation of the Airline Deregulation Act.[9] Lesser savings are cited by the former DOT secretary (and deregulation proponent) James Burnley, who said that inflation-adjusted fares fell 13 percent from 1978 to 1988.[10] The same 13 percent decline in average inflation-adjusted fares from 1979 to 1988 is cited in a recently released DOT study on competition in the airline industry.[11]

According to the Air Transport Association, real yields (revenues per passenger mile) have fallen 22 percent since 1978 and 28 percent since 1977, when Kahn took over at the CAB and began to allow more flexible

Table 21.1
Yield and Fuel Price Indices (1978 = 100)

	Real Yield (revenue per passenger mile)	Real Fuel Prices	Fuel Adjusted Real Yields
1967	129.2	55.9	143.8
1968	123.5	54.1	138.0
1969	121.7	50.9	137.1
1970	117.3	47.0	133.4
1971	117.7	46.2	134.2
1972	114.3	46.3	130.3
1973	112.3	47.7	127.6
1974	116.2	82.1	120.9
1975	111.0	90.2	113.3
1976	110.1	92.5	111.9
1977	109.3	99.3	109.5
1978	100.0	100.0	100.0
1979	94.2	131.7	88.0
1980	104.9	180.1	90.0
1981	106.9	189.8	90.3
1982	95.9	168.6	84.0
1983	91.9	148.0	83.3
1984	91.8	135.7	84.9
1985	85.4	124.0	80.7
1986	77.5	140.7	79.1
1987	76.5	81.6	78.4
1988	78.4	74.9	81.4
Growth Rates:			
1967-77	-1.7	5.9	-2.7
1978-88	-2.4	-2.8	-2.0

Source: Air Transport Association and author's calculations—see Appendix

pricing by the airlines (see column 1 of table 21.1). This seems like an impressive achievement indeed, until it is compared with the historical record before deregulation, on the one hand, and the behavior of the crucially important price of jet fuel, on the other. This sobering comparison, which clearly shows the emptiness of the attempt to attribute the reduction in real fares since 1977–78 to deregulation, is displayed in table 21.1.

In the first place, it should be noted that real yields fell in the period before deregulation as well. From 1967 to 1977, they fell at an annual average rate of 1.7 percent a year, compared with the post-deregulation (1978–88) rate of decline of 2.4 percent per year. On the surface, it appears as if deregulation may have at best sped up the rate of decline; attributing the *entire* decline to deregulation ignores the preexisting downward trend. (Melvin Brenner has made this point for the eight years before and after deregulation.[12] Another source[13] points out that airfares have been declining at about the same rate for more than 40 years—a long-term trend preceding deregulation by several decades.)

But even this more moderate claim of an accelerated rate of decline in prices after deregulation is put in doubt by the figures presented in column 2 of table 21.1. Here we see that real yields (prices) fell in the 10-year period before deregulation despite *a doubling of the real cost of fuel,* whereas the somewhat higher rate of decline after deregulation occurred in the context of a 25 percent *decline* in the real price of fuel. During the period as a whole, fuel constituted anywhere from 12 percent of costs in the early 1960s to 30 percent after the second oil shock in 1979 and back down to 15–16 percent in recent years, according to data from the Air Transport Association.[14] Thus, between 12 and 30 percent (depending on the year) of the percentage change in real airfares that occurs during a given period has absolutely nothing to do with whether the industry is regulated or deregulated.

The third column of table 21.1 accounts for fuel prices by taking out of the real yield series the changes that were solely attributable to changing real fuel prices (calculated for a given year as the product of the fuel share of all cash expenses in the previous year and the contemporaneous percentage change in real fuel costs—see the appendix to this chapter for details). This shows that *holding fuel prices constant,* the real price of air travel fell more rapidly (an annual average percentage decline of 2.7 percent) in the period before deregulation than after deregulation (2.0 percent). Roughly, real yields would have fallen 1 percentage point more a year (17 percent of 5.9) had it not been for the average 5.9 percent increase in real fuel prices during the period from 1967 to 1977; real yields would have fallen 0.4 percent less per year (about 14 percent of 2.8) had it not been for a totally gratuitous 2.8 percent annual *decline* in real fuel prices during the 1978–88 period.

The fuel-adjusted series dramatically shows what the person on the street senses about deregulation but what the unadjusted data obscure, namely the enormous front-loading of the gains from deregulation. From 1977 to 1978 and from 1978 to 1979, fuel-adjusted real yields fell 10 and 12 percent respectively; it then took from 1979 to 1988 for fuel-adjusted real yields to fall another 10 percent—an annual average percentage decline of only 0.9 percent! The unadjusted data obscure this by making the first few years of deregulation, which coincided with the second oil shock, look worse than they were, whereas the latter part of the period, when real fuel prices plummeted, looks much better than it actually was. Note too, that from 1985 to 1988, fuel-adjusted real fares actually rose—the only three-year period during 21 years when this was so.

The apparent difference in the rate of decline of real revenues per passenger mile before and after deregulation was tested for statistical significance using regression techniques (see the appendix for detailed results). The *unadjusted* real yield series falls significantly faster after deregulation than before (at a continuously compounded annual rate of 3.1 percent from 1978 to 1988, compared with 1.5 percent from 1967 to 1977.)[15] The fuel-adjusted series, on the other hand, falls at a significantly faster rate *before* deregulation—at 2.7 percent from 1967 to 1977 versus 1.9

Figure 21.1
Fuel-Adjusted Real Yields 1967–88 (Actual vs. Pre-deregulation Trend)

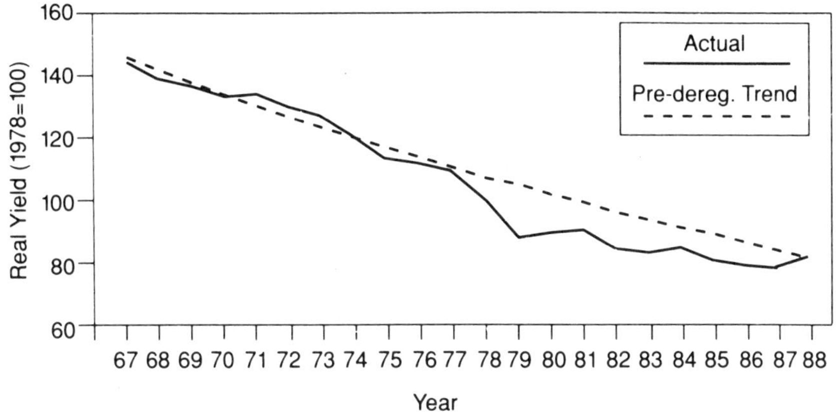

Source: Air Transport Association and author's calculations

percent from 1978 to 1988.[16] Instead of falling twice as quickly after de-regulation—as the unadjusted numbers would suggest—real airline yields per passenger mile fell at a 30 percent slower rate *after* deregulation.

The regressions also suggest that deregulation was responsible for a one-time reduction in fares on the order of 13 percent; as figure 21.1 shows, however, by 1988—due to the slower rate of decline of real fares—all the gains of this one-time shift had been dissipated. By 1988, that, is consumers were paying "net" prices (net of the effects of fuel) exactly equal to what they would have paid had pre-deregulation trends continued. By contrast, figure 21.2 shows the pre-deregulation trend compared with actual when only "gross" prices—unadjusted for fuel cost changes—are examined. Again this is dramatically misleading as an indicator of consumer gains—attributing to deregulation what is really a result of lower oil prices. The case for a gain for consumers from deregulation, based on the 28 percent fall in unadjusted real yields since 1977, is entirely vacuous, to put it charitably.

The industry's use of revenue per passenger mile as a measure of consumer prices also presents a significant methodological distortion. Consumers who in 1988 were paying, in real per passenger mile terms (net of fuel), exactly what they paid before deregulation were in general flying more miles to make the same trip after deregulation than before. Thus, a decline in revenue per passenger mile may represent only an increase in miles for making the same trip, with no reduction—or even an increase—in the actual price of the trip!

Hub and spoking has significantly increased circuity in air travel, thereby lengthening the distance between origin and destination. Many (if not most)

Figure 21.2
Real Yields, 1967–1988 (Actual vs. Pre-deregulation Trend)

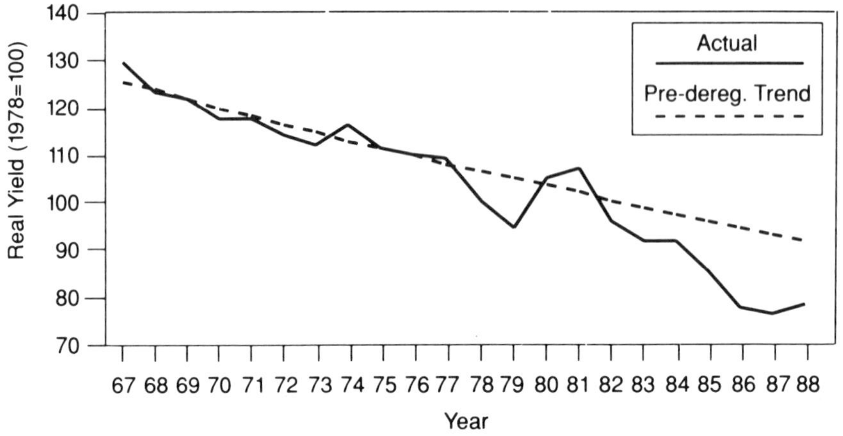

Source: Air Transport Association and author's calculations

passengers who do not begin or end their trip in a hub airport have to fly more miles to get to their destination than before deregulation, with estimates of this effect ranging from 4 percent to 30 percent for the average trip.[17] For example, the loss of pre-deregulation Boston–San Francisco nonstops means that some travelers in the market have no choice but to fly through a hub (through, for example, Minneapolis on Northwest, through Atlanta on Delta, through St. Louis on TWA, through Dallas on American, through Chicago or Denver on United, and so on). As Robert Kuttner has noted, the pre-deregulation Boston–San Francisco passenger yield was for fewer miles (2,429 to be exact) than the post-deregulation Boston–Dallas–San Francisco trip (which is 3,024 miles, or 24 percent more, and takes about four hours longer).[18] In its frequent flyer magazine, American Airlines' CEO Robert Crandall used the Wichita–Dallas–LaGuardia example to show the benefits of hub-and-spoke operations. If it existed, a Wichita–LaGuardia nonstop would be 1277 miles. Routed through Dallas, the trip consumes 1715 miles, or 34 percent more miles. *Due to the greater circuity, then, consumers paid more in 1988 than they would have paid projecting the pre-deregulation trend*—the same net price per passenger mile amounted to a higher charge to go from point A to point B. Quantitatively, this effect would mean that the price of a trip in 1988 would be higher by some 4 to 30 percent—the range reflecting the wide range of estimates of increased circuity noted above.

We have argued that the purported 28 percent fall in real yields that has occurred since deregulation would have occurred as well under regulation—given the same fall in real fuel prices as occurred under deregulation and projecting the pre-deregulation trend behavior of real yields net of fuel costs. A widely cited study of deregulation by Steven Morrison and Clifford Winston of the Brookings Institution[19] alleges that deregulation is responsible for a 30 percent real fare reduction, a reduction that we have argued cannot properly be attributed to deregulation. They claim, however, to be doing a "counter-factual" analysis to come up with their estimate—asking what deregulation did to fares, holding all other factors constant. If they had in fact done so, their estimate would not be subject to the argument we have made here. However, as is argued in the appendix, their estimate does *not* hold all other factors constant. In particular it does not hold time constant—a crucial consideration in industries that become more efficient over time and in which a time trend proxies the secular gain in efficiency. The airline industry is such a progressive sector.

Thus we claim that both the naive attribution of the actual reduction in real fares since 1977 to deregulation and the more sophisticated "counter-factual" analysis of Morrison and Winston are misleading, that the *average real fare per mile was not lower in 1988 (and is estimated to be some 2.6 percent higher in 1989) as a result of deregulation* but that the *real fare per trip was actually higher* (perhaps by as much as 30 percent) due to the greater circuity attributable to hub and spoking, and that the volatility and associated transactions costs were higher as well. More recently, Morrison and Winston have employed the CAB's archaic Standard Industry Fare Level (SIFL) methodology to calculate what fares might be today if still regulated. The SIFL computes fares based on industry costs. But Morrison and Winston fail to account for the fact that with hubbing-and-spoking, a practice that has proliferated under deregulation, industry costs have risen significantly. Moreover, the CAB changed its pricing methodology periodically, and well might have abandoned SIFL by now.

In addition, the good we were buying before deregulation is not the same good we buy today—it is significantly lower in quality along many dimensions, adding insult to injury. We are paying more for less, on average, despite gains for some consumers, particularly pleasure travelers on long-haul routes between large cities.

CROSS-SUBSIDIZATION AND PRICING DISCRIMINATION

Before deregulation, there was some amount of cross-subsidization within the transportation industry. Although carriers were allowed to serve specified lucrative routes, they were also required to serve less lucrative markets in the geographic territory designated by their operating certificates. Carriers were expected to cross-subsidize losses or meager profits earned from serving small communities with healthier revenues earned from dense, lucrative markets and to provide just and reasonable rates to both. Dereg-

Table 21.2
Average Unadjusted Fares by FAA Hub Class

Hub Class	Average Fare ($) 1979	1988	Percent Change 1979–1988
Large (27 cities)	$97.41	$134.69	+38.3%
Medium (29 cities)	$95.24	$132.28	+38.9%
Small (56 cities)	$90.22	$143.81	+59.4%
Nonhub (362 cities)*	$96.36	$155.49	+61.4%
Total	$96.19	$134.50	+41.5%

*represents the number of cities still receiving service in 1988

Source: Department of Transportation, Secretary's Task Force on Competition in the U.S. Domestic Airline Industry, February 1990, Pricing Section, 1: 53.

ulation was designed to end this internal cross-subsidy on the grounds that such wealth redistribution created allocative inefficiency.

Actually, cross-subsidization appears merely to have been reversed in direction, rather than eliminated. Under deregulation, carriers began to impose higher rates in their monopoly and oligopoly markets to cross-subsidize the losses they incurred as a result of the intensive competitive battles they waged for market share in dense traffic lanes.[20] For example, recently the airline rate from Dubuque to Chicago was $1 per seat mile, whereas the fare from New York to Los Angeles was 3.3¢ per seat mile.[21] In 1987, a round-trip coach ticket between International Falls, Minnesota and Minneapolis/St. Paul was 86¢ per seat mile; between Washington, D.C., and Minneapolis/St. Paul, the fare was 27¢ per seat mile. The trip from Madison, Wisconsin, to St. Louis cost $225 one way, whereas a ticket from New York to Los Angeles via St. Louis was only $199.[22]

Generally, fares in larger, more competitive markets have been lower than fares in smaller, less competitive markets. In a recently published study, the DOT analyzed average unadjusted fares over the 1979–88 period for each of the four hub-size class categories: large, medium, small, and non-hub (see table 21.2).[23] Average unadjusted fares in the small and nonhub categories have increased more than fares in the large and medium hub categories over the 1979–88 period.

Further evidence from small community travelers attests to increased price discrimination. Senator Ernest Hollings (D-S.C.), chairman of the Senate Commerce Committee, said it now costs him $510 to fly from Charleston, South Carolina, to Washington, D.C., compared with $120 in 1977.[24] Senator Larry Pressler (R-S.D.) said it costs $269 to fly from Aberdeen, South Dakota, to Rapid City, South Dakota, whereas a Washington–to–Los Angeles ticket can cost $239.[25] Simply put, constituents from many smaller communities have not benefited from fare reductions during deregulation.[26]

Table 21.3
Airfare Changes under Deregulation

Year	Number of Fare Changes	Net Price Changes
1982	4,611,888	−4%
1983	6,532,728	−2%
1984	6,090,834	+4%
1985	10,624,574	−3%
1986	20,255,405	−7%
1987	49,369,278	+2%
1988	48,241,972*	+7%**

*annualized
**estimate

Source: FORTUNE, Dec. 19, 1988, at 9.

The complete disconnection of relative prices from relative costs is apparent in cases such as Delta's flights from Oakland to Salt Lake City versus Oakland to Phoenix. The latter flight stops in Salt Lake, Delta's hub, but costs much less than the former. Obviously, unless the leg from Salt Lake to Phoenix has negative costs, the lower unit costs of flying longer distances are not the explanation. The level of competition in the Oakland-to-Phoenix market (comparatively high) versus Oakland to Salt Lake (low) is the explanation. (Unfortunately for Delta, they have not figured out a way to stop Salt Lake City–bound travelers from buying tickets to Phoenix and getting off in Salt Lake, throwing away the unused coupon.)

Moreover, deregulation appears to have brought us a roller-coaster ride of high and low fares—fares that change on an hourly basis. This instability of the rate structure is reflected in table 21.3. Today, two travelers sitting next to one another on the same plane traveling on the same route can be paying dramatically different fares. For example, a round-trip coach seat between Boston and Dallas on short notice costs $1,020; a nonrefundable, 14-day advance purchase, Saturday-night stayover ticket costs $318.[27] Fares change so often that *when* you buy your ticket (and how flexible your travel plans are) has become more important than *where* your ticket takes you. Thus, in the deregulated air travel market, prices for the same service can be unbelievably, and discriminatorily, different.

The choice among a bewildering array of fares has undoubtedly made the acquisition of information for consumers more difficult and more costly. Transactions costs for both producers and consumers appear to have grown sharply under deregulation. And those (largest) airlines with control of computer reservations systems have been in the most advantageous position regarding these costs.

THE EMERGING OLIGOPOLY

The price benefits many consumers enjoyed under deregulation were a short-term phenomenon.[28] As noted above, the trend of deregulation is an oligopoly of megacarriers. Holding fuel prices constant, airfares fell sharply during the first several years of deregulation, a reflection of the downward pricing spiral of head-to-head, destructive competition. As carriers became adept at seizing monopoly market opportunities by merging (there were a rash of mergers in 1985–86) and creating hub dominance, and as weaker rivals dropped from the skies into bankruptcy, prices began to rise.

In 1989, the General Accounting Office compared fares at 15 concentrated hub airports—those where one or two airlines dominate the traffic—with fares at 38 unconcentrated airports and found average fares 27 percent higher at the concentrated hubs.[29] Furthermore, the rate of fare increase at concentrated airports rose much faster from 1985 to 1988 than at unconcentrated airports.[30] What this means is that monopoly power is manifesting itself at "fortress hub" airports in the form of higher fares for those passengers departing from or arriving at the hub. Travelers living in hub cities, who at one time were basking in the benefits of increased non-stop service, now find themselves at the mercy of a megacarrier and have little choice but to pay whatever is charged.

The GAO study on airline concentration was at least partly corroborated by the recently released DOT study on competition in the airline industry.[31] This study found that the average fare per mile at the eight most concentrated hubs was 18.7 percent higher than at other airports (see figure 21.3). Thus, even the DOT has clearly acknowledged the relationship between hub concentration and higher fares for those passengers traveling from or to such cities.

Since the beginning of 1988, coach fares in many markets have increased by more than 50 percent.[32] Between September 1988 and February 1989, the largest carriers announced four fare increases, with several more after Eastern's bankruptcy in March 1989.[33]

The data on revenue per passenger mile for 1989 imply an estimated rise in inflation-adjusted yields of 2.4 percent for the year. At the same time, real fuel prices rose by 4.1 percent, and the fuel share of costs was about 14.5 percent. Thus, adjusting for fuel price increases still puts the increase in fuel-adjusted real yields at 1.8 percent[34]—an historical outlier (see figure 21.1). Since, as we saw above, real fares net of fuel trended downward before deregulation at 2.7 percent a year, and consumers in 1988 paid the same real yield net-of-fuel that they would have paid had the pre-deregulation trend continued, it follows that *in 1989 consumers are estimated to be paying roughly 2.6 percent more than they would be paying under the pre-deregulation trend per mile* (and thus anywhere from 10 to 36 percent more per trip, given the range of estimates for the effect of deregulation on circuity).

Even without the estimated 1989 data, the effect of consolidation in the industry shows up in the behavior of the annual percentage change in fuel-adjusted real yields before and after deregulation, graphed in figure 21.4.

Figure 21.3
Average Fare per Mile at Concentrated Hubs (Compared with Domestic Average)

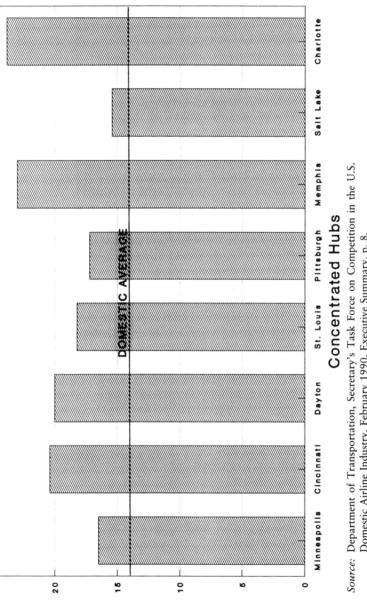

Cents/Passenger-Mile

Concentrated Hubs

Source: Department of Transportation, Secretary's Task Force on Competition in the U.S. Domestic Airline Industry, February 1990, Executive Summary, p. 8.

Figure 21.4
Percent Change in Real Fuel-Adjusted Fare, 1967–1988

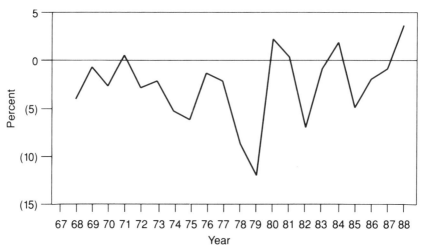

Source: Air Transport Association and author's calculations

Before deregulation, real yields were decelerating slightly (but not significantly). Deregulation—after a one-time drop in the rate of change—has imparted a significant upward trend on the series, with the percentage increase in yields going up by 0.84 percent after deregulation, instead of falling by 0.13 percent, as was true before deregulation.[35]

QUALITY DISINTEGRATION: TICKET RESTRICTION AND DELAYS

It is widely recognized that the average fare reductions we have seen during deregulation are a reflection not of lower unrestricted first-class or coach fares but of the enormous increase in discounting (from 48.2 percent of revenue passenger miles in 1979 to 91 percent in 1988).[36] But the discount fare category differs (lower-quality) in many respects from the undiscounted version—due to various time restrictions, advance-purchase requirements, nonrefundability, etc. If instead of looking at the average fare paid regardless of quality, one treated each fare category as a different good—which goes too far in the other direction but is instructive nonetheless—the behavior of fares appears *dramatically* worse under deregulation. Indeed, *full* fares have risen 156 percent since 1978, more than double the rate of growth of the Consumer Price Index. As Melvin Brenner has noted, "Getting a 50 percent discount is no bargain, when it's calculated from a list price that was first raised 200 percent or more."[37]

The Bureau of Labor Statistics (BLS), in the air transport component of

Figure 21.5
Real Airfares, 1967–1988

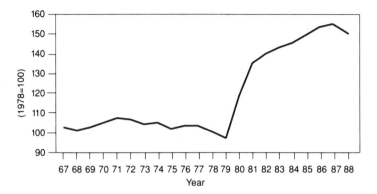

Source: Bureau of Labor Statistics.

the Consumer Price Index (CPI), prices a fixed bundle of fares in different fare categories—first-class, discount first-class, coach and discount coach—to construct an index of air fares.[38] Figures 21.5 and 21.6 show the behavior of this index over the period 1967–88 after adjusting for inflation and—for figure 21.6 only—changes in real fuel prices. The index rises dramatically after deregulation in both cases. Before deregulation, this index of airfares was either flat or falling, depending on whether the measure is adjusted for fuel price changes. In either case, however, *real fares rose some 50 percent after deregulation.* Given that the mix of discounted ver-

Figure 21.6
Fuel-Adjusted Real Airfares, 1967–1988

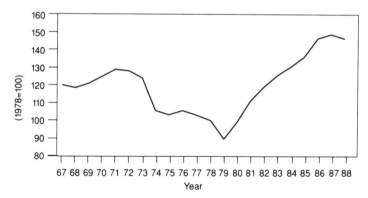

Source: Bureau of Labor Statistics.

sus undiscounted traffic has remained roughly flat (at 90 percent) in recent years,[39] the post-deregulation behavior of this mix-held-constant measure of airfares does not augur well for consumers in the future.

Thus, the changing mix of air travel toward discounted fare categories entails some deterioration in quality. But perhaps more important is the increase in delays and schedule uncertainty that pertain to flying—in any category—in the brave new world of deregulation. The opportunity cost of air travel—the time we lose stranded at airports, imprisoned in aircraft, or routed through circuitous hub connections—seems to have increased substantially under deregulation.

The widely acclaimed Brookings Institution study on airline deregulation by Steven Morrison and Clifford Winston alleged that consumers save $6 billion annually as a result of deregulation, a sum composed of fare discounts and opportunity cost savings realized as a result of "improved service convenience [to business travelers] attributable to the accelerated development of hub-and-spoke operations and to frequency improvements in low-density markets."[40] Of the $6 billion, approximately $4 billion is attributable to opportunity cost *savings* while the remaining $2 billion comes from savings in fares.[41] (We have already seen reason to doubt that consumers have saved *anything*, let alone $2 billion, from lower fares.) The overall import of the study was that airline service had not declined since deregulation began but, because of additional frequencies, had actually improved. Ostensibly, business travelers save time because they have more frequencies from which to choose. It is fair to say that most business travelers, if polled, would find such an assumption implausible.

By focusing on the number of flights in larger markets as the dominant measure of airline service, the Brookings study appears to have missed what most real-world flyers see. Whatever the improvements in the rate structure since deregulation, the consensus of most of what is written about airlines in this environment is that service has declined significantly. Although consistently measured data on delays over a long time period are not available, the epidemic of delays that pervades the airline industry seems actually to have imposed significant opportunity costs, not benefits.

Other sources maintain that the United States has suffered billions of dollars in opportunity costs as a result of air travel delays. Travel delays in 1986 alone cost airlines $1.8 billion in extra operating expenses and cost consumers $3.2 billion in lost time.[42] Travelers at 17 of our nation's most congested airports collectively lose more than 20,000 hours per year.[43] Because of the undependability of airline schedules, many business travelers find they must arrive in a city the evening before a business meeting in order to be sure they will be at the meeting.[44] Moreover, the delays experienced at congested airports constitute the other side of the coin of the frequency improvements in thin markets noted by Morrison and Winston;

both result, arguably, from the same phenomenon—the move to hub and spoking. As Brenner notes:

The very increase in hub-and-spoke frequencies which played so large a part in the study's calculation has been an important contributor to the congestion and delays which by 1987 had become a matter of widespread concern. While reducing the time interval between published departure times, the increased hub-and-spoke frequencies have increased the actual delay time at the gate, and in runway queues— a form of lost time that is especially costly to business traveler productivity.[45]

In 1988, many airlines amended their schedules to incorporate anticipated delays. Initially, this brought an "improvement" in on-time performance by airlines, as measured by the FAA (which counts only nonmechanical delays of more than 15 minutes). Despite creative-accounting methodology, delay figures in late 1988 were significantly higher than the year before.[46] Moreover, delays for the first nine months of 1989 were 22 percent higher than in the same period the preceding year.[47]

Note too that, even accounting for lost time, for which there is some equivalent dollar measure, we do not take into account the other, less measurable costs to society of deregulation. The aggravation and anxiety many travelers suffer because of delays, congestion, and a narrower margin of safety cannot easily be calculated. The Brookings study in fact explicitly omitted the psychic costs to the actual business traveler; the authors' measure encompasses only the monetary "savings" to the *businesses* that employ the increasingly harried travelers[48] (see chapter 23 for more on the Brookings study).

In the 1990s, the principal opportunities for low prices will be for discretionary travelers on long-haul (greater than 1000 miles), one-stop flights (via hubs) between large cities at off-peak times.[49] But the *average* air passenger in 1989 is paying roughly 2.6 percent more per mile than he or she would have paid without deregulation, and the differential is growing. We are flying more miles than we would have flown before deregulation; we are flying in fare categories with more restrictions; and we seem to be experiencing more actual delays. In short, we are paying more and enjoying it less. In addition, as the next chapter argues, service has deteriorated along many dimensions.

APPENDIX

Data, Methods, and a Note on Morrison and Winston's[50] "Counter-factual" Analysis of the Effects of Airline Deregulation on Fares

Annual data on nominal yields (revenue per passenger mile) came from the Air Transport Association. The annual average value of the Consumer Price Index (all items) was used as a deflator to construct an index of real yields. Annual average fuel costs and the share of fuel in all expenses came from the Air Transport Association's Airline Cost Index for the years from 1970 on. For the earlier years included in the analysis (1967–69), both fuel costs and the fuel share of expenses were estimated, as noted below, since these data were generally unavailable. Real fuel costs were constructed using the CPI as a deflator.

The fuel-adjusted real yield was constructed as follows. Starting from an arbitrary level at the beginning of the period, the percentage change in the index in each year is computed as the difference between the percentage change in the unadjusted real yield, on the one hand, and the product of the percentage change in real fuel prices and the fuel share of costs in the previous year (and thus at the beginning of the current year), on the other. For example, if for some year real yields rose 10 percent, real fuel costs rose 20 percent, and fuel costs in the prior year were 20 percent of costs, then the calculation of the percentage change in fuel-adjusted real yields would be: $10 - .2(20) = 6$ percent. The reported trend differences before and after deregulation were obtained by regressing the natural logarithm of unadjusted and fuel-adjusted real yields, in turn, on time (with $1967 = 0$), a time/deregulation dummy to capture any change in trend after deregulation and a deregulation dummy to capture any one-time shift after deregulation. The first two columns of table 21.A1 present these results.

The coefficient on time gives the estimated trend rate of increase in real yields before 1978. The sum of the coefficients on time and DTime gives the estimated trend rate of increase in real yields after 1978. Thus a statistically significant coefficient on DTime indicates a statistically significant *difference* in the trend before and after deregulation. In column 1, with the *unadjusted* real yield as the dependent variable, the trend rate of increase is significantly lower (a bigger rate of decay) after as compared with before deregulation. As column 2 shows, however, the *fuel-adjusted* real yield grew significantly faster (a slower decay rate) after deregulation. The estimated one-time percentage shift in real yields for which deregulation is responsible can be computed as the antilogarithm of the difference between the absolute value of the coefficient on DInt, on the one hand, and the product of 11 (the value of Time in 1978) and the coefficient of DTime, on the other. For the fuel-adjusted real yield, this computation yields 12.8 percent as the one-time decline in yield due to deregulation.

Fuel costs for the years 1967–69 were predicted by kerosene prices, based on the regression of fuel costs on kerosene prices (obtained from the Bureau of Labor Statistics, Producer Price Index) in those years (1970 and after) when both variables were available. The estimate of the fuel share of expenses in 1967–69 was based on the regression of fuel share on the natural logarithm of real fuel costs for 1970 and after. Finally, for computing the estimated real yield in 1989, only year-to-date data on the yield for domestic services were available (from the Air Transport Association). The estimate for the yield on all services was made using the predicted value based on the regression of the yield for all services on the yield for domestic service alone in the years (1978 and after) when both were available. In each case, the variables used to make the estimate were highly correlated with the variable to be estimated (R^2 in the regression of fuel costs on kerosene prices was .998; for the regression of fuel share of expenses on real fuel costs, R^2 was .960;

Table 21.A1
Regression Results

| | Dependent Variable | | | |
| | (1) | (2) | (3) | (4) |
Coefficient on:	Ln. of Real Yield	Ln. of Fuel- Adjusted Real Yield	Ln. of Real Yield	Percentage Change in Fuel-Adj. Real Yield
Time	−.0149	−.0275	−.0355	−.1314
	(.0043)	(.0029)	(.0121)	(.3863)
DTime*	−.0165	.0085	.0153	.9701
	(.0060)	(.0040)	(.0164)	(.5110)
DInt**	.1772	−.2298	.24474	−14.0162
	(.0740)	(.0496)	(.1887)	(5.9598)
Ln. Real Fuel	—	—	.1620	—
			(.0368)	
Ln. Real Labor	—	—	.1319	—
			(.3373)	
R^2	.9300	.9839	.9647	.2737
Time Period	1967-88	1967-88	1970-88	1968-88
D.F.	18	18	13	17

(Standard errors in parentheses)

* DTime takes on the value of time in 1978 and after, and 0 for prior years.

** DInt takes on the value of unity in 1978 and after, and 0 for prior years.

and for the regression of the yield on all service on the domestic yield only, R^2 .989).

The third column of table 21.A1 reports the results of a regression that would appear to throw some doubt on the methodology employed by Morrison and Winston in their study of the effects of airline deregulation, in which they claim that deregulation was responsible for "an overall reduction in fares of nearly 30 percent."[51] This estimate is based on a "counter-factual" methodology in which they ask what fares would have been in 1977 if deregulation had been in effect then and compare the result to *actual* fares in 1977. The estimate of what fares would have been in 1977 under deregulation is based on the relationship between input costs—chiefly fuel and labor—and revenues per passenger in the period 1980–82 for major carriers. The regression of fares on input costs for this period allows them to predict 1977 deregulated fares based on 1977 input costs.

This method is highly problematic once the secular downward trend in real yields— even holding input costs constant—is appreciated. The third column of table 21.A1 illustrates this trend. Even holding real labor and fuel costs constant, real yields fell by 3.5 percent per year, indicating a secular increase in productivity before deregulation, which if anything, has been adversely affected by deregulation (the coefficient on DTime is positive but not statistically significant).

Given such a secular trend, a substantial part of the difference between Morrison and Winston's "1977 deregulated yield" and the actual yield simply reflects the passage of time and the correlated productivity improvements between 1977 and 1981–82, when Morrison and Winston estimated their fare-cost relationship. This trend has nothing to do with deregulation or whether anything was slowed down by deregulation. Half of the effect they find (14 of 28 percent—they round up the latter to 30) might well be spurious for this reason.

Second, if disaggregated data were to confirm the lower rate of productivity improvement (lower rate of price decline, holding costs constant) after deregulation, which is weakly supported in the aggregate annual data used for the regression reported above, then Morrison and Winston would be telling us only about a one-time shift, which would eventually be dissipated. In fact, coincidentally, the 14 percent shift that their results, properly interpreted, can attribute to deregulation is almost identical to the one-time shift (12.8 percent) that our descriptive trend analysis finds in the data. Thus, our conclusion that consumers paid in 1988 what they would have paid without deregulation—and actually paid 2.6 percent more in 1989—is not inconsistent with Morrison and Winston's finding of a 30 percent fare *reduction*, using their faulty methodology. Before their estimates of "gains" to consumers in form of lower fares can be taken seriously, it behooves the authors to use a method that allows for the trend decline in fares, holding costs constant, a trend that the industry, deregulated or not, has historically exhibited.

NOTES

1. C.R. McConnell, Economics, 10th Edition, 49 (1987).

2. *See* E. Bailey, D. Graham, & D. Kaplan, Deregulating the Airlines (1985); Graham, Kaplan, & Sibley, *Efficiency and Competition in the Airline Industry,* 14 Bell J. Econ. 118 (1983); Call & Keeler, *Airline Deregulation, Fares, and Market Behavior: Some Empirical Evidence,* in Analytical Studies in Transport Economics (A. Daugherty, ed., 1985); Moore, *U.S. Airline Deregulation: Its Effect on Passengers, Capital, and Labor,* 29 J. L. & Econ. 1 (1986); Morrison & Winston, *Empirical Implications and Tests of the Contestability Hypothesis,* 30 J. L. & Econ. 53 (1987).

3. A recent General Accounting Office study shows that average yields are higher and have increased faster from 1985 to 1988 at concentrated hub airports than at unconcentrated airports. *See* General Accounting Office, Air Fares and Service at Concentrated Airports (1989). Low fares have stimulated new traffic in the past decade, mostly for vacation travelers flying between large cities served by more than a single carrier. In 1977, the airlines carried 240 million passengers; in 1987, they carried 447 million passengers. Ott, *Industry Officials Praise Deregulation, but Cite Flaws,* Aviation Week and Space Technology, Oct. 31, 1988, at 89. But business travelers and others unwilling to sleep in strange

cities on Saturday nights, individuals flying to small towns or people who, at the last minute, have to fly home for funerals or other emergencies, are ineligible for these discounts. Full fares have increased 156 percent since 1978, more than double the growth in the Consumer Price Index. So when deregulation's proponents emphasize that 90 percent of passengers travel today on discounts and that the average discount is 60 percent, remember that the full fare from which such "bargains" are discounted has risen sharply.

4. The new low fares that are offered in larger, competitive markets during the last decade have stimulated significant new passenger demand. People who had never flown took advantage of the bargain-basement fares of the early 1980s and became addicted to flying. And as demographics shifted the median age upward, more people than ever entered the 40–60 age bracket of affluent Americans who can afford to fly.

5. *See* Wagner & Dean, *A Prospective View toward Deregulation of Motor Common Carrier Entry,* 48 ICC PRAC. J. 406, 413 (1981). *See also* Wagner, *Exit of Entry Controls for Motor Common Carriers; Rationale Reassessment,* 50 ICC PRAC. J. 163, 172–73 (1983).

6. McGinley, *Bad Air Service Prompts Call for Changes,* WALL ST. J., Nov. 9, 1987, at 28.

7. *Ex-Official Suggests Lid on Air Fares,* ROCKY MOUNTAIN NEWS, Nov. 5, 1987, at 100. Said Kahn: "I don't reject the idea as a matter of principle. If price gouging gets bad enough, consumers will demand and deserve protection." Hamburger, *Fighting Back Begins as Costs Go Up, Up and Away,* MINNEAPOLIS STAR TRIBUNE, Dec. 24, 1988, at 6A. *See also* A. Kahn, *Surprises of Airline Deregulation,* 78 AEA PAPERS AND PROCEEDINGS (1988), at 320.

8. Carroll, *Higher Fares, Better Service Are Forecast,* USA TODAY, Oct. 24, 1988, at B1.

9. *See* A. Kahn, *Transportation Deregulation . . . and All That,* ECONOMIC DEVELOPMENT QUARTERLY (1987), at 96.

10. *Airline Deregulation under Fire,* DENVER POST, Sept. 23, 1988, at 4A. "Average fares, adjusted for inflation, have declined 13% in the ten years of free-market travel." Burnley, *Soaring Air Travel in Unfettered Skies,* WASHINGTON TIMES, Oct. 31, 1988, at 4-D.

11. Department of Transportation, "Secretary's Task Force on Competition in the U.S. Domestic Airline Industry," February 1990, Pricing Section, Vol. 1, pp. 52–53.

12. Brenner, *Rejoinder to Comments by Alfred Kahn,* 16 TRANSP. L. J. 253, 255 (1988).

13. *Better Buys, Crowded Skies,* U.S. NEWS AND WORLD REP., Oct. 31, 1988, at 50, 52.

14. AIR TRANSPORT ASSOCIATION, AIRLINE COST INDEX, September 18, 1989.

15. The rates reported here differ from those in table 21.1 because they are least-squares growth rates, first, and because they are continuously compounded annual growth rates, second.

16. See previous note.

17. In R. GORDON, PRODUCTIVITY IN THE TRANSPORTATION SECTOR (1991). It is alleged that circuity is only 4 percent, although the source cited provides no independent means of verification. In S. MORRISON & C. WINSTON, ECONOMIC

EFFECTS OF AIRLINE DEREGULATION 22 (1986), 5.4 percent is given as an estimate of increased *travel time* after deregulation, which the authors attribute partly to increased circuity, partly to increased congestion. The percentage increase in *mileage* due to circuity may be very different, of course. Since so much of travel time is takeoff and landing, increased mileage per trip will not add proportionately to travel time. But hub and spoking also increases takeoffs and landings per trip, which goes in the other direction. Taking 5.4 percent as an estimate of increased mileage amounts to treating these opposing factors—arbitrarily—as a wash. The 30 percent is an estimate of increased trip distance after deregulation by airline consultant Ted Harris, cited in Henwood, *Deregulation and Beyond*, ECONOMIC NOTES, November–December 1989.

18. Kuttner, *Plane Truth*, NEW REPUBLIC, July 17, 1989, at 21, 22.

19. S. MORRISON & C. WINSTON, *supra* note 17.

20. Brenner, *Are Airlines Off Course?*, WALL ST. J., Feb. 1, 1989, at A13.

21. *Effect of Airline Deregulation on the Rural Economy, Hearings before the Subcomm. on Rural Economy and Family Farming of the Sen. Comm. on Small Business*, 100th Cong., 1st Sess. (1987) (testimony of John J. Nance).

22. Dempsey, *Fear of Flying Frequently*, NEWSWEEK, Oct. 5, 1987, at 12.

23. DOT, Secretary's Task Force on Competition in the U.S. Domestic Airline Industry, February 1990. The FAA has defined hub size classes as follows: a large hub is a city and its metropolitan area that enplanes at least 1.00 percent of the total number of U.S. certificated airline passengers per year; a medium hub enplanes between 0.25 and 0.999 percent; a small hub enplanes between 0.05 and 0.249 percent; and a nonhub enplanes less than 0.05 percent. FEDERAL AVIATION ADMINISTRATION AND RESEARCH AND SPECIAL PROGRAMS ADMINISTRATION, AIRPORT ACTIVITY STATISTICS OF CERTIFICATED ROUTE AIR CARRIERS iv (1987).

24. *Airline Deregulation under Fire, supra* note 10, at 4-A.

25. *Id.*

26. J. F. Molloy, Jr., in his book on the U.S. commuter airline industry, has commented on service and fares to small communities: "The average smaller community in the U.S. air transport system has generally had the worst of all worlds since airline deregulation. Departures have decreased, seats have decreased, passengers have decreased, flights have decreased, and the fares have increased at a rapid rate." J. F. MOLLOY, JR., THE U.S. COMMUTER AIRLINE INDUSTRY 62 (1985).

27. Kuttner, *supra* note 18, at 21.

28. Paradoxically, higher ticket prices are a mixed blessing. The unprecedented economic anemia created by deregulation deprived the airline industry of sufficient resources to replace aging aircraft. As the skin peeled off the aging jets, a chilling realization swept over the industry—that it may be time to retire them. Higher profits will enhance the industry's ability to retire old aircraft and spend more on maintenance. Although there is no guarantee that airline executives will so invest their healthier profits, if we are not to have a series of aviation catastrophes, the industry needs to do both.

29. *See* GENERAL ACCOUNTING OFFICE, *supra* note 3, at 2.

30. *Id.* at 23.

31. Department of Transportation, Secretary's Task Force on Competition in the U.S. Airline Industry, February 1990, Executive Summary, 8.

32. Rose & Dahl, *Skies Are Deregulated, But Just Try Starting a Sizeable New Airline*, WALL ST. J., July 19, 1989, at 1.

33. *Happiness Is a Cheap Seat*, ECONOMIST, Feb. 2, 1989, at 68.

34. Estimate based on the percentage change, 1988–89, in the yield for domestic service only, from ATA. See the Appendix for details.

35. See equation 4 in table 21.A1.

36. AIR TRANSPORT ASSOCIATION, AIR TRANSPORT 1989 (1989).

37. Brenner, *supra* note 20.

38. Doug Henwood, editor of the LEFT BUSINESS OBSERVER, has been a voice crying in the wilderness about the disparity between this data and the usually cited industry data. The authors owe their knowledge of the existence and significance of the BLS data to a phone conversation with him. See Henwood, *Deregulation and Beyond*, ECONOMIC NOTES, November/December 1989, and forthcoming article in HARPER'S. The authors also wish to thank Dale Smith of the Bureau of Labor Statistics, who kindly provided the data and patiently explained the methodology employed to get it.

39. AIR TRANSPORT ASSOCIATION, *supra* note 36. The preparation of revenue passenger miles flown on discounted tickets is on pp. 79–88.

40. S. MORRISON & C. WINSTON, *supra* note 17, at 33.

41. *Id.*

42. *Gridlock!*, TIME, Sept. 12, 1988, at 52, 55.

43. Another source estimates that delays of more than 15 minutes have increased 117 percent since 1976, creating productivity losses of more than $3 billion annually. *Better Buys, Crowded Skies*, U.S. NEWS & WORLD REP., Oct. 31, 1988, at 50.

44. *Gridlock!*, *supra* note 42, at 55.

45. Brenner, *Airline Deregulation—A Case Study in Public Policy Failure*, 16 TRANSP. L.J. 179, 223 (1988).

46. Dahl & Valente, *Airline Delays Rise Sharply after Earlier Improvement*, WALL ST. J., Nov. 23, 1988, at B1.

47. Winans, *Flight Delays Surge; Airlines Blamed by FAA*, WALL ST. J., Nov. 3, 1989, at B1.

48. *See* S. MORRISON & C. WINSTON, *supra* note 17, at 16. In their model, "The mode choice by pleasure travelers is based on utility-maximizing behavior; the choice by business travelers on *cost-minimization behavior by their firms.*"

49. Pelline, *Bumpy Ride under Deregulation*, SAN FRANCISCO CHRONICLE, Oct. 28, 1988, at 21.

50. *See* S. MORRISON & C. WINSTON, *supra* note 17.

51. MORRISON & WINSTON, *Airline Deregulation and Public Policy*, SCIENCE 18, August 1989, at 708.

22

SERVICE

The concept of service has numerous dimensions. Air service can be measured in a strictly quantitative sense, such as the number and type of flight departures per city over a given time period or the number of flight delays over time. Service can also be assessed more qualitatively, by focusing on its more intangible elements such as convenience, comfort, reliability, and consumer satisfaction.

In this chapter, we will evaluate both the quantitative and qualitative aspects of service. We will first examine changes in the frequency and type of air service for both large and small communities. Next, we review the question of delays and on-time performance in the air transportation system as a whole. Finally, other, more qualitative aspects of service will be addressed.

SERVICE FREQUENCY

Under regulation, trunk and local-service air carriers were required to possess CAB certificates of "public convenience and necessity" for every route they served. The CAB generally awarded routes to carriers so as to divide markets into systems.[1] Airlines were awarded mixtures of higher- and lower-density routes to allow for cross-subsidization of the less profitable routes with profits earned from more lucrative markets.[2] Because of the need for beyond-segment passenger flow to increase load factors on nonstop flights, what emerged were patterns of predominantly linear route structures.[3]

With deregulation, the CAB lost the authority to award routes, so that now all carriers that are "fit, willing, and able" are allowed to serve any domestic route.[4] Airlines adapted to the new environment by developing hub-and-spoke route structures in order to (1) accommodate larger vol-

Table 22.1
Locations of Airline Hub Facilities and Traffic Growth

		Aircraft Departures, 1980-87	
Hub Cities	Airlines	Absolute Change	% Change
Atlanta	Delta, Texas Air	+15,049	+ 5.64
Baltimore	USAir Group	+28,673	+ 80.98
Charlotte	USAir Group	+61,228	+189.71
Chicago	United, American	+87,442	+ 30.32
Cincinnati	Delta	+32,416	+111.82
Dallas	American, Delta	+67,952	+ 34.20
Dayton	USAir Group	+18,433	+ 92.45
Denver	United, Texas Air	+43,091	+ 27.25
Detroit	Northwest	+47,585	+ 53.21
Houston	Southwest, Texas Air	+57,200	+ 55.54
Memphis	Northwest	+51,455	+ 92.15
Minneapolis	Northwest	+32,325	+ 41.54
Nashville	American	+28,156	+ 99.44
Newark	Texas Air	+79,713	+140.27
Pittsburgh	USAir Group	+23,950	+ 24.97
Raleigh-Durham	American	+26,731	+132.05
St. Louis	TWA	+37,582	+ 37.33
Salt Lake City	Delta	+31,860	+ 75.04
Syracuse	USAir	+19,497	+101.49
Washington, D.C.	United	+67,247	+ 56.41

Sources: GAO, "Air Fares and Service at Concentrated Airports," June 1989; Brenner, 16 Transp. L. J. 190 (1988); FAA, "Airport Activity Statistics," 1987, 1980.

umes of traffic from an increased number of city-pairs, (2) avoid destructive competition, and (3) establish market power opportunities.

Hub systems have a multiplier effect on the number of origins and destinations a carrier serves. Each major airline developed hub facilities at strategic points in its air service networks so that passengers from many origins could be funneled into a hub and then flown out to their desired destinations. Thus, those cities with hub operations became centers of very high concentrations of passenger traffic and flight frequencies (see table 22.1).

In particular, cities such as Chicago, Newark, Dallas, and Washington, D.C., had the largest absolute increases in flight frequencies among all cities from 1980 to 1987. The largest percentage increases over the same time period were for cities such as Charlotte, Newark, Raleigh, and Cincinnati. The increased number of flights arriving at and departing from the major hub cities has been a mixed blessing for people who reside in these places. On the positive side, the increased number of flight departures has meant more frequent nonstop service from hubs to more different destinations—a direct improvement in hub accessibility. For travelers living in cities such as Atlanta, Chicago, and Dallas, there are now more air travel options from which to choose. The cities themselves have benefited from increases in business activity attributable to the availability of frequent and extensive air service.

However, the increased traffic coming through the hub airports has con-

Table 22.2

Changes in Frequency of Air Service by Size Classification, June 1978–June 1987

Size Class	Flights/Week 1978	Flights/Week 1987	Percent Change	Seats/Week (000) 1978	Seats/Week (000) 1987	Percent Change
Large	63,484	103,063	+62.3%	7,104	12,132	+70.8%
Medium	19,731	30,712	+55.7%	1,953	3,031	+55.2%
Small	13,256	18,806	+41.9%	1,112	1,405	+26.3%
Nonhub	29,543	29,271	− 0.9%	1,175	971	−17.4%
Total	126,014	181,852	+44.3%	11,344	17,539	+54.6%

Source: Brenner, 16 Transp. L. J. 211 (1988).

tributed to severe problems of congestion and delays. Some of the benefits of increased frequency of service have been offset by the increases in delays. Furthermore, as these hub airports became more firmly dominated by one or two airlines, the fares for travelers originating at or destined to a hub have increased significantly.[5]

Other problems related to the huge increases in traffic to the major hub airports include the increased noise pollution around the airports. With more flights arriving and departing around the clock, nearby residents have been bombarded by noise. Increased pressure for limiting use and/or expanding capacity at the most heavily utilized airports has been exerted. The nature of hub-and-spoke systems creates a situation of very high peaks of demand at certain times at only a few concentrated places across the country. The demand for certain "slots" at the most heavily congested airports has become intense. At four of the nation's busiest airports (Chicago O'Hare, Washington National, and New York's Kennedy and La-Guardia), the FAA has had to limit use at peak hours of demand by constraining the number of landing and takeoff slots while turning a partially deaf ear to pleas for building more infrastructure capacity.

Along with the hubs, larger and intermediate-size cities have generally experienced higher volumes of service frequency since deregulation began. Under the FAA size-classification scheme,[6] the large and medium-size hubs have had more substantial increases in the number of weekly departures and weekly seats than either the small or nonhub classes from 1978 through 1987 (see table 22.2).

Also noteworthy is the increase in enplaned passengers during the deregulatory period. Mirroring trends in service frequency, passenger enplanements for the major hub cities have exhibited skyrocketing growth, much of it due to the increased number of connecting passengers these airports now are handling (see table 22.3). Overall, passenger totals have increased substantially, with most of the growth occurring in the large hub category. Both the medium and the small hub groups also experienced healthy passenger growth, but the nonhubs showed little, if any, increase in passenger enplanements (see table 22.4).

Why has the number of passenger enplanements increased so much? One

Table 22.3
Locations of Airline Hub Facilities and Passenger Growth

| | | Enplaned Passengers, 1980-87 | |
Hub Cities	Airlines	Absolute Change	% Change
Atlanta	Delta, Texas Air	+2,655,320	+ 13.28
Baltimore	USAir Group	+2,357,286	+142.65
Charlotte	USAir Group	+4,540,317	+306.62
Chicago	United, American	+9,253,425	+ 47.65
Cincinnati	Delta	+1,872,984	+134.59
Dallas	American, Delta	+9,564,977	+ 74.87
Dayton	USAir Group	+1,277,512	+143.70
Denver	United, Texas Air	+5,977,798	+ 62.17
Detroit	Northwest	+4,203,738	+ 83.23
Houston	Southwest, Texas Air	+4,052,166	+ 59.53
Memphis	Northwest	+2,874,317	+133.77
Minneapolis	Northwest	+3,925,507	+ 89.53
Nashville	American	+1,865,149	+166.22
Newark	Texas Air	+7,082,930	+168.40
Pittsburgh	USAir Group	+2,774,356	+ 51.55
Raleigh-Durham	American	+1,450,204	+167.46
St. Louis	TWA	+4,407,759	+ 82.86
Salt Lake City	Delta	+2,731,889	+136.82
Syracuse	USAir	+ 705,315	+ 88.80
Washington, D.C.	United	+4,273,958	+ 55.10

Sources: GAO, "Air Fares and Service at Concentrated Airports," June 1989; Brenner, 16 Transp. L. J. 190 (1988); FAA, "Airport Activity Statistics," 1987, 1980.

immediate answer is that the rise in airline passengers is part of a long-term trend that extends throughout the history of aviation. With or without deregulation, the number of passengers flying would have increased substantially simply on the basis of these long-term trends.

Nonetheless, the "fare wars" of the early 1980s undoubtedly contributed to increased passenger demand for air travel. But as we have already seen, these intense fare wars were a fleeting phenomenon, as the industry has once again become highly concentrated and fares are trending sharply upward. Still, the positive public relations image of the early fare wars period has contributed to a continuing large air passenger demand.

Also contributing to that demand is the increasing absence of practical

Table 22.4
Changes in Number of Enplaned Passengers by Size Classification, 1980–1987

Size Class	Enplaned Passengers 1980	1987	Absolute Change	Percent Change
Large	197,679,376	316,041,613	+118,362,237	+59.88
Medium	51,664,627	70,646,403	+ 18,981,776	+36.74
Small	23,393,324	30,300,694	+ 6,907,370	+29.53
Nonhub	9,339,408	9,806,176	+ 466,768	+ 5.00
Total	282,076,735	426,794,886	+144,718,151	+51.30

Source: FAA, "Airport Activity Statistics," 1987, 1980.

alternatives. Both passenger rail and bus transportation have continued their declines in intercity service and have not been strong competitors for the airline industry. The United States is the only G-7 industrialized nation without a high-speed rail alternative. Finally, a large segment of the U.S. population—the "baby boom" generation—has been entering those affluent age categories where the propensity to travel begins to increase. This is true for both business trips and vacation trips as increasing disposable incomes generally suggest a greater likelihood of increased air travel.

Moreover, although the overall number of passenger boardings has grown since 1978, it grew every year after 1938 when economic regulation was first imposed. In fact, the percentage of growth was slower in the 1978–87 decade than in any other previous decade.[7]

Before deregulation, one of the chief concerns of the proposed legislation was the expected impact on service to small communities. Many observers feared that free exit might result in a total loss of service to many small towns. The CAB had protected small-town service in the past by the use of subsidies under Section 406 of the Federal Aviation Act and through stringent exit requirements.[8]

To appease these apprehensions, the Airline Deregulation Act included Section 419, providing for 10 years of guaranteed Essential Air Service (EAS)[9] to all communities on the 1978 certificated route network. The ADA's policy statement specifies that it is the purpose of the new legislation to maintain "a comprehensive and convenient system of continuous scheduled airline service for small communities and isolated areas . . . with direct Federal assistance where appropriate."[10]

Before the promulgation of Section 419, many proponents of deregulation had discounted the likelihood that it would lead to any significant service reductions for small communities. It was predicted that if a large trunk carrier pulled out, a smaller commuter or local-service carrier would fill the void, in most cases without a subsidy.[11] Thus, it was maintained that small communities had little to fear from deregulation.

Constituents from small communities were more skeptical. Even though the CAB had the authority to prohibit those service abandonments it deemed inconsistent with the public interest, between 1960 and 1975 it had approved the abandonment of 173 communities (or an average of 9.6 per year).[12] Without CAB oversight, many feared an even greater acceleration in abandonments.

In the first year of deregulation, 70 of the communities that were receiving service lost all of it[13] while 260 cities suffered some deterioration in air service.[14] In the first two years of deregulation, more than 100 communities lost all scheduled service.[15] So, whereas abandonments averaged 9.6 per year before deregulation, they soared to more than 50 per year during deregulation's first two years.

More recent statistics express similar trends. Although the rate at which abandonments are occurring appears to have slowed, they continue to result in reductions of more than 10 communities per year, further shrinking the air service network. Of the 514 nonhub communities receiving air service in 1978, by 1987, 312 (60.7%) had experienced declines in flight frequency; 143 (27.8%) had lost all service, and only 32 (6.2%) enjoyed the initiation of new service.[16] In terms of both flight departures and seating capacities, nonhubs have experienced an absolute decline (see table 22.2). These results tend to suggest that optimistic expectations for continued small-community service under deregulation have not materialized.

However, Alfred Kahn insists that small communities have not suffered under deregulation. He points out that not a single community receiving certificated service in 1978 has completely lost all service.[17] True, the Essential Air Service program has assured subsidies to these points, although the Department of Transportation recently announced its intention to drop a number of these cities from the EAS program.[18] But in fact, the existence of the subsidies itself mercifully dulls the impact of deregulation, which would likely deprive most of these communities of all service.

Furthermore, cities not previously certificated were ineligible for the subsidies. The program may have hastened abandonment of the small towns served only by the noncertificated commuter airlines because as the large carriers left the certificated small cities for denser markets, the commuter airlines shifted their operations to take advantage of the new subsidies, exiting towns not eligible for them.

Other proponents of deregulation, when assessing its impact on smaller communities, tend to emphasize the increased number of departures many have enjoyed.[19] As a group, nonhub communities had experienced a 9.3 percent increase in the number of departures per week from June 1978 to June 1984.[20] Much of that increase resulted from the replacement of larger-capacity jet aircraft by more frequent, but smaller, commuter aircraft.[21] Since 1984, the number of nonhub departures per week has returned to pre-deregulation levels, but the mix now favors commuter aircraft to an even larger extent.[22]

The shift to commuter aircraft represents a substantial reduction in service quality. The smaller, unpressurized aircraft flown by commuter and local-service carriers are considered to be less comfortable by most passengers. They are also less safe.[23] Depending on how it is measured, commuter airlines have a safety record of between 3 and 37 times worse than established jet airlines.[24] Passengers also appear to be less satisfied with the service schedules and flight delays of commuter airlines.[25] Service in small communities is also highly unstable, as carriers drop into bankruptcy, with service suspended until a replacement carrier can be found.[26]

Small towns lie remotely scattered under the dark and cloudy skies of deregulation, where not enough sunlight falls to give passengers a glimpse of the super-saver discounts prevalent in major markets.[27] With the airline industry becoming an oligopoly, passengers in small towns find their service reduced to a single airline, providing circuitous connections out of a major hub and charging whatever the market will bear.[28]

Even the deregulation proponent Thomas Gale Moore admits that 40 percent of small communities have suffered both a loss of air service and a disproportionate increase in ticket prices since deregulation began.[29] Similarly, Professor Abdussalam Addus has observed, "As a result of airline deregulation . . . fares for traveling between small points have increased rapidly; and commuter air carrier fares are reported to be particularly high in most cases."[30] Assessing the quantitative and qualitative impacts, observers have noted that "smaller communities are receiving markedly worse air service than existed prior to deregulation."[31] J. Molloy, Jr., has flatly stated, "The average smaller community in the U.S. air transport system has generally had the worst of all worlds since airline deregulation."[32]

The loss of service has an unhealthy ripple effect throughout the economy of each of these communities. As one commentator has noted, "Besides increasing transportation costs for companies already doing business in many small communities, the impact of deregulation is decreasing the attractiveness of locating new businesses in these communities."[33] A survey of executives of the 500 largest American corporations reveals that 80 percent would not locate in an area that did not have reasonably available scheduled airline service.[34]

A decade has elapsed since the federal government launched its grand experiment in transportation deregulation. The outlines of a consistent trend are becoming visible in all deregulated transport modes—airlines, railroads, trucking, and bus companies. Although deregulation has created a class of beneficiaries, consumers in small towns and rural communities are not among them. Today, they pay higher prices for poorer service.[35]

ON-TIME PERFORMANCE

Another dimension of air service that is more directly quantifiable is on-time performance. Before 1978, the Civil Aeronautics Board collected data on airline delays. With deregulation, the airlines were exempted from submitting delay data. This situation continued until September 1987, when the Department of Transportation, in response to an unprecedented number of consumer complaints about delays, required that airlines once again make public their on-time performance records.

Before regulation, on-time performance was not a major issue, since passengers generally expected and received reasonably punctual air service. Under deregulation, the airlines have almost universally adopted hub-and-spoke route structures, making congestion and delays inevitable, especially at the largest hub airports.

The very nature of hub-and-spoke systems requires that airlines concentrate as many incoming and outgoing flights in as narrow a window of time and space as possible in order to maximize the total number of city-pair combinations that can be effectively served through a hub airport. Before the DOT required publication of delay statistics, carriers also had an incentive to engage in unrealistic scheduling of the shortest possible origin and destination times so as to include their flights on the first page of the computer reservations system (CRS) screen display, where most flights are sold by travel agents. Hence, clustering of arrivals and departures may bear no correlation to an airport's or air traffic control system's capacity to handle them.

Because each airline independently schedules these hub-and-spoke flights, even the largest airports have been overtaxed. For example, in 1987, at Atlanta's Hartsfield Airport, the 9:00 A.M. crunch was illustrated by the airlines' scheduling of 32 arrivals in 15 minutes, whereas the optimum capacity was 21 arrivals in 15 minutes.[36] As a result of such saturation, delays occur and ripple throughout the entire national system of air transport and are exacerbated by inclement weather.

Delays have been rising since the beginning of deregulation, but some of the most recent years have been the worst. Delays in 1986 increased by 25 percent over 1985 at the nation's large hub airports and increased another 13 percent in 1987.[37] The traveling public tolerated delays through the early years of deregulation, perhaps accepting some degree of trade-off between service quality and lower fares. By 1987, however, the situation had grown so intolerable that a consumer backlash occurred, exemplified by an unprecedented number of complaints.[38]

In response to growing public disenchantment with the airlines' on-time performance, and an angry Congress poised to promulgate a legislative solution, the DOT in late 1987 began requiring airlines to disclose their on-time performance records so that consumers could begin to choose among airlines on that basis. The DOT stated:

A flight is counted as "on time" if it is operated less than 15 minutes after the scheduled time shown in the carriers' Computerized Reservation Systems. . . . Because of our concern that our rule not penalize carriers for conscientious safety practices, a delay is not reported to DOT if it results from a mechanical problem that is required to be reported to the Federal Aviation Administration.[39]

Despite these stated criteria, confusion and misunderstandings persist about how delays actually are recorded and classified. For instance, it is unclear whether the clock starts when the airplane leaves the gate or when it actually takes off. Likewise, there is ambiguity about when a flight actually arrives. Not all airlines report their delays in exactly the same way, so that consumers still have to be wary of footnotes and asterisks.[40]

Table 22.5
On-Time Arrival Performances of Airlines and Airports, 1987–1988*

	SEP 87		DEC 87		FEB 88		APR 88	
Carriers	%	Rank	%	Rank	%	Rank	%	Rank
Alaska	79.8	(6)	59.2	(13)	76.9	(6)	77.7	(12)
American	84.5	(1)	73.1	(4)	80.8	(4)	85.0	(5)
America West	73.4	(10)	76.4	(2)	88.7	(2)	90.8	(2)
Continental	81.1	(3)	60.5	(12)	67.7	(13)	81.5	(8)
Delta	72.3	(11)	61.8	(11)	73.6	(9)	85.6	(4)
Eastern	80.4	(4)	69.5	(6)	70.6	(11)	75.5	(14)
Northwest	69.0	(13)	63.3	(9)	61.7	(14)	84.2	(6)
Pacific SW	70.5	(12)	57.6	(14)	90.6	(1)	91.1	(1)
Pan American	74.3	(9)	77.3	(1)	80.1	(5)	76.5	(13)
Piedmont	80.3	(5)	67.2	(7)	75.0	(7)	81.0	(10)
Southwest	82.4	(2)	74.2	(3)	88.5	(3)	90.3	(3)
TWA	78.4	(8)	63.5	(8)	69.4	(12)	81.5	(8)
United	79.2	(7)	62.6	(10)	73.2	(10)	81.8	(7)
US Air	67.4	(14)	71.9	(5)	74.8	(8)	77.9	(11)
Total	77.1		66.4		74.7		82.6	
Airports								
Atlanta	82.4	(3)	70.1	(11)	75.3	(11)	83.6	(10)
Boston	69.5	(22)	71.2	(10)	70.9	(17)	67.7	(26)
Charlotte	85.1	(1)	71.9	(7)	80.7	(3)	84.9	(7)
Wash.-National	74.1	(16)	73.7	(4)	75.7	(9)	78.3	(23)
Denver	81.9	(5)	52.4	(25)	66.6	(24)	82.0	(13)
Dallas	84.5	(2)	67.4	(16)	79.1	(4)	85.9	(4)
Detroit	69.3	(23)	68.8	(13)	64.0	(26)	83.6	(11)
Newark	76.2	(13)	69.2	(12)	72.1	(15)	76.4	(24)
Hous.-Intercont.	80.6	(8)	56.3	(23)	68.9	(20)	78.5	(22)
New York-JFK	68.8	(24)	74.3	(2)	74.9	(12)	67.6	(27)
Las Vegas	76.7	(10)	68.2	(14)	82.9	(1)	84.3	(8)
Los Angeles	70.4	(21)	53.1	(24)	76.2	(7)	81.9	(14)
NY-LaGuardia	75.1	(14)	74.7	(1)	76.0	(8)	79.3	(20)
Orlando	74.0	(18)	72.4	(5)	62.7	(27)	79.9	(19)
Memphis	78.4	(9)	68.2	(15)	66.8	(23)	86.7	(2)
Miami	74.1	(17)	74.0	(3)	71.6	(16)	82.1	(12)
Minn./St. Paul	74.8	(15)	61.1	(20)	66.2	(25)	85.7	(5)
Chicago-O'Hare	80.9	(7)	66.5	(17)	68.7	(21)	80.6	(18)
Philadelphia	68.5	(25)	72.0	(6)	74.2	(13)	79.1	(21)
Phoenix	72.2	(19)	65.6	(18)	82.8	(2)	86.7	(3)
Pittsburgh	67.3	(26)	71.4	(9)	77.2	(6)	81.6	(16)
San Diego	71.7	(20)	57.1	(22)	77.9	(5)	81.8	(15)
Seattle	76.5	(12)	51.5	(26)	69.3	(18)	80.9	(17)
San Francisco	65.3	(27)	41.8	(27)	75.6	(10)	68.4	(25)
Salt Lake City	81.5	(6)	60.0	(21)	74.1	(14)	90.6	(1)
St. Louis	82.0	(4)	64.8	(19)	69.1	(19)	85.0	(6)
Tampa	76.7	(11)	71.7	(8)	67.8	(22)	84.2	(9)

* The DOT counts only nonmechanical delays greater than 15 mins.
Source: U.S. Dept. of Transportation, Air Travel Consumer Reports. 1987; 1988.

Since the records were made public, on-time performances have generally improved (see table 22.5). Most of the recent improvements, however, may be attributed to the carriers' practice of adding time to schedules rather than shaving actual transit time. More passengers are arriving "on-time" even though they are not arriving at their destinations earlier.[41]

The Federal Aviation Administration (FAA) has been tracking delays since 1982, although the criteria it uses differ from those used by the DOT. The FAA considers a flight late if it takes off more than 15 minutes after the plane leaves the gate or the pilot requests takeoff.[42] According to this definition, the FAA reported a 22 percent increase in delays during the first quarter of 1989, the highest number of delays for a first quarter since the FAA started keeping track.[43] What this seems to suggest is that an increasing number of planes are "stacking up" on runways, waiting to take off.

The capacities of our major hub airports are being saturated and often overloaded. With more flights being scheduled through a more constrictive air traffic control system network, it's no wonder that it now takes more time to get from point A to point B. Yet because of the way the airlines are recording delays, on-time performances appear to have improved. This scenario can only be described as being misleadingly rosy.

SERVICE QUALITY

In addition to service frequency and on-time performance, other aspects of service quality need to be addressed. Earlier, the issue of small-community service was discussed, along with the fact that service quality has generally declined. Much of that discussion could also be applied to the national system of air travel as a whole.

Even though some travelers are enjoying reduced fares, the quality of the product that they buy today is inferior to that they could have purchased before deregulation.[44] To pose an analogy, a decade ago we traveled by Carnival Cruise Line. It was luxury service, but at a premium price. Today we travel by slave ship. It is crowded and uncomfortable, but for some, it's dirt cheap.

A recent editorial in the *Washington Post* summed up what many firmly believe to be the results of deregulation: "Airline Service Has Gone to Hell."[45] Why? One authority on services marketing said: "It's one of those terrible debt spirals. Without profit, there can be no service and no safety."[46]

Flying has become a rather unpleasant experience. The planes are filthy, delayed, cancelled, and overbooked; our luggage disappears; the food is processed cardboard. Chronic delays, missed connections, near misses, and circuitous routing all are products of hub-and-spoking, adopted by every major airline. Too often, we find ourselves stranded in airports or imprisoned in aircraft, waiting endlessly to get to our destinations.

A recent survey of consumers reveals that almost 50 percent believe that airline service had declined since deregulation; less than 20 percent said service had improved. Among the complaints: late departures, crowded seating, long lines at check-in, unappetizing food, overbooked aircraft, and an unacceptably long wait for baggage.[47] Another survey, this one of 15,000 frequent flyers, found even more negative attitudes of the impact of deregulation on air service. Approximately 68 percent said that deregulated air service was "less convenient and enjoyable," whereas only 19 percent thought it more convenient and enjoyable.[48] Still another survey, this one of 461 members of the Executive Committee (a group of corporate presidents and chief executives), revealed that 36 percent had lost job efficiency because of air travel delays.[49]

These results parallel those of the U.S. Department of Transportation. DOT data reveal that consumer complaints about airline delays, congestion, overbooking, bumping, missed connections, lost baggage, cancellations, and deteriorating food have soared in recent years.[50] From a low of 7,326 in 1983, complaints filed with DOT against U.S. airlines skyrocketed to a high of 40,985 in 1987.[51] Consumer complaints have since receded, reflecting perhaps the public's acclimation to poorer service and their lowered expectations.

Consumer abuses do not stop with miserable service. Under deregulation, management philosophy in the airline industry is dominated by the philosophy of P. T. Barnum: "There's a sucker born every minute." Without government oversight, airlines freely engage in imaginative forms of consumer fraud, including bait-and-switch advertising, deliberate overbooking, unrealistic scheduling, and demand-based flight cancellations.[52]

Admittedly, some consumers are paying less for air service than they did before deregulation. Those who have benefited most are vacation (discretionary) travelers in densely traveled markets served by several carriers, often via alternative hubs. Business travelers flying between small towns served by only a single carrier generally have not benefited from fare reductions. (However, large corporations, with more than half a million dollars in annual travel, can negotiate a rate nearly as low as that offered the discretionary traveler, but without the restrictions). And today, both the vacation and the business traveler are often routed through a circuitous hub connection, causing the traveler to consume more time in both aircraft and airports, and a decidedly less pleasurable consumption of time, than before deregulation. For many, opportunity costs have increased since deregulation began. And for passengers who begin or end their trip at a concentrated hub, air travel is significantly more expensive vis-à-vis those who connect at the hub.

Why has the unregulated market not corrected this deterioration in service? Some suggest that service deterioration is attributable to the decline in firms' profitability caused by the "destructive competition" unleashed by deregulation.[53] Hence, during deregulation's first decade, carriers did not have sufficient resources to staff flights with more flight attendants than FAA minimums, to staff ticket counters or baggage areas adequately, to provide better food, to avoid deliberate overbooking or unrealistic scheduling, to buy new aircraft or even to clean them properly. While some airlines are worse than others, the decline appears to be nearly universal.

Another explanation of the market's failure may be reflected in the nature of the item being sold. When a consumer purchases a manufactured product, he can examine it in a retail store before he spends his money—pull it off the shelf and turn it over and make some assessment of its quality. But when a consumer buys a service, like transportation, its definition beyond a mere description of "the movement of my body from A to B" is more amorphous. Air transportation is a "credence good," one that cannot be examined before consumption. This enables unscrupulous providers to exploit unwary consumers.

When booking a flight, most consumers do some price shopping. Where competitive alternatives exist, there has been some measure of pricing competition under deregulation. Those who price shop usually opt for the lower fare (although, as noted earlier, frequent-flyer mileage programs and travel agent commissions militate against the lowest price). Travelers who have been through the ordeal of a hub connection may ask for a nonstop if one is available, or, if not, a one-stop. Some may also shop for a convenient departure, although published schedules are today unreliable.

But beyond that, how many consumers ask the following questions? What kind of aircraft is being flown, how old is it, and when was it last overhauled and cleaned? How often is this flight late and by how much, on average? By what percentage of passengers do you usually overbook the flight? What percentage of bags are usually lost on the flight, and if you don't lose them, how long will I have to wait at the destination for my bags? How many flight attendants are on board, and will I be offered a magazine, pillow, cup of coffee, or bag of peanuts? What's for dinner, and how tasty is it? What's the average wait in the line at the airport? How crowded is the flight and the waiting lounge at the gate? How much knee and leg room do you give me between seats? How comfortable is the seat? Because most of these questions are not asked by consumers before they purchase their ticket (nor would the question probably be answered even if asked), the market has not responded to consumer desires for better service.[54]

In the next chapter, the issues of service and pricing will be explored in greater depth through an analysis and critique of a particular study on the economic effects of airline deregulation.

NOTES

1. E. Bailey, D. Graham, & D. Kaplan, Deregulating the Airlines 11–12 (1985).

2. Dempsey, *The Rise and Fall of the Civil Aeronautics Board—Opening Wide the Floodgates of Entry*, 11 Transp. L. J. 112–113 (1979).

3. M. Brenner, J. Leet, & E. Schott, Airline Deregulation 75 (1985).

4. Entry in most international markets, however, continues to be regulated. See P. Dempsey, Law & Foreign Policy in International Aviation 65–69 (1987).

5. GAO Testimony, *Air Fares and Service at Concentrated Airports,* June 7, 1989.

6. Hub size classes were defined by the CAB and the FAA as follows: a large hub is a city and its metropolitan area that enplanes more than 1.00 percent of the total number of U.S. certificated airline passengers per year, a medium hub between 0.25 and 0.999 percent, small hubs between 0.05 and 0.249 percent, and nonhubs less than 0.05 percent.

7. The increase in boardings rose 65% from 1978 to 1988. It had risen 89% in the preceding decade, 201% in the decade before that, and 240% from 1948–58. W. Pogue, Airline Deregulation, Before and After: What Next? (1991).

8. Beitel, *CAB Rules for Essential Air Service,* Airport Services Mgmt, June 1980, at 15, 16.

9. The CAB defined EAS as being at least two daily round-trips, five days a week. *Id.*

10. U.S.C. sec. 1302 (a)(8) (1979).

11. E. Bailey, D. Graham, & D. Kaplan, *supra* note 1, at 33.

12. Civil Aeronautics Board, Competition and the Airlines: An Evaluation of Deregulation 135 (1982).

13. Meyer, *Section 419 of the Airline Deregulation Act: What Has Been the Effect on Air Service to Small Communities?,* 47 J. Air L. & Com. 151, 181 (1981).

14. *See* Civil Aeronautics Board, Report on Airline Service 43–50 (1979).

15. Havens & Heymsfeld, *Small Community Air Service under the Airline Deregulation Act of 1978,* 46 J. Air L. & Com. 641, 673 (1981).

16. Goetz & Dempsey, *Airline Deregulation Ten Years After: Something Foul in the Air,* J. Air L. & Com. 947 (1989).

17. Testimony of Alfred Kahn Before the California Public Utilities Commission 6247-48 (Jan. 31, 1989).

18. Under Section 419 of the Airline Deregulation Act of 1978, small-community subsidies were to last until 1988, when they were extended by Congress. In 1985, 142 cities were receiving subsidized service under the program. General Accounting Office, Deregulation: Increased Competition Is Making Airlines More Efficient and Responsive to Consumers 31–32 (1985). In 1989, DOT announced its intention to eliminate subsidies to several cities. No doubt, most will lose air service altogether if federal economic subsidies dry up.

19. *See* E. Bailey, D. Graham, & D. Kaplan, *supra* note 1, and R. Noll & B. Owen, The Political Economy of Deregulation: Interest Groups in the Regulatory Process (1983).

20. Civil Aeronautics Board, Report on Airline Service 68 (1984).

21. *See* Civil Aeronautics Board, Report on Airline Service, Fares, Traffic, Load Factors, and Market Shares 32 (Sept. 1, 1984).

22. Since these data are no longer reported by the CAB, it has been updated to June 1987 from information reported in the Official Airline Guide (June 1, 1987); *see also* Goetz & Dempsey, *supra* note 16, at 947; and A. Goetz, The Effect of Airline Deregulation on Air Service to Small and Medium-Sized Communities: Case Studies in Northeastern Ohio 35 (1987) (Ph.D. Dissertation).

23. Oster, Jr., & Zorn, *Deregulation and Commuter Airline Safety,* 49 J. AIR L. & COM. 315, 316 (1984).

24. *See* Oster, Jr., & Zorn, *Airline Deregulation, Commuter Safety, and Regional Air Transportation,* 14 GROWTH AND CHANGE 3, 7 (1984). Author John Nance summarized the reasons for the deterioration of safety resulting from the substitution of commuter carrier service for scheduled airlines:

> The aircraft [commuter airlines] that fly are usually less sophisticated, largely unpressurized, and much smaller than mainstream jetliners. Many are devoid of not only restrooms, they are also devoid of radar, devoid of decent cockpit communications, devoid of sophisticated flight instruments, devoid of those elements that are part of the safety buffer which all of us as Americans have come to expect of our air transportation system, whether we are boarding in a rural area or not.
>
> In addition [most] of these aircraft . . . fly at altitudes most vulnerable to weather hazards and potential mid-air collisions. They are maintained by less sophisticated maintenance departments, they are flown by less experienced pilots, usually the first airline job of their career.

Effect of Airline Deregulation on the Rural Economy, Hearings before the Subcomm. on Rural Economy and Family Farming of the Sen. Comm. on Small Business, 100th Cong., 1st Sess., 81–82 (testimony of John J. Nance).

25. *See* Ahmed, *Air Transportation to Small Communities: Passenger Characteristics and Perceptions of Service Attributes,* 38 TRANSP. Q. 15, 21 (1984).

26. M. Kihl, *The Impacts of Deregulation on Passenger Transportation in Small Towns,* 42 TRANSP. Q. 243, 248 (1988).

27. P. Dempsey, *Life since Deregulation: It Means Paying Much More for Much Less,* DES MOINES REGISTER, Dec. 30, 1987.

28. P. Dempsey, *Fear of Flying Frequently,* NEWSWEEK, Oct. 5, 1987. Kahn has also admitted that pricing for small community service has become discriminatory under deregulation. In his book, he wrote: "Experienced travelers can readily cite examples of seemingly outrageous geographic price discriminations against relatively thin routes. These doubtless reflect the lesser effectiveness of competition in thin than in dense markets." Kahn, *Transportation, Deregulation . . . and All That,* ECON. DEVELOPMENT 91, 94 (1987). A. KAHN, THE ECONOMICS OF REGULATION xix (1988). In recent testimony before the Senate Commerce Committee, he said much the same: "I must admit, of course, that the benefits of this price competition have been unevenly distributed geographically. Yields are definitely higher on thinly traveled than on densely traveled routes." Safety and Re-Regulation of the Airline Industry, Hearings Before the Senate Comm. on Commerce, Science and Technology, 100th Cong., 1st Sess. 155 (1987). In fact, Kahn admitted that the national air system, the national rail system, and the national bus system have all suffered a shrinkage in the number of communities served under deregulation. Kahn Oral Testimony, *supra* note 17, at 6300–1.

29. T. G. Moore, *U.S. Airline Deregulation: Its Effects on Passengers, Capital, and Labor,* 24 J. L. & ECON. 1, 15, 18 (1986).

30. A. Addus, *Subsidizing Air Service to Small Communities,* 39 TRANSP. Q. 537, 548 (1985).

31. Meyer, *supra* note 13, at 184.

32. J. MOLLOY, JR., THE U.S. COMMUTER AIRLINE INDUSTRY 62 (1985).

33. Meyer, *supra* note 13, at 175.

34. Dempsey, *The Dark Side of Deregulation: Its Impact on Small Communities*, 39 ADMIN. L. REV. 458 (1987).

35. After promulgation of the Bus Regulatory Reform Act of 1982, more than 4,500 small towns lost service while fewer than 900 gained it. After promulgation of the Staggers Rail Act of 1980, more than 1,200 communities lost all rail service. *Id.* at 455. Even Kahn saw a need for economic regulation to protect service to small communities, saying, "I'm not sure I would ever have deregulated the buses because the bus is the lifeline of many small communities for people just to get to the doctor or to the Social Security office." Kahn Oral Testimony, *supra* note 17, at 6337.

36. Morganthau, *Year of the Near Miss*, NEWSWEEK, July 27, 1987, at 20, 24.

37. Brenner, *Airline Deregulation—A Case Study in Public Policy Failure*, 16 TRANSP. L.J. 212 (1988).

38. *See* the next section on service quality for a discussion of consumer complaints to the DOT.

39. OFFICE OF CONSUMER AFFAIRS, U.S. DEPARTMENT OF TRANSPORTATION, AIR TRAVEL CONSUMER REPORTS ii (Nov. 1987).

40. Brown & Dahl, *New Data on Airline Performance May End Up Misleading Travelers*, WALL ST. J., Nov. 23, 1987, at 29.

41. Dahl, *Why On-Time Flights Take Forever*, WALL ST. J., April 26, 1989, at B1.

42. *Id.*

43. *Id.*

44. A. KAHN, *supra* note 28, at xxii. Kahn has acknowledged that "deregulation and competition have clearly contributed to a substantial deterioration . . . in the quality of service." *Airline Deregulation, Hearings Before the Subcomm. on Antitrust of the Senate Judiciary Comm.*, 100th Cong., 1st Sess. 62 (1987). Kahn has been inconsistent on this subject, saying in 1986, "In most instances the quality—and especially the variety [of service]—has sharply improved [under deregulation]." A. Kahn, *The Theory and Application of Regulation*, 55 ANTITRUST LAW JOURNAL 179 (1986). Kahn also criticized the "widespread but nevertheless erroneous popular supposition" that the quality of service had deteriorated. *All That, supra* note 28, at 97.

45. Rowen, *Airline Service Has Gone to Hell*, WASHINGTON POST, July 23, 1987, at A21. *See also* Dempsey, *Consumers Pay More to Receive a Lot Less*, USA TODAY, July 16, 1987, at 8A.

46. Coleman, *No Silver Lining Expected to Brighten Airlines' Stormy Skies*, MARKETING NEWS, Sept. 25, 1987.

47. *The Big Trouble with Air Travel*, CONSUMER REPORTS, June 1988, at 362, 363.

48. Brenner, *supra* note 37, at 223.

49. *Gridlock!*, TIME, Sept. 12, 1988, at 55. Many said they took the precaution of arriving in a city the night before an appointment rather than risk flight delays or cancellations; they thereby saddled their firms with the cost of a hotel room. *Id.*

50. Brenner, *supra* note 37, at 223.

51. Civil Aeronautics Board Consumer Complaint Report 13b (1982); DOT Air Travel Monthly Consumer Complaint Report 1 (1988); Coleman, *supra* note 46, at 1. The top 10 complaints, in order of number registered, were as follows:

- *Flight Problems:* Cancellations, delays, or any other deviation from schedule
- *Baggage:* Claims for lost, damaged, or delayed baggage; charges for excess baggage; carry-on problems; difficulties with airline claim procedures
- *Refunds:* Problems in obtaining refunds for unused or lost tickets or fare adjustments
- *Customer Service:* Rude or unhelpful employees; inadequate meals or cabin service; poor treatment of delayed passengers.
- *Reservations, Ticketing, and Boarding:* Airline or travel agent mistakes in reservations and ticketing; problems in making reservations and obtaining tickets due to busy phone lines or waiting in line; delays in mailing tickets; problems boarding the aircraft (except over-sales)
- *Oversales:* All bumping problems, whether or not the airline complied with DOT oversale regulations
- *Other:* Cargo problems, security, airport facilities, claims for bodily injury, and other miscellaneous problems
- *Fares:* Incorrect or incomplete information about fares; discount fare conditions and availability; overcharges; fare increases; level of fares in general
- *Smoking:* Inadequate segregation of smokers from nonsmokers; failure of the airline to enforce no-smoking rules; objections to the rules
- *Advertising:* Ads that are unfair, misleading, or offensive to consumers

Id.

52. As the *Wall Street Journal* observed:

Complaints about service are at an all-time high, with flight delays and cancellations provoking protest chants and even violence among angry passengers. The alarming rise in reported midair near-collisions has sharpened demands for improved safety. Meanwhile, mergers have given some carriers so much market clout that fliers are seeing the consumer benefits of deregulation eroded.

McGinley, *Bad Air Service Prompts Call for Changes,* WALL ST., J., Nov. 9, 1987, at 28.

53. *See* Brenner, *supra* note 37.

54. The U.S. Department of Transportation has the authority to protect consumers from many of these evils, including deliberate overbooking, unrealistic scheduling, and fraudulent (bait-and-switch) advertising. But the Reagan administration's DOT was reticent to do much of anything to correct market failure.

Another consideration that increasingly influences both service and fare levels is the level of industry concentration that has emerged under deregulation. With fewer carriers, with some traffic lanes and hubs now a monopoly or oligopoly, and with no government agency to protect consumers, it is quite likely that as time passes, prices will rise and service will continue to decline.

23

THE ECONOMIC EFFECTS OF DEREGULATION: THE $6-BILLION MYTH

Of all the studies that have examined the economic effects of airline deregulation, the most widely quoted has been a Brookings Institution study published in 1986.[1] The authors, Steven Morrison and Clifford Winston, allege that as a result of airline deregulation, there has been "at least a $6 billion (in 1977 dollars) annual improvement in the welfare of travelers."[2] This $6 billion (now $10 billion, adjusted for inflation) has been repeatedly cited by proponents of deregulation in political hearings, in newspapers, and on television as evidence of the overwhelmingly positive economic impact of airline deregulation.[3]

During this same time period, there have been only a few studies that have seriously questioned the methodology or assumptions on which the Morrison and Winston study was based.[4] Most reviews of the book have been very favorable, although salient questions were raised in those reviews.[5] Certainly, the study has been one of the most thorough and academically rigorous of those published since the beginning of deregulation. Yet, there are some serious questions about the findings, questions that merit exploration.

This chapter will first focus on the Morrison and Winston study itself, describing its assumptions, methodology, and findings. A brief analysis of results will follow, and the remainder of the chapter will be a critique of the study, showing why the $6-billion benefit figure is misleading and ultimately inaccurate.

THE BROOKINGS STUDY

Morrison and Winston adopted a "counter-factual" approach to compare 1977 regulated levels of service with what would have been levels of service in 1977 under deregulation. They use 1983 service and fare data

Table 23.1
Estimates for Air Travel Derived from Disaggregate Logit Intercity Demand
Model

	Pleasure	Business
Value of Travel Time (as fraction of hourly wage)	1.49	0.85
Value of Time Between Departures (as fraction of hourly wage)	0.23	1.44

Source: S. Morrison and C. Winston, THE ECONOMIC EFFECTS OF AIRLINE DEREGULATION (1986), p. 17.

as the basis for their deregulation scenario and use a fare deflator to esti-mate what deregulated fares would have been in 1977. This counter-fac-tual approach was used by Morrison and Winston so as to control for macroeconomic changes between the years 1977 and 1983, specifically recession-related events of the early 1980s.

From the standpoint of consumers of air travel, the major objective of Morrison and Winston's study was to determine the difference in travelers' welfare between regulation and deregulation. The change in travelers' wel-fare is due to changes in fares, in travel time, and in time between depar-tures and is measured by calculating travelers' compensating variations (CVs)—that is, how much money travelers would have to be given after a price or service quality change to be as well off after the change as they were before the change.[6]

The welfare calculations are based on results from the calibration of disaggregate logit models of mode choice for intercity pleasure and busi-ness travelers, models that were developed in a previous study by the au-thors.[7] The parameter estimates of the mode choice models were used to calculate estimates of travelers' elasticities, value of travel time, and value of time between departures (see table 23.1).[8]

The values of travel time and time between departures for both business and pleasure travelers were used to calculate CVs based on a sample of 812 city-pairs of all hub classes that received air service both before and after deregulation.[9] The 812 city-pairs represent a considerable mixture of large, medium, small, and nonhub cities (see table 23.2).[10]

A quick scan of the weighted-average percentage changes in fares, travel times, and frequencies between 1977 regulation and 1977 counter-factual deregulation by hub classification reveals some interesting trends. Coach fares for all hub classes have increased during deregulation. Perhaps sur-prisingly, discount fares have also increased for almost all route categories. Only the large hub–large hub and medium hub–large hub route categories (granted, the most heavily traveled city-pair routes) exhibited decreases in discount fares.

According to Morrison and Winston, travel times generally have not

Table 23.2
City-Pair Sample by Hub Classification: Weighted-Average Percentage Change in
Fares, Travel Time, and Frequency between Regulation and Counter-factual
Deregulation (in 1977)

Category of Route	Number of City Pairs	Coach Fare	Discount Fare	Travel Time	Freq.
Nonhub-nonhub	51	21.2	22.1	-4.1	33.9
Nonhub-small hub	52	22.5	12.3	-0.8	1.4
Nonhub-medium hub	45	5.4	-0.4	-5.7	24.3
Nonhub-large hub	53	16.3	9.1	-1.5	28.7
Small hub-small hub	50	15.3	11.3	-5.1	33.9
Small hub-medium hub	69	18.7	10.4	-4.8	20.8
Small hub-large hub	57	25.0	8.1	10.1	19.2
Medium hub-medium hub	69	15.6	2.0	-4.5	-4.3
Medium hub-large hub	161	17.4	-6.8	12.7	14.4
Large hub-large hub	205	8.6	-17.6	4.2	-3.5
Total	812				

Source: S. Morrison and C. Winston, THE ECONOMIC EFFECTS OF AIRLINE DEREGULATION
   (1986), p. 23.

changed as much as the other measures, although categories involving large
hubs exhibited significant increases in travel time. This is probably due to
the reduction in nonstop flights between large hubs and slightly smaller
hubs and the increase in one-stop hub-and-spoke connections during deregulation.

Finally, according to the 1983 schedule data, the weighted-average percentage change in frequencies has increased dramatically for almost all
route categories. The only exceptions are for large hub–large hub and medium hub–medium hub traffic.

For each of the 812 city-pairs, CVs were calculated based on air schedule and fare data from the June 1977 and June 1983 editions of the *Official Airline Guide*.[11] Travel time and time between departures were based
on the published scheduled (not actual) departure and arrival times of all
feasible flight alternatives (some of which may have involved connections).[12]

With regard to fares, Morrison and Winston used the median discount
fare for all pleasure travelers under both regulation and deregulation. They
further assumed that business travelers paid only coach fares (the lowest
Y-class fare) under regulation but that one-half paid the coach fare and
one-half paid the median discount fare under deregulation. These fare assumptions were based on the Air Transport Association's "Monthly Discount Reports," which contain aggregate information on the percentage of
all domestic air travel that occurs at coach fares (based on the Y fare
classification) and discount fares for business and pleasure travelers.[13]

By applying the values of travel time and time between departures to the changes in those measures between the regulation and deregulation scenarios, as well as including changes in fares, Morrison and Winston calculated the CV, or the mean annual dollar welfare change per person, for each of the 812 city-pairs. The dollar welfare changes were then aggregated and weighted by hub classification, so that a weighted-average welfare change per traveler was calculated at $10.62 per round-trip.[14] This figure was then multiplied by the estimated number of intercity passenger trips during 1977 to yield an estimated total annual benefit to travelers of $5.7 billion, or $6 billion after rounding.[15]

ANALYSIS OF STUDY RESULTS

Where exactly does this $6 billion come from and who has benefitted from it? According to Morrison and Winston, "The greatest net benefit [is] going to business travelers from increased flight frequency".[16] Roughly $4 billion (or two-thirds of the total) is directly attributable to this.[17] The authors contend that because business travelers place a very high value on greater frequencies of airline service, and because frequencies have increased during the first five years of deregulation, there has been an estimated annual total benefit to travelers of $4 billion.

This conclusion assumes, however, that it is possible to place a monetary value on increases in frequencies and furthermore that the authors' estimation of that monetary value is accurate. Other air transportation studies have dealt at length with the concept of the value of time, but almost exclusively in the context of travel time.[18]

George Douglas and James Miller III[19] were the first to ascribe a monetary value to the concept of "schedule delay," that is, the average time difference between all travelers' preferred departure times and their actual departure times.[20] According to Douglas and Miller, schedule delay equals the sum of frequency delay (the difference between the traveler's desired departure time and the closest scheduled departure) and stochastic delay (the additional delay encountered if a seat is not available on the best scheduled flight).[21] In other words, Douglas and Miller felt that the "full cost" of travel should include not only direct price or monetary cost and trip time but also some measure of schedule convenience.

Morrison and Winston address the issue of schedule convenience by using the average time between scheduled departures as a surrogate for schedule delay. They admit that this approach does not completely capture the full effect of schedule delay on traveler behavior, but they maintain that all of the effect of frequency delay and a part of the effect of stochastic delay are captured. Without information regarding load factors on specific flights, it would be difficult to capture the full effect of stochastic delay. The authors contend, however, that their findings are insensitive to this remaining component of stochastic delay.[22]

The other $2 billion of benefits is attributed primarily to fare savings for pleasure travelers in large and medium hub markets. During the first five years of deregulation, many low discount fares were available, mainly in those highly competitive, high-density city-pair markets. Meanwhile, both discount fares in smaller markets and coach fares in all markets have risen sharply. Given the post-deregulation assumption that all pleasure travelers use a discount fare and that business travelers are split between discount and coach fares, it then follows that some business travelers and pleasure travelers in large markets have benefited from lower fares, whereas pleasure travelers in smaller markets and the rest of the business travelers have not.

CRITIQUE

There are a number of serious problems in the assumptions and methodology of the Morrison and Winston study on the economic effects of airline deregulation. Each will be discussed in turn.

Exaggerated Dollar Benefits

The most disturbing aspect of the study from our point of view is the rather large benefit, for business travelers, attributed to greater flight frequencies. Roughly two-thirds of the $6-billion total benefit to consumers is allegedly derived from this form of savings in time.

As a result of the calibration of Morrison and Winston's disaggregate logit model of intercity travel demand, the value of time between departures for business travelers was estimated to be 1.44 times the value of the average business traveler's hourly wage (see table 23.1). For pleasure travelers, it was estimated to be only 0.23 times the value of the average pleasure traveler's hourly wage.[23] Morrison and Winston explain this as follows,

The value of time between departures reflects travelers' value of the inconvenience involved in schedule delay, manifested in their valuation of waiting time both at their home (or hotel) or business and in the terminal. Because pleasure travelers generally plan air trips reasonably far in advance, they are not likely either to experience significant schedule delay, or, as found here, to place a particularly high value on the wait time, which will be largely incurred at their residence.

The estimated high value placed on time between departures by business travelers reflects the high disutility to them of adjusting departure times to the schedule and capacity constraints of the air carriers. It is to be expected that the value of time between departures for business travelers is significantly greater than it is for pleasure travelers.[24]

This line of reasoning appears correct; business travelers should value departure frequencies more highly than pleasure travelers. But is it plausible to expect that the value of time between departures for business travelers is *greater* than their value of travel time and that it is *greater* than their average hourly wage rate?

The value of travel time for business travelers was estimated to be 0.85 times the hourly wage, which seems reasonable given that business travelers could use some of the time for work while in flight.[25] But a value of time between departures of 1.44 times the wage rate seems incredibly high.

Douglas and Miller, architects of the concept of schedule delay, took the opposite view of the relationship between the value of time in flight and on the ground. They had this to say about the value of time attributed to schedule delay,

We should emphasize again at this point the distinction between the traveler's value of time en route and his value of timing as represented by schedule delay. The former almost certainly exceeds the latter, since the time represented by schedule delay has alternative uses; it need not be spent in waiting,* whereas the alternative uses of time en route are quite limited. Thus, one would expect to assign a higher value to time en route than to schedule delay.

*Exceptions might include emergency travel, where the arrival at or before a specific time is of great value.[26]

The Douglas and Miller interpretation seems intuitively more appropriate; businesspeople can get more work done on the ground than in the air. Despite the footnote, which explains extreme circumstances where schedule delay might possibly be valued more highly, they clearly state that it would be unrealistic to expect a higher value of time between departures than for travel time.

Likewise, it seems equally unrealistic that a business traveler would value time between departures more highly than his hourly wage. Morrison and Winston explain further: "Business travelers generally do not plan their travel in advance. They are also likely to experience significant disutility from schedule delay, both because their wait is spent inconveniently (for example, in a hotel) and because it delays important business at their destination."[27]

It is true that fewer frequencies should translate into more inconvenience, in terms of both opportunity costs and schedule delay. But the time attributed to schedule delay is not completely lost time. The business traveler most certainly would be using the extra time at his destination in some productive capacity. In hotels, business travelers can place business telephone calls, fax documents, read the documents brought with them, or draft memoranda and letters. At the airport, business travelers can open their briefcases and pull out documents to read. Even if the business traveler was not using that time, why should it be valued even higher than his or her hourly wage rate?

Most business travelers do not have the type of business schedule that characterizes individuals like Donald Trump or Ted Turner. The opportunity costs of not getting the most desirable departure times might be quite costly to these jet-set corporate entreprenuers (in this case, they would have private jet service anyway). To the average business traveler, opportunity costs simply are not that high.

Also, some clarification is necessary about how far in advance business travelers plan their travel. If the business traveler is using a coach fare, increased frequencies do allow him much more flexibility, and the business traveler in need of a short-notice, emergency-type flight would probably value more frequencies quite highly. But according to the assumptions of Morrison and Winston, fully one-half of all business travelers since deregulation are assumed to be traveling under the median discount fare.

Given the nature of restrictions on most discount fares, flights must be reserved and paid for well in advance, which would obviously mitigate against the high savings attributed to the flexibility afforded by more frequencies. In other words, business travelers flying on discount fares should have a value of time between departures more akin to that of pleasure travelers, given that their schedules would be determined well in advance of the flight. They would thus be able to plan around their flight schedule and use the time attributed to schedule delay much more efficiently.

The business traveler flying on a discount fare is essentially "locked in" to the scheduled flight because of airline limitations on discount fares in the form of penalties assessed for changing or cancelling flights after they have been reserved and paid for. The penalties for cancellation or change of departure have grown very large. Their imposition must of necessity dilute or destroy any benefits ascribed to opportunity cost savings. Thus, a business traveler flying on a discount fare cannot realistically change flights at the last minute. For business travelers, this greatly decreases the benefits of greater flexibility attributed to more frequencies.

Only those business travelers still flying on coach (or higher) fares are afforded the luxury of being able to change flights at the last minute without any penalties. Given, however, that fewer business travelers have been flying on coach (or higher) fares under deregulation, and therefore cannot avail themselves of an alternative departure without incurring a significant penalty, the benefits attributed to more flexibility from increased flight frequencies are overstated.

Furthermore, in the earlier Morrison and Winston study,[28] there seems to be an implicit assumption in their interpretation of the value of time between departures; they seem to assume that all air travelers have unlimited flexibility to change flights at the last minute. They state: "With respect to stochastic delay, while flights are often unavailable due to capacity constraints, the high frequency of service indicates that the delay in waiting for the next available departure is likely to be relatively small. In addition, it is not expected that a traveler would perceive that waiting in an airline terminal is a particularly onerous experience."[29]

The scenario just described seems more applicable to stand-by or shuttle service rather than scheduled, discount-fare service. The extra flights during the rest of the day mean very little to the discount-fare traveler once that traveler has already booked and paid for the flight. If a traveler has scheduled a discount flight, he or she would have done it well in advance and would be able to plan approximately when to arrive at the airport in order to minimize waiting time, thus substantially mitigating the value of time between departures.

With respect to waiting time in airline terminals, when travelers are in control over how much waiting time they endure (i.e., by planning their arrival time at the airport), waiting is probably not a "particularly onerous experience." But when travelers are not in control over waiting time (i.e., because of actual delays caused by overcrowding, overbooking, and unrealistic flight scheduling), the level of "onerousness" probably increases exponentially (see the next section for more on actual delays).

As an example of how the "benefits" of reduced time between departures can add up, consider two different scenarios. Scenario A is four round-trip flights per day between two cities, whereas scenario B is five round-trip flights per day between the same two cities. The time between departures for scenario A is calculated at 5 hours (20 hours/day divided by four flights/day); for scenario B, it is 4 hours (20 hours/day divided by five flights/day). Thus, the difference in the time between departures for the two scenarios is 1 hour. At a value of time between departures of 1.44 times the wage rate, and an average hourly business wage rate of $14.39 (in 1977 dollars),[30] the welfare benefit attributed to one more frequency would be $20.72 (in 1977 dollars) per business traveler. Aggregating across all business travelers would result in a rather large figure. Hence, this is the origin of most of the $6 billion of the purported benefits of airline deregulation.

A further note about the practical validity of the concept of time between departures has been raised by Melvin Brenner:

If a new schedule is added at mid-day on a route which previously had only morning and evening schedules, there would surely be a potential gain in business traveler productivity. But if, on the other hand, a new schedule is added at 3 p.m. on a route which already had a 5 p.m. nonstop, it is not clear that this would translate into a meaningful gain in a business traveler's productive time. (This new departure option, while reducing the business time available at the origin city, would provide an arrival too late to add meaningfully to the business day at the arrival city.)[31]

According to the Morrison and Winston methodology, the dollar benefits accruing to fewer hours of time between departures would occur even if some of the departures were bunched together in time. Although contributing to a potential reduction in stochastic delay, extra frequencies bunched together do not give the business traveler that much more flexibility. More would need to be known about the timing of the extra departures before such a hefty dollar benefit could be proclaimed.

To summarize this first point, we can levy specific criticisms in one or all of the following three areas. First, too much reliance has been placed on a quantitative measure (value of time between departures) that has not been theoretically or empirically well developed in the transportation literature. Douglas and Miller in 1974 were the first to develop the concept of schedule delay in air transportation service, but even they clearly pointed out the lesser value that it should attain. Since that time, no other study has used the value of timing as represented by schedule delay (or its surrogate, value of time between departures) in such a conspicuous manner.

Second, the conceptual development of the methodology concerning value of time between departures is flawed. Even if one assumes the validity of the general concept of the value of time between departures, the way in which it is used in this study seems exaggerated. It has been given as much (if not more) conceptual weight as the seemingly more important values of travel time and fares, whereas other service variables (such as actual delay time) are not even considered. Also, more practical matters, such as the problem associated with the grouping of flights or the problem associated with the balance between assumptions of fare levels and values of time, have not been adequately accounted for in the methodology.

And finally, even if the first two arguments are ignored, the authors have not satisfactorily explained why the value of time between departures for business travelers is so high. In light of the Douglas and Miller statement, as well as intuitive observations about business travelers' perceptions of the quality of airline service since the beginning of deregulation, the authors need to provide a more convincing argument as to why business travelers would place such an incredibly high value on time between departures.

Actual Delay Time

The question of real delay time is an issue that has been raised earlier in this book (see chapter 22) as well as by Brenner, as follows:

There is a more serious drawback to the premise of the Brookings study. The very increase in hub-and-spoke frequencies which played so large a part in the study's calculations has been an important contributor to the congestion and delays which by 1987 had become a matter of widespread concern. While reducing the time interval between published departure times, the increased hub-and-spoke frequencies have increased the actual delay time at the gate, and in runway queues—a form of lost time that is especially costly to business traveler productivity.[32]

In their study, Morrison and Winston used only published departure times and did not take into account actual delay time at the gate or on the runway. Part of the problem lies in the unavailability of delay data during the time period under study. In spite of this limitation, though, Morrison and Winston do not even mention the issue of real delays but instead focus almost entirely on the value of greater frequency of service. But, as Brenner suggests, what good are more frequencies if they in turn cause increased delays and thus force travelers to miss scheduled appointments? The benefits of increased frequencies become substantially mitigated by the lost time in actual delays.

As was suggested in chapter 21, increased delays have cost both the airlines and consumers dearly. It has been estimated that in 1986 alone, travel delays cost airlines $1.8 billion in extra operating expenses and cost consumers $3.2 billion in lost time.[33] The already inflated dollar benefits accruing to greater flight frequencies would be more than offset by these dollar losses.

Furthermore, it is more believable that travelers (especially business travelers) would place a much higher value on time lost due to actual delays than due to schedule delay. Schedule delay is known; actual delays are unknown. The level of travelers' anxiety and tension increases as they are informed, just before scheduled departure at the gate, of delays that they can do nothing about. This kind of time is really lost time, and because of the last-minute, unexpected nature of actual delays, this time would probably be valued well beyond that of time between departures, actual travel time, or the average hourly wage rate.

This uncertainty surrounding actual delays makes scheduling a particular flight a risky proposition. With increased frequencies and increased actual delays, there is a higher probability that the preferred flight will cause the traveler to miss the scheduled appointment or meeting at the traveler's destination. Therefore, the traveler is forced to choose a "safer" flight, that is, one that leaves well before the preferred departure time, just because of the high probability of incurring actual delays.

This uncertainty has led many travelers to arrive in their destination city the *night before* an important meeting, whereas before deregulation they could rely on the airline to get them to their destination the *day of* the meeting.[34] In this way, effective schedule delay is not decreased by additional frequencies.

The 1977–1983 Period

Morrison and Winston's study used data that covered only the first five years of airline deregulation. The authors can hardly be criticized for this, given that when the study was published in 1986, they were using what was, at that time, fairly recent data. Thus, this point is more of an external comment than an internal criticism, but it is very important nonetheless.

During the first decade of airline deregulation, we witnessed a great deal of volatility and turbulence in the airline industry.[35] As noted in chapter 20, there have been very important macrostructural changes in the industry during these ten years. The first five years of deregulation (1978–83) were strikingly different from the last nine years (1983–92). Assessments made on the basis of 1983 data most surely do not reflect what has happened in the airline industry in the recent period. Also, the year 1983 was the trough of the deepest recession since the Great Depression, a circumstance that further clouds the data analysis.

A related point concerns the use of 1977 data to calibrate the disaggregate logit model of intercity travel demand that was used to derive the values of time in the study. Morrison and Winston used data from the 1977 U.S. Census of Transportation National Travel Survey to calibrate their intercity demand model.[36] Those data reflect individual travelers' preferences for air, auto, rail, and bus modes as of 1977. This was before airlines and bus companies were deregulated and thus before service quality began to decline. The preferences for air travel manifested in the 1977 data, then, reflect the assumption of the type of service and fares that existed before deregulation.

Passenger assumptions about the nature of air travel in 1977 are much different than they are today. For example, travelers in 1977 would probably have assumed air travel to be faster, more comfortable, more convenient, more costly, and of higher quality in general than would travelers in 1983 or 1992. Many travelers from smaller communities might have assumed air travel to be of jet-quality service in 1977, whereas in 1983 or 1992, more would probably associate air travel with turboprop-quality service. These differences could have dramatic effects on individual travelers' preferences regarding each of the modes and consequently on travelers' value of time for each of the modes.

Decline in Service Quality

Another related point involves consideration of aspects of service quality other than time and cost. David Graham raised this issue in his review of the Morrison and Winston study:

Net benefits are exaggerated because the assessment omits several aspects of air service quality, such as in-flight amenities and schedule reliability, that have declined under deregulation. . . . a careful analysis of these [quality-of-service] variables would provide a welcome perspective on the current concerns over the quality and reliability of airline service.[37]

Admittedly, time and cost variables are the most important, and reliable, measures of transportation service. Aspects of service quality, such as comfort, convenience, and perceived safety, are more amorphous and much more difficult to quantify. Yet, certain other quality-of-service variables (e.g., type of aircraft, type of service [nonstop vs. one-stop], provision of meals) are more easily quantifiable and could have been addressed in some manner in the Morrison and Winston study.

Problematic Fare Data

One of the areas of greatest change in the deregulated airline environment has been the proliferation of different fare categories and the consequent inability to accurately determine how many people are flying on which fares. Morrison and Winston assumed that all pleasure travelers use median discount fares both before and after deregulation, that business travelers use the lowest coach fares before deregulation, and that of business travelers after deregulation, one-half use the lowest coach fare and the other half use the median discount fare.[38] These assumptions are based on the Air Transport Association's "Monthly Discount Reports," which contain aggregate information on the percentage of travelers using coach and discount fares.[39]

The $2-billion remainder of the $6-billion benefits attributed to airline deregulation (after "savings" from reduced time between departures are subtracted) is mostly attributed to savings in fares for pleasure travelers. According to Morrison and Winston's findings (see table 23.2), only discount fares in large hub–large hub and medium hub–large hub route categories exhibited substantial declines between regulation and counter-factual deregulation.

Once again, the question of the volatility of the airline industry, especially with respect to airfares, becomes germane. Brenner addressed Morrison and Winston's fare analysis methodology as follows:

This approach treated the fares up to 1983 as representing a sustainable level. It thus failed to consider the widespread impact of "fare wars" which reflected below-cost pricing. . . . In other words, the base which the Brookings study used to represent deregulated pricing was abnormally depressed, and therefore the use of that base overstated the fare savings that could be counted upon as an ongoing benefit.[40]

Brenner contended that by using 1983 fare data (deflated to 1977 dollars as part of Morrison and Winston's counter-factual approach), the authors used an abnormally low base year for fares as the standard for deregulated fares. This criticism has become increasingly more valid as most fares (in constant dollars) since 1983 have risen sharply (see chapter 21). Thus, the $2-billion savings in fares probably represents an upper bound of accuracy applicable only in the early 1980s.

SUMMARY

This chapter has critically examined the Morrison and Winston study on the economic effects of airline deregulation. The most significant criticism of the supposed $6 billion worth of annual savings accruing to travelers as a result of deregulation is the exaggerated $4-billion benefit for business travelers attributed to increased flight frequencies. Other problems of the study include the lack of consideration of actual delay time, the use of 1977–83 data only, the omission of quality-of-service measures other than time and cost, and the problematic nature of making inferences from deregulated fare data.

Together, these criticisms cast a rather large shadow on the results of the Morrison and Winston study, and largely invalidate the often-quoted "$6 billion (now $10 billion) annual economic benefit" of airline deregulation. The question of the economic effect of airline deregulation is much more problematical than Morrison and Winston have suggested. Despite these deficiencies, Morrison and Winston should nonetheless be credited with at least attempting to quantify the effects of deregulation.

NOTES

1. S. MORRISON & C. WINSTON, THE ECONOMIC EFFECTS OF AIRLINE DEREGULATION (1986).

2. *Id.* at 1–2. In a more recent study, Morrison and Winston have updated the benefit figure from $5.7 billion (measured in 1977 dollars) to $10.4 billion (measured in 1988 dollars). This is simply an accounting for inflation over the 11-year period and does *not* represent a recalculation using updated data from the 1983–88 period. *See* Morrison & Winston, *Airline Deregulation and Public Policy*, 245 SCIENCE 707–11 (1989).

3. Now the figure being cited is $10 billion a year, or $100 billion over 10 years. *See* THE ECONOMIST, Feb. 4, 1989, at 16; and FEDERAL TRADE COMMISSION, THE DEREGULATED AIRLINE INDUSTRY: A REVIEW OF THE EVIDENCE x (1988).

4. See, for example, Brenner, *Airline Deregulation—A Case Study in Public Policy Failure*, 16 TRANSP. L.J. 222–25 (1988); and Goetz and Dempsey, *Airline Deregulation Ten Years After: Something Foul in the Air*, 54 J. AIR LAW & COM. 927 (1989).

5. *See*, for example, Graham, 26 J. ECON. LIT. 131–33 (March 1988).

6. S. MORRISON & C. WINSTON, *supra* note 1, at 15–16.

7. See Morrison & Winston, *An Econometric Analysis of the Demand for Intercity Passenger Transportation*, 2 RESEARCH IN TRANSP. ECON. 213–37 (1985).

8. S. Morrison & C. Winston, *supra* note 1, at 17.

9. *Id.* at 20.

10. Hub size classes were defined by the CAB and the FAA as follows: a large hub is a city and its metropolitan area that enplanes more than 1.00 percent of the total number of U.S. certificated airline passengers per year, a medium hub between 0.25 and 0.999 percent, small hubs between 0.05 and 0.249 percent, and nonhubs less than 0.05 percent.

11. S. Morrison & C. Winston, *supra* note 1, at 21.

12. *Id.* at 22.

13. *Id.* at 21.

14. *Id.* at 31.

15. *Id.* at 31–32.

16. *Id.* at 2.

17. Brenner, *supra* note 4, at 222.

18. See, for example, R. Gronau, The Value of Time in Passenger Transportation: The Demand for Air Travel (1970); and De Vany, *The Revealed Value of Time in Air Travel,* 56 Rev. of Econ. & Statistics 77–82 (1974).

19. G. Douglas & J. Miller III, Economic Regulation of Domestic Air Transport: Theory and Policy (1974).

20. S. Morrison & C. Winston, *supra* note 1, at 5.

21. *Id.* at 5.

22. *Id.* at 18.

23. "Based on our sample of travelers used in the demand model, the average wages of pleasure travelers and business travelers in 1977 were $10.30 an hour and $14.39 an hour, respectively." *Id.* at 70.

24. *Id.* at 18.

25.

Business travelers' value of [travel] time actually captures their employer's valuation of their time, because the employer is ultimately responsible for the mode choice. Because business travelers can work on airplanes, it is not unreasonable to find that employers value their employees' air travel time somewhat below the wage. The high value of travel time for pleasure travelers largely reflects their valuation of the opportunity costs of not being at their destination as opposed to their disutility from airline travel.

Id. at 17.

26. Douglas & Miller, *supra* note 19, at 32–33.

27. *Id.* at 18.

28. See Morrison & Winston, *supra* note 7.

29. *Id.* at 225.

30. *Id.* at 70.

31. Brenner, *supra* note 4, at 222.

32. *Id.* at 223.

33. *Gridlock!,* Time, Sept. 12, 1988, at 52, 55.

34. *Id.* at 55.

35. Goetz & Dempsey, *Airline Deregulation Ten Years After: Something Foul in the Air,* 54 J. Air Law & Com. 927 (1989).

36. Morrison & Winston, *supra* note 7, at 217–18.

37. Graham, 26 J. Econ. Lit. 132–33 (1988).
38. Morrison & Winston, *supra* note 1, at 21.
39. *Id.*
40. Brenner, *supra* note 4, at 223–24.

24

SAFETY

A Gallup poll reveals that two-thirds of Americans have less confidence in the safety and reliability of airlines than they did in the early years of deregulation. A *Wall Street Journal* poll reveals that Americans have less confidence in airlines overall than in any other industry.

Aside from natural disasters like hurricanes and earthquakes, individual airline crashes are the deadliest domestic phenomena. Often, the critical findings of the National Transportation Safety Board (NTSB) are not released until many months after a crash. Only after examining the broken and twisted metal and the plane's maintenance records, interviewing the witnesses and survivors, reviewing the digital flight data recorder, and listening to the cockpit voice recorder (the little black box), can the NTSB get a complete picture of what happened.

The three principal causes of aviation catastrophes appear to be pilot or ground control (human) error, faulty aircraft (equipment failure), and acts of God (weather).

Pilot or ground control error may be attributed to inadequate training, insufficient rest, or just plain old bad judgment. A NASA psychologist, Clay Foushee, said: "The reasons [for a pilot error] are probably as extensive as the reason any person makes a mistake—a distraction, fatigue, task overload or stress. It can be complacency. It can be inexperience." Fatigue, stress, and inexperience all appear to have worsened under deregulation.

Faulty aircraft may be a product of sloppy maintenance or inadequate engineering or construction. As we shall see, the U.S. airline industry now flies the oldest fleet of aircraft in the developed world, and maintenance also seems to have declined.

Acts of God include, of course, various forms of inclement weather, including ice, snow, and wind shear. Actually, adverse weather and other natural disasters should be blamed on the Devil, not God; they should be called acts of Satan.

In one form or another, fear of flying stalks about one in every six Americans. Twenty-five million Americans are afraid to fly, and another 15 million fly only when they must.

For some, fear of flying manifests itself in no more serious way than compelling them to grip their armrests tightly with sweaty hands on take-offs and landings. For others, the fear surpasses anxiety to the point of phobia—aviaphobia—and many of these tormented souls avoid air travel altogether. The former National Football League head coach John Madden took a bus or a train rather than fly to Oakland Raiders' games around the country.

Crowding people together and propelling them in a metal cylinder several hundred miles an hour through the clouds at an altitude of several miles above sea level creates discomfort, even fear, among some. Fear of flying is fear of falling plus fear of drowning or burning alive plus claustrophobia and sheer helplessness if anything should go wrong. Not a pretty sight. For the record, statistics prove that flying in a commercial aircraft is significantly safer than driving an automobile; and safety is not the worst problem of deregulation (the other chapters in this book describe those problems).

Because of the destructive competition unleashed by deregulation, overall industry financial performance during the first decade of deregulation declined to the point of inadequacy. Poor or nonexistent profits create a natural tendency of management to curtail costs. Among those that can be significantly diminished are maintenance costs, including mechanics' wages, spare or replacement parts, and idle aircraft time lost during inspections and maintenance. A decade of economic anemia has, quite naturally, deprived carriers of the resources to reequip with new aircraft or maintain the wide margin of safety the public previously enjoyed.

Since deregulation, the average age of our nation's aircraft fleet has increased sharply. Expenditures for maintenance and the number of mechanics per aircraft have been reduced. The number of near misses has soared. The average age of cockpit crew members is the lowest since deregulation began; the standards for hiring, and the duration and quality of crew training, have declined.[1]

The father of airline deregulation, Alfred Kahn, admits that the margin of safety has "possibly" narrowed since 1978, although fatality statistics do not yet reflect it.[2] True, more people died in crashes in 1985 than in any other year since 1977; there were more aircraft accidents in 1987 than in any other year since 1974; and there were more fatal incidents involving U.S. airlines in 1989 than in any other year since 1968.[3] But statistics prove the fatality *rate* has fallen under deregulation (they also fell during the four decades of regulation, flattening out in the 1970s). Of course, if

Table 24.1
Accidental Deaths, 1985

45,901	Motor vehicle
12,001	Fall
4,938	Fire
4,407	Drowning
3,612	Drug or medication
1,663	Food inhalation or ingestion
1,649	Firearms
1,428	AIR TRAVEL
1,288	Machinery
903	Struck by falling object
802	Electric current
305	Alcohol poisoning
85	Lightening
49	Venomous bite or sting
15	Dog bite
11	Fireworks

Source: TIME, Mar. 27, 1989, at 27.

the body count were the only measure of victory, we would have won the war in Vietnam.[4]

As table 24.1 reveals, there are many ways to lose your life other than in air travel. Although passenger fatalities have not ascended to the levels one would expect in such an environment, every other measure of safety paints a different picture. The economic imperatives of market Darwinism collided with safety objectives and brought about a comprehensive deterioration in the age of aircraft, maintenance, near misses, and the age, training, and qualifications of pilots.

The average age of cockpit crew members is the lowest since deregulation began; the standards for hiring, and the duration and quality of training, have declined.[5] For example, in 1983, a prospective pilot needed 2,300 hours of flight time and uncorrected 20/20 vision to be hired by one of the major airlines. Today, one needs only 800 hours of flight time and (for all but one airline) correctable vision to be hired by a major carrier and merely 300 hours to be hired by a commuter carrier. The number of pilots with fewer than 2,000 hours of flight time soared from 2 percent in 1983 to 14 percent in 1988.[6] Among commuter airlines, pilots with less than 2,000 hours of flight time rose from 20 percent in 1985 to nearly 40 percent in 1990; some are hired with as little as 300 hours of flight time, just 50 hours above the FAA minimum.

The economic anemia unleashed by deregulation has caused management to push pilots to fly more hours with less rest. Although working longer for less pay may increase productivity, it can induce fatigue, which has a negative impact on safety. Between 1982 and 1988, fatigue was responsible for two operational errors per week—errors such as pilots falling asleep in the cockpit, landing on the wrong runway, or wandering out of assigned flight paths.[7]

Ninety-seven percent of airline pilots believe that deregulation has had an adverse impact on airline safety.[8] Among the problems identified are "lagging and inadequate maintenance; pressure to avoid delays; lowered hiring and experience standards for new pilots; increased use of waivers and exemptions from safety rules; increased flying hours for pilots; [and] the profusion of new, inexperienced airlines."[9] One out of every five pilots has been involved in a near miss during a recent two-year period, although only 25 percent of those were reported to the FAA.[10]

Between 1978 and 1987, departures for major airlines increased by 27 percent.[11] With airlines funneling their flights into hub-and-choke bottlenecks and scheduling takeoffs and landings through a narrow window of time and space, near misses are soaring.[12] Thus, the flight paths of the nation's major airports are heavily congested during peak periods. There were 311 near misses during 1982, 475 in 1983, 589 in 1984, 758 in 1985, 840 in 1986, and 1,058 during 1987 (although they have declined since).[13] The number of near misses increased under deregulation, both in absolute numbers and in rates per 100,000 flight hours.[14]

Near misses is an inaccurate term for what really are near hits. At no other time in the history of aviation have so many aircraft passed so uncomfortably close, with so little a margin for error. By funneling their flights into constipated bottlenecks, and scheduling takeoffs and landings through a narrow window of time and space, airlines have sent their fleets down arteries of coagulation.

All of this has placed serious strains on the air traffic control system at a time when it is least capable of handling the surge in demand. In 1981, President Ronald Reagan fired 11,000 members of the Professional Air Traffic Controllers Organization (PATCO) for striking, leaving only a third of the work force, and the Federal Aviation Administration has yet to replace them all.[15] Not only is the system understaffed, but many airports and navigational facilities are equipped with obsolete and aging equip-

Table 24.2
Average Age of Boeing 727s and 747s (in years)

	727s	747s
U.S.	16.47	14.99
Europe	16.17	10.33
Asia/Pacific	13.48	8.17
Africa	13.46	10.75
Middle East	11.93	10.17
Canada	11.56	14.90

Source: CHICAGO TRIBUNE, Mar. 16, 1989.

ment. Operational errors, or mistakes by controllers, increased by 20 percent during the first half of 1987 over the same period one year earlier.[16]

Legitimate concerns have also been raised over the problem of the age and poor maintenance of jets flown by unhealthy airlines, which lack the financial resources to reequip with modern aircraft or properly maintain their aging fleets.[17] This is particularly a concern in the commuter airline industry, seemingly plagued by endless bankruptcies, where used, recycled aircraft dominate the fleets of the smaller carriers.[18] Professor Frederick Thayer reminds us that "safety always has suffered when airlines were largely unregulated."[19]

The intensive competition unleashed by deregulation has deprived many carriers of the resources to replace their aging fleets of aircraft. The average age of the airline industry's domestic fleet grew from 7.5 years in 1975 to 12.5 years in 1989.[20] Today, more than half of the 2,767 jets in service are 16 years old or older.[21] Table 24.2 shows how the U.S. fleet compares with those of other regions of the world.

Of course, some airlines have older fleets than others. Table 24.3 is an airline-by-airline breakdown of the average ages of U.S. carriers' aircraft.[22] By December 1991, when Pan Am ceased operations, the average age of its aircraft exceeded 18 years. The "economic design goal" of an aircraft is the length of time it can be flown before it becomes too costly to be flown safely. Traditionally, the economic life of Boeing 727s, 737s, and 747s has been about 20 years. Thirty-one percent of U.S. carrier commercial aircraft now exceeds the economic design goals established by the manufacturers.

Boeing has produced most of the world's commercial jet aircraft. It manufactured about 6,000 of the world's 8,800 jetliners, including approximately 2,000 that are 15 years or older. As of this writing, United

Table 24.3
Average Ages of Carrier Fleets

Carrier	Age
TWA	14.3
Northwest	14.1
Eastern	13.8
United	13.6
Pan Am	12.8
Continental	11.0
American	9.4
USAir	9.0
Delta	8.7

Source: WALL STREET JOURNAL, Mar. 31, 1989, at B1.

Airlines flies the first Boeing 727, a plane nearly 30 years old. As of 1988, United flew 10 Boeing 727s that were built in 1963 and 29 DC-8s, which are almost 20 years old.

Before entering bankruptcy Pan Am flew the oldest Boeing 747 in commercial operations, the Clipper Juan Trippe—more than 20 years old. The plane was named after Juan Trippe, the buccaneer who built Pan Am into what was once the world's premiere international airline, which deregulation destroyed.

The Clipper Juan Trippe holds all the records for commercial distance, time, and endurance. The plane averaged approximately 50 flights a year (about 400 flying hours) for two decades. It flew more than 34 million miles, carried more than 3 million passengers, and had nearly 30,000 take-offs and landings. The aircraft had nearly 2,000 tires and more than 100 engines replaced and has gulped more than 200 million gallons of aviation fuel. Galleys were changed twice, the lavatories and passenger compartment four times.

The Clipper Juan Trippe initially seated some 360 passengers. Deregulation caused all airlines to jam the seats closer together to try to squeeze out more revenue; it now seats 414. The plane originally cost $22 million; today, a new 747 would cost around $130 million. Thus, renovation of old aircraft is often cheaper than buying new ones.

Before the top peeled off the second most heavily flown Boeing 737, over Hawaii in 1988, Aloha Airlines flew the two 737s that had seen the most service.

How old can planes get before they start to fall apart? Traditionally, the U.S. airlines shed themselves of old aircraft by selling them off to Third World carriers. Fifty-year-old DC-3s are still taking off and landing on dirt strips in Latin America.

There are those who argue that planes can fly forever—that planes never need be retired. An Air Transport Association spokesman claimed, "We have seen nothing in our service history that implies there is a quantitative limit to an aircraft's useful life."

The notion of aircraft immortality is an interesting, if not remarkable, concept. It is said that humans fear only time and that time fears only the Great Pyramids at Giza. Now we know. Time also fears the geriatric aircraft, which can fly forever.

Fortunately, the healthier airlines (there are a few) have ordered new planes. The largest carrier, American Airlines, is taking delivery of one new plane about every five days.

The airline industry has $150 billion in new aircraft on order or option. To give that figure some perspective, note that the largest debtor nation in Latin America, Brazil, carries a debt of only $114 billion. But many of the geriatric jets in our U.S. fleet will not be retired until 1999, when Congress has mandated retirement of older, Stage 2 aircraft. Only one in four of new aircraft will replace retiring jets; the other three are for fleet expansion.

But most of the orders are being placed by the healthier airlines, not the anemic ones. Debt-ridden carriers nearing bankruptcy have little alternative but to defer new aircraft purchases, trim maintenance, and coerce flight crews to fly more and rest less. Profit drives everything in the deregulated airline industry, including the margin of safety.

As aircraft become older, airlines should spend 2 percent more on maintenance per year, on average.[23] But rather than spending *more* on aircraft maintenance, U.S. airlines are spending *less* while the fleet has grown steadily older. Resources devoted by commercial airlines to aircraft maintenance fell 30 percent during deregulation's first six years.[24] More recent data indicate that airline spending on maintenance fell from nearly 13 percent of operating expenses in 1977 to 8 percent in 1982, but partially recovered to 11 percent in 1988.[25] A survey of commercial airline pilots reveals that almost half believe that their companies defer maintenance for an excessive period of time.[26] Table 24.4 shows what has happened to aircraft maintenance under deregulation. The number of mechanics per aircraft fell at all but three airlines during this period. This occurred at a time when the U.S. fleet grew to be the oldest in the world—when you would think the industry would devote *more* resources to maintenance, not less.

Before deregulation, the FAA minimum safety standards were a floor below which airlines would not go. Regulation encouraged a healthy margin of safety above these standards.

Table 24.4
Number of Mechanics per Aircraft

	1987	1982
Pan Am	28.2	27.4
TWA	25.7	30.9
United	21.2	17.8
Eastern	16.9	22.1
American	15.6	16.6
Delta	14.9	21.3
Continental	13.0	14.6
Northwest	12.4	11.6
USAir	11.8	12.4

Source: WALL STREET JOURNAL, July 18, 1988, at 25.

The unprecedented criminal indictments handed down in 1990 against Eastern Airlines suggest that Eastern put the bottom line ahead of the life line. Despite massive fines previously levied by the FAA (one totaled $9.5 million for more than 78,000 safety violations), Eastern allegedly installed defective and untested equipment, skipped repairs, and doctored records. Stripped of assets and laden with debt by the corporate pirate Frank Lorenzo, ailing Eastern pinched pennies until it collapsed into bankruptcy.

It is doubtful that the economic health of airlines can ever be divorced from safety. Carriers nearing bankruptcy simply cannot afford to take defective aircraft out of service. Every day, indeed every hour, that a plane sits on the ground is lost revenue.

The level of public and media concern over the trimmed margin of safety has turned up the heat on the Federal Aviation Administration, causing it to become more vigilant in enforcing its safety-regulation mandate—something it was lethargic in doing during the early years of the Reagan administration. Toward the end of the Reagan administration, significant fines were levied on the major airlines.[27]

The Federal Aviation Administration discovered 63,191 safety violations by airlines in 1987, compared with only 28,864 in 1984.[28] Nonetheless, the FAA recently came under fire in a report prepared by the Office of Technology Assessment (OTA).[29] The OTA found the FAA understaffed in the number of inspectors, controllers, and technicians employed and stated that the FAA maintains inadequate programs to improve the performance of aircraft crews, air-traffic controllers, and mechanics. It urged the

FAA to continue surprise inspections and, in particular, to engage in intensive and extensive oversight of the commuter airline industry "during the shakeout expected over the next few years."[30]

The OTA also had a few words of criticism for the airline industry. It found that although all airlines profess adherence to high safety standards, there are significant variations in corporate cultures and maintenance procedures. According to the OTA, safety means one thing to a financially well-off carrier and quite another to a financially strapped airline forced to choose between discretionary aircraft maintenance and the purchase of needed facilities.[31] The OTA concluded, "While airline officials are concerned about safety, financial considerations drive many industry decisions and will continue to do so as strong competition exists among the airlines."[32] Further, "many airlines have lowered hiring standards, [and] increased pilot and mechanic duty time."[33] In a more recent study, the U.S. General Accounting Office found that some airlines are not complying with government orders to correct critical safety defects.

The economic strains created by the intensive pricing competition unleashed by deregulation have had a deleterious effect on carrier safety.[34] Why, then, haven't fatality rates been worse? Two reasons.

First, we can thank the people with the slide rules at Boeing and McDonnell-Douglas. They build very sturdy aircraft, designed to fly even when falling apart. Every generation of aircraft is better engineered than the last to withstand the elements. The redundancies built into over-engineered jets have made it possible for aircraft to stay aloft even when poorly maintained. If one system fails, a backup will usually do the job.

The truth is the accident rate is lower *despite* deregulation, not because of it. Many airlines are slashing expenditures on such things as maintenance and new aircraft purchases. Nonetheless, no matter how badly airlines abuse their aging jets, the aircraft usually hold together. They are as structurally sound as cutting-edge technology can make them.

Good pilots made safe landings in Hawaii when an Aloha Airlines 737 became a convertible and when a United Airlines 747 popped a "Kahn door." People died in both accidents, but the carnage could have been far worse had the pilots and planes not held together.

Second, pilots know that maintenance has taken a beating as profit margins have plummeted. This has mandated a keen level of vigilance in the cockpit. Few of life's ordeals are as sobering as the fear of death. Pilots know to keep a sharp eye out, or else a near miss could become an actual hit.

Add to that the sluggishness of the FAA to restaff the air traffic control system to pre-PATCO strike levels, replace obsolete equipment (the agency is reputed to be the largest user of vacuum tubes in the world), or impose peak-period landing fees or limit general aviation access, and you have a prescription for disaster. One investigatory agency found that the FAA "moved at glacial speed, slowed by inadequate system planning, technology development difficulties and administration and congressional budget decisions."

Many deregulation advocates insist that economic regulation had nothing whatever to do with safety. They pretend that the Civil Aeronautics Board, which regulated the airline industry for four decades, played no role in maintaining airline safety; that was the FAA's job, they say.

In fact, the CAB insured airline safety in three principal ways. First, the CAB employed its licensing powers as a safety filter. The Federal Aviation Act requires that, before a carrier be given a license to operate, it prove itself "fit, willing and able" to provide the service safely. A grossly undercapitalized applicant could not demonstrate financial fitness and therefore was denied permission to fly.

Second, if several applicants sought authority to enter a lucrative market, the CAB would often favor those most in need of economic assistance. Thus, weaker airlines were bolstered so that they wouldn't be tempted to scrimp on safety.

Third, like the banking regulatory agencies, the CAB would encourage a healthy suitor to acquire a company approaching the precipice of bankruptcy. This was how Delta Air Lines acquired ailing Northeast Airlines in the early 1970s.

These policies allowed the airline industry to afford a comfortable margin of safety well above FAA minimums. But today, for some carriers, those standards have become the ceiling, rather than the floor.

The Airline Deregulation Act of 1978 gutted much of the CAB's jurisdiction and caused the agency's sunset in 1985, when its regulatory responsibilities were transferred to the U.S. Department of Transportation. Nonetheless, the act's first two provisions reminded both agencies that the promotion of safety should be a policy imperative of the highest priority and that deterioration in safety should be prevented.

Imaginative arguments have been advanced to assuage the fear of flying nowadays. One of the most cold-blooded is that the market will ultimately solve the safety problem. No airline can afford to have too many crashes, or else the public will shy away from it. Remember the collision of that Air Florida jet with the 14th Street Bridge in Washington, D.C.? The picture of the Air Florida jet's tail sticking out of the icy Potomac dominated the week's news, and the carrier was soon submerged in bankruptcy.

No one can quarrel with the basic economic principle that if an airline kills off its passengers, it won't have any. But the men and women who lost their lives in the Air Florida disaster would likely have preferred government protection before they ascended into heaven. No doubt, their souls gain little comfort in knowing that Adam Smith's "invisible hands" vindicated their deaths by strangling the economic life out of the unsafe airline.[35]

NOTES

1. Thomas & McGinley, *Airlines' Growth, Pilot Shortage Produce Least Experienced Crews in Nine Years,* WALL ST. J., Nov. 20, 1987, at 28. Welling, *The Airline's Dilemma: No Cash to Buy Fuel-Efficient Jets,* BUS. WEEK, Sept. 27, 1982, at 65. P. DEMPSEY, LAW & FOREIGN POLICY IN INTERNATIONAL AVIATION 90 (1987).

2. Kahn, *Airline Deregulation—A Mixed Bag, But a Clear Success Nevertheless,* 16. TRANSP. L.J. 229, 251 (1988); and Kahn, *Transportation Deregulation . . . and All That,* ECONOMIC DEVELOPMENT QUARTERLY (1987), at 98, where he noted that the pressure created by competition exerted "on carriers to reduce prices and costs, may be inducing them also to cut corners on safety." In response to a question on whether the margin of safety had narrowed under deregulation, Kahn conceded, "No one can deny that, under the pressure of competition, we may be walking on a thinner margin." *Interview with Alfred Kahn,* USA TODAY, Oct. 5, 1988, at 13A.

3. *Air Safety Record Worst Since '74,* CHICAGO TRIBUNE, Jan. 13, 1988, at 5.

4. Dempsey, *Cross Your Fingers, Hope Not to Die,* CHICAGO TRIBUNE, Aug. 28, 1987, at 27.

5. Thomas & McGinley, *supra* note 1.

6. Valente, *United's Flight 811 Showed How Vital Capable Pilots Can Be,* WALL ST. J., Mar. 1, 1989, at 1, col. 1. The Continental Airlines DC-9 crash in Denver during a takeoff in a snowstorm in 1987 was piloted by a 26-year-old individual with less than 37 hours of flight time in a DC-9, whereas the captain had only 33 hours. Knox, *Policy Shift Silent Factor in Crash?,* ROCKY MOUNTAIN NEWS, Oct. 4, 1988, at 1-B, 2-B. Before deregulation, 80 percent of pilots had experience as a military pilot; today only half do.

7. *Fatigue Blamed for Dangerous Pilot Errors,* DENVER POST, Sept. 12, 1988, at 3A, col. 1.

8. Duffy, *View from Cockpit Is Clearly Negative,* DENVER POST, Dec. 7, 1987, at 2E.

9. *Id.*

10. *1 in 5 Airline Pilots Has Had Near Collision,* DENVER POST, Dec. 24, 1987, at 2, col. 1. *See also Increasing Near-Midair Incidents Spur Drive to Improve ATC Performance,* AV. WEEK & SPACE TECH. 127 (1987).

11. *Skies Safe Today, But Turbulence Is Brewing,* ROCKY MOUNTAIN NEWS, May 4, 1988, at 37.

12. Dempsey, *supra* note 4, at 28.

13. *Increasing Near-Midair, supra* note 10, updated by DOT Total Near Midair Collision (NMAC) Reports (Sept. 30, 1988).

14. In 1981, there were 317 hazardous midair incidents (or 0.66 per 100,000 flight hours), 85 critical incidents (0.18 per 100,000 hours), and 230 potential incidents (0.48 per 100,000 hours). The corresponding figures for 1986 were 642 (1.46), 163 (0.37), and 473 (1.07). *Focus: A Decade of Deregulation,* TRAFFIC WORLD, Dec. 5, 1988, at Supp. C.

15. Morganthau, *Year of the Near Miss,* NEWSWEEK, July 27, 1987, at 20.

16. Molinari, *How Safe Is the Air Traffic Control System?,* USA TODAY, Nov. 1987, at 12, 13.

17. Welling, *supra* note 1; P. DEMPSEY, *supra* note 1.

308   The Results of Deregulation

18. Dempsey, *Transportation Deregulation—On a Collision Course,* 13 TRANSP. L.J. 329, 354 n. 100 (1984).

19. Rowen, *Airline Deregulation Doesn't Work,* WASHINGTON POST, Apr. 8, 1982, at A27.

20. Valente, Harris, Jr., & McGinley, *Should Airlines Scrap Their Oldest Planes for Sake of Safety?,* WALL ST. J., May 6, 1988, at 1.

21. *Id.*

22. Another source indicates that average ages of airline fleets are as follows:

American	11.14
Continental	11.96
Eastern	14.49
Delta	9.76
Northwest	14.54
Pan Am	14.67
TWA	15.14
United	14.43
USAir	11.58
Piedmont	10.25

Aging Jets Problem Discussed Years Ago, ROCKY MOUNTAIN NEWS, May 8, 1988, at 32.

23. Valente, Harris, Jr., & McGinley, *supra* note 20, at 12, col. 3.

24. Fischetti & Perry, *Our Burdened Skies,* 23 IEEE SPECTRUM 36, 79 (1986).

25. Knox, *supra* note 6, at 2-B.

26. Duffy, *supra* note 8.

27. McGinley, *Fifteen Airlines Face FAA Fines Totaling about $6.5 Million for Alleged Violations,* WALL ST. J., May 12, 1988, at 4.

28. Ott, *Industry Officials Praise Deregulation, But Cite Flaws,* AV. WEEK & SPACE TECH., Oct. 31, 1988, at 89.

29. OFFICE OF TECHNOLOGY ASSESSMENT, SAFE SKIES FOR TOMORROW, SUMMARY 89 (1988).

30. *Id.* at 13.

31. McGinley, *Congressional Report Warns Air Safety May Be Imperiled without Swift Action,* WALL ST. J., July 28, 1988, at 35.

32. OTA REPORT ON AIRLINE SAFETY, *supra* note 29, at 11.

33. *Id.* at 12.

34. *Collision Course, supra* note 18, at 352.

35. Dempsey, *Deregulation Has Spawned Abuses in Air Transport,* AV. WEEK & SPACE TECH., Nov. 21, 1988, at 147.

AIRLINE SURVIVAL AND MARKET DARWINISM: DAWN OF THE GLOBAL MEGACARRIERS

THE DISINTEGRATION OF THE U.S. AIRLINE INDUSTRY

The airline industry is immersed in an unprecedented crisis, one that was not entirely unforeseen but one that was nonetheless unfortunate and avoidable. As 1991 dawned, six major airlines found themselves in some stage of liquidation, desperately selling off operating assets to raise enough cash to stay aloft. Five of them stumbled into bankruptcy, and by the end of 1991, three of those died.

Before deregulation, many industry analysts warned that after a binge of destructive competition, only a handful of airlines would survive.[1] These warnings were dismissed by deregulation proponents who saw nearly textbook levels of competition everywhere they looked.[2] Alfred Kahn, the architect of airline deregulation, recently confessed, "We thought an airplane was nothing but a marginal cost with wings."[3]

Deregulation was supposed to produce lots of new airlines. Congress was told that barriers to entry and economies of scale were insignificant; new entrants would emerge to prevent the industry from becoming concentrated.[4] In deregulation's inaugural years, new airlines appeared; but most couldn't survive. Many, like People Express, were consumed in mergers and acquisitions or, like Air Florida and Midway and 150 other airlines, fell into the abyss of bankruptcy. Although they sent ticket prices spiraling downward, new entrants never accounted for more than about 5 percent of the passenger market. Only one of the 176 airlines to which deregulation gave birth is, as of this writing, still alive and it, America West, languishes in bankruptcy. Today, new entry is highly unlikely.

The magnitude of the crisis with which the airlines are now confronted is unparalleled in the history of commercial aviation. In 1991, after prolonged illnesses, Eastern Airlines, Pan American World Airways, and Midway Airlines were laid to rest. Eastern's tragedy could be dismissed as an aberration were it not for the fact that other major U.S. airlines are also liquidating major operating assets to stay aloft. Continental and America West have also stumbled into bankruptcy, Continental for the second time (some call it Chapter 22 bankruptcy). More will likely follow. Take a closer look at the disintegrating airlines.

A year after closing its Kansas City hub, *Eastern* entered bankruptcy and sold its Washington–New York–Boston shuttle (to Donald Trump for $365 million), as well as the Latin American routes it had picked up a few years earlier at Braniff's fire sale (to American Airlines for $310 million). After running out of cash, it ceased operations in January 1991. Delta and United were the highest bidders in the Eastern liquidation of gates, landing slots, and routes.[5]

Pan Am sold its transatlantic routes to London and beyond to United for $400 million.[6] Pan Am sold its Washington–New York–Boston shuttle and remaining European authority to Delta. The 1980s was a decade of dismemberment for anemic Pan Am, a time during which it sold off its transpacific routes (again to United, for $750 million), its Intercontinental Hotel chain, and the Manhattan skyscraper that still bears its name. It ceased operations in December 1991.

TWA is selling off international routes, gates, and landing slots at Chicago and Washington, D.C. American is spending $445 million for TWA's Heathrow authority as well as other domestic airport and landing slot assets.[7]

Midway sold its Philadelphia gates, which it had picked up at Eastern's fire sale, to USAir at a $32-million loss, then entered bankruptcy.[8] It ceased operations in 1991.

In bankruptcy for the second time in a decade, *Continental* sold its lucrative Seattle-Tokyo route to American Airlines for $150 million.[9] As of this writing, it too is in bankruptcy.

Other U.S. airlines are having serious problems. USAir lost about half a billion dollars during 1990, and was projected to lose as much in 1991. It has tightened its belt significantly by reducing flights, withdrawing from markets, and furloughing thousands of workers.[10]

Of course, a few gargantuan airlines will survive. The healthiest three—United, American, and Delta—already control more than half the market.[11] All three are on a buying binge, gobbling up the dismembered parts of the disintegrating airlines.

The airline industry suffered recessions and sharply increased fuel costs before deregulation. Fuel prices shot up 300 percent in the 1970s after the Arab oil embargo of 1973, and there was a recession in the early 1970s as well.[12] But never before have major airlines collapsed.

All the world's airlines are paying the sharply higher fuel prices inspired by the Persian Gulf crisis, and all are suffering from the early pangs of global recession. But only America's are in bankruptcy, only America's have died, and only America's are selling off operating assets—despite the fact that international aviation fuel costs more than domestic fuel. Why?

FORMER DOT SECRETARY SKINNER'S OBSERVATIONS ON THE CONTEMPORARY CRISIS IN THE AIRLINE INDUSTRY

Former Secretary of Transportation (now White House Chief of Staff) Samuel Skinner recently delivered a speech before the National Press Club and testimony before two congressional committees, in which he addressed the contemporary crisis in the airline industry. Distilled to their essence, Skinner made the following points:

1. The contemporary shakeout will leave air passenger transportation dominated by "more than three and less than seven" airlines over the next few years, and as a consequence, "some of the lowest fares will disappear."[13]
2. The deregulation experiment is not the cause of the industry's problems. Deregulation is instead a profound success, and the deregulation debate is proclaimed over.
3. Although deregulation is not the cause of the industry's problems, labor costs are.[14]
4. Foreign ownership is the cure for the industry's ills.[15]

Only Skinner's first conclusion is correct. The industry will achieve levels of concentration even higher than the unprecedented levels it has already reached. Before deregulation, the eight largest airlines controlled 80 percent of the domestic passenger market. They now control 95 percent. The five disintegrating airlines accounted for about 25 percent of the domestic market, which, if Skinner is right, will likely be distributed among the surviving airlines. The three largest airlines (American, United, and Delta) already account for more than half the domestic market. The four largest (these three, plus Northwest) control more than two-thirds of the market.

A growing number of industry experts and concerned citizens dispute Skinner's second point. Marty Shugrue, formerly Eastern Airlines trustee, observed: "Deregulation is simply not working out as anticipated. There are far fewer airlines than when deregulation began. Of the remaining carriers, more than half are struggling and several may well go the way of Eastern."[16]

Aviation fuel costs soared 300 percent during the 1970s, and the industry was plagued by recession then as well; but not a single airline folded, entered bankruptcy, or liquidated operating assets. Then, of course, the industry was regulated; today it is not.

Today, aviation fuel is cheaper than before Saddam Hussein invaded

Kuwait. Although fuel costs rose significantly during the crisis, they were nonetheless *lower* in actual and real terms than they were a decade ago. Between 1981 and 1984, the actual cost per gallon of aviation fuel ranged between $0.79 and $1.04 per gallon, ranging in real terms (adjusted for inflation) between $1.04 and $1.47. In contrast, aviation fuel sold in 1990 for only $0.80 per gallon.[17] Despite the fact that fuel is cheaper, several major U.S. airlines are liquidating operating assets.

The first decade of deregulation produced a bloodbath of ruinous competition. The industry as a whole enjoyed an average profit margin of less than 1 percent during the 1980s (compared with an average of between 3 percent and 6 percent for manufacturers).[18] Excessive losses produced 150 bankruptcies and 50 mergers during deregulation's first decade. DOT never met a merger it didn't like, approving all 21 submitted to it.[19] Deregulation also freed corporate raiders like Carl Icahn and Frank Lorenzo to strip airlines of assets. Debt service is now crushing the operating profits of the disintegrating airlines. DOT could have stopped the carnage, but chose not to intervene.

The economic anemia unleashed by deregulation forced airlines to defer new equipment purchases. Sadly, U.S. airlines today fly the oldest fleet of aircraft in the developed world. The geriatric jets burn more fuel.[20] They are also less safe.

Deregulation created the fuel-guzzling hub-and-spoke phenomenon, which requires passengers to fly more miles, with more takeoffs and landings, and creates more airway congestion than before. Flying older jets more miles necessarily consumes more fuel. So when fuel costs rise even modestly, as they did during the Persian Gulf crisis, the profit margin disappears.

Skinner is therefore wrong. Deregulation must shoulder at least part of the blame for the industry's disintegration and unprecedented concentration. The same is true in the savings and loan industry and the trucking and bus industries.

We will address Skinner's other conclusions in greater detail below. First, let us examine the principal survival characteristics of airlines in these unfriendly skies.

SURVIVAL CHARACTERISTICS OF U.S. AIRLINES

After more than a decade of deregulation, several characteristics appear essential for the survival of airlines. Listed below are nine.[21] They are neither listed in order of importance, nor are they of equal value. But generally speaking, the more of them an airline possesses, the better its chances for survival.

Figure 25.1
Single-Carrier Market Share at Concentrated Airports

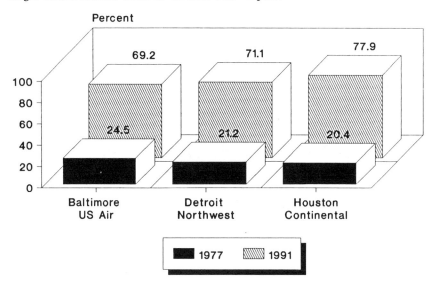

Multiple Hubs, Strategically Located

Before deregulation, although Atlanta (for Delta) and Pittsburgh (for Allegheny, now USAir) were moderately concentrated, no airline dominated more than 50 percent of the market (measured by gates, passengers, or takeoffs and landings) at any major airport in the nation. Today, dominant airlines control more than 60 percent of the market (sometimes more than 90 percent) at about 18 major airports. The infrastructure of gates and landing slots at the major airports has been consumed by the megacarriers, leaving little room for new entry.[22] Figures 25.1 through 25.3 reveal the growth in concentration at several of the nation's largest airports.

Strategically located hubs are designated to allow the carriers to blanket the nation with service. For example, United has hubs at Chicago, Denver, San Francisco, and Washington, D.C. (Dulles). American Airlines has expanded its traditional hubs at Chicago and Dallas/Ft. Worth and established new ones at San Jose, Nashville, Raleigh/Durham, and San Juan. Delta has hubs at Atlanta, Dallas/Ft. Worth, Salt Lake City, and Cincinnati.

In contrast, TWA has a domestic hub only at St. Louis (and an international hub at New York–Kennedy), and Pan Am dominated no domestic airport. Among the troubled airlines, only Continental has multiple

Figure 25.2
Single-Carrier Market Share at Concentrated Airports

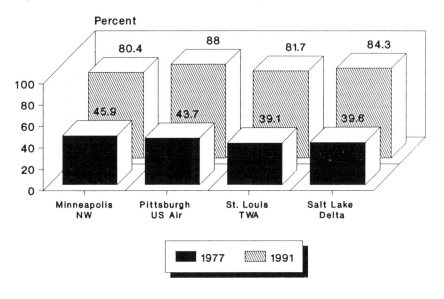

Figure 25.3
Single-Carrier Market Share at Concentrated Airports

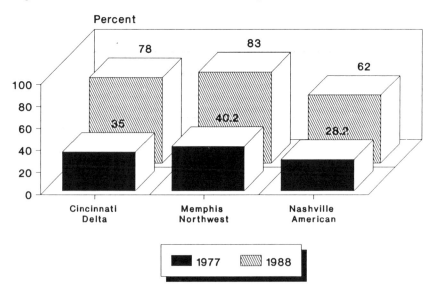

strategically located hubs—at Houston, Denver, Cleveland, and Newark (the latter it acquired from People Express on its deathbed).

Moreover, consumption of airport infrastructure can translate into higher yields. Yields at concentrated airports are 27 percent higher per mile for passengers who begin or end their trips there than at unconcentrated airports.[23] Airlines with more gates, takeoff and landing slots (at capacity-constrained airports), and/or code-sharing agreements charge significantly higher prices than those without, according to the U.S. General Accounting Office.

For example, as of 1988, the eight largest airlines owned 96 percent of the landing and takeoff slots at the four slot-constrained airports (Chicago O'Hare, Washington National, and New York's Kennedy and LaGuardia). In 1985, before the Department of Transportation decreed slots could be bought and sold in the market, the eight largest airlines controlled only 70 percent of the slots.[24] Fares are 7 percent higher, on average, at slot-constrained airports.[25] Moreover, an airline that doubles the number of its gates enjoys a 3.5 percent increase in fares.[26]

Frequent-Flyer Programs

The widespread service permitted by multiple hubs allows airlines to enjoy economies of density and to better market their product to the most lucrative customer, the business traveler. For example, United Airlines serves all 50 states, not because each is profitable but because United can thus offer to fill all the geographic needs of business travelers.

Airlines offer to fill business needs while luring business travelers with rewards of free travel to exotic destinations. In essence, they encourage business fraud. Suppose, for example, a distributor of copying paper offered to sell a business executive paper at a price 25 percent higher than his competitors but promised the executive two free first-class airline tickets to Hawaii if he bought the distributor's paper all year long. Wouldn't the business executive be defrauding his company if he purchased the higher-priced paper? Yet that is precisely the type of inducement that airlines offer business travelers addicted to the frequent-flyer programs. Once addicted, many business travelers select (and bill their companies for) the higher-priced flight on the airline, satiating their desire for free travel. Indeed, 75 percent of travel agents report that business customers chose to fly a particular airline more than half the time because of their membership in a frequent-flyer program.[27]

Computer Reservations Systems

Eighty percent of flights are booked through travel agents, and 95 percent of agents use one of the airline-owned computer reservations systems.[28] According to the GAO, an airline that owns its own computer reservations system stands between a 13 to 18 percent better chance of selling its product through its system than does a competitor.[29] American Airlines pioneered such systems, with SABRE. United owns APOLLO. Continental owns SYSTEM ONE, which it took from Eastern for a good deal less than fair market value. TWA, Northwest, and Delta share the combination of PARS and DATAS II (now named WORLDSPAN).

Computer reservations systems have created a sophisticated and expedient means of exchanging pricing proposals and have facilitated implicit price-fixing.[30] They also produce extraordinary profits for their owners, far beyond the rents that could be exacted in a fully competitive market.

Sophisticated Yield Management

Airlines have learned that by watching passenger demand carefully, they can shrewdly manipulate the number of seats for which restricted discounts are offered on an hourly basis and can fill seats with passengers paying the maximum price. That explains the phenomenon of tens of thousands (40,000 to 100,000) of rate changes each day.

Consumer groups complain that by offering cut-rate fares for only a relatively small number of seats, airlines are engaging in "bait-and-switch" advertising.[31] The bewildering array of fares has also increased transactions costs for consumers.

Fuel-Efficient Fleet of Standardized Aircraft

The economic anemia created by the destructive competition unleashed by deregulation left airlines with inadequate resources to buy new planes, causing the U.S. fleet to degenerate into the oldest in the developed world. Thirty-one percent of the U.S. fleet now exceeds the economic design goals originally set by the manufacturers.[32] Older-generation aircraft gulp more fuel. TWA flies the oldest jets in our geriatric U.S. fleet.

Merged airlines have been forced to deal with the problems of consolidating huge fleets of aircraft of inconsistent types and several manufacturers, problems that increase the cost of maintenance and require multiple inventories of spare parts. Deregulation led to an unprecedented number of mergers and acquisitions during its first decade (see table 3.6).

As a consequence, Continental, which flies the fleets of former carriers like Texas International, New York Air, People Express, and Frontier, experiences this problem. Northwest flies the fleets of North Central, Southern, and Hughes Airwest, all of which merged to form Republic, which Northwest acquired. In contrast, airlines that grow from within (such as, for the most part, American and United) save maintenance cost and aircraft downtime by incrementally growing with relatively standardized fleets.

United has placed orders for new aircraft—which will expand its fleet by between 40 percent and 90 percent—all with a single manufacturer, Boeing, "promoting commonality within the fleet which assures significant long-term operational efficiencies."[33]

Incidentally, the largest airlines now control the order books at the major aircraft manufacturers. Both American and United are taking delivery of new jets every week (and will through the middle of this decade), whereas the collapsing airlines are not. As noted above, newer-generation aircraft are relatively fuel-efficient. This will matter more as the decade proceeds toward the statutory retirement of Stage 2 aircraft on January 2, 1999. As of May 1990, the airlines with the highest percentage of aging Stage 2 aircraft were Eastern (70%), Northwest (65%), Pan Am (58%), USAir (55%), TWA (55%), Continental (52%), and Midway (85%).[34] In contrast, only 31 percent of American's fleet consists of Stage 2 aircraft.[35]

As noted above, deregulation also produced the fuel-guzzling hub-and-spoke phenomenon—the dominant megatrend on the deregulation landscape. Hubbing requires that airlines fly passengers more miles in smaller aircraft with more takeoffs and landings. Indeed, hubbing led many airlines to cancel orders for widebody aircraft in the early 1980s and either fly their existing jets or place orders for narrowbodied planes. The average seat mile costs for a widebodied aircraft like a Boeing 747 are significantly lower than those of a narrowbodied plane like a Boeing 737 or 727. Figure 25.4 reveals the pre-deregulation trend toward larger-capacity (and lower seat mile cost) aircraft, compared with the reversal of the trend in the post-deregulation period. Funneling passengers through constipated hub-and-choke bottlenecks not only squanders billions of dollars of business travelers' time and productivity but also burns fuel wastefully. Smaller, older jets flying more miles with more takeoffs and landings necessarily cause their airlines to suffer increased costs during a period of ascending fuel prices. That plus the higher per passenger labor costs created by hubbing has caused productivity to turn flat under deregulation.

Low Debt (Conservative Growth)

The operating losses engendered by deregulation created enormous debt. Despite reduced wages, airline operating expenses increased 94 percent during deregulation's first six years.[36] During deregulation's first decade, the industry suffered a 74 percent decline in its profit margin, to a mere .6 percent—until now, the worst financial period in the industry's history.[37] The industry became an economic basket case, prompting the rash of mergers in the mid-1980s.

Deregulation also freed corporate raiders, like Frank Lorenzo (at Continental and Eastern) and Carl Icahn (at TWA), to loot airlines, leaving

Figure 25.4
Average Seats per Aircraft, Fiscal Years 1969–1989

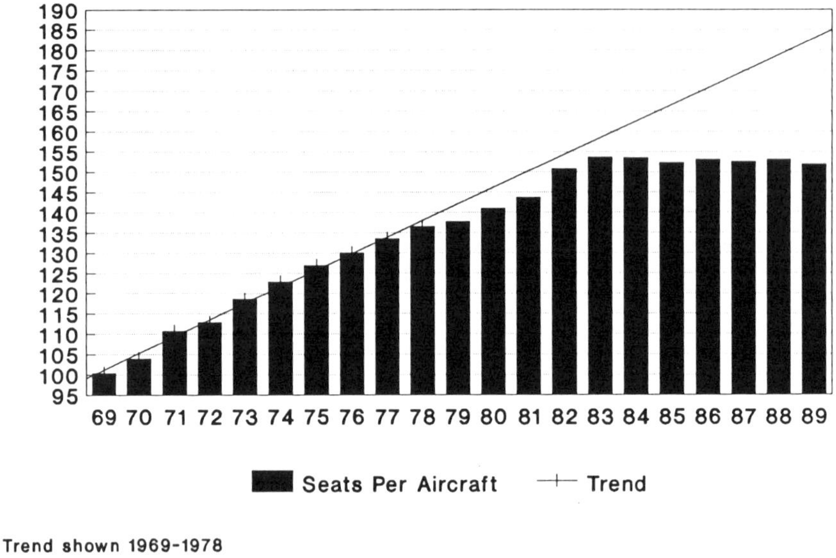

Seats Per Aircraft   ─┼─ Trend

Trend shown 1969-1978

them with suffocating debt. Frank Lorenzo is the only man in history to have bankrupted two airlines (one of them twice).

TWA owes $3.2 billion in long-term debt, lease obligations, and unfunded pension liability.[38] Continental suffers from about $2.2 billion in debt.[39] Eastern's collapse could expose its parent, Continental, to an additional billion dollars of liability for Eastern's unfunded pension obligations and the transfer of assets into the Texas Air empire at less than fair market value. Interest payments recently exceeded 8 percent of operating expenses at both TWA and Eastern—the highest in the industry.[40]

As a percentage of total capitalization, Pan Am's debt soared from 62 percent in 1980 to 273 percent in 1989.[41] Pan Am has $3 billion in long-term debt, lease obligations, and unfunded pension liability.[42] Eastern's debt climbed from 79 percent of total capitalization in 1980 to 473 percent in 1988, its last year before bankruptcy.[43] TWA's debt soared from 62 percent in 1980 to 115 percent in 1989.[44] Continental's rose from 62 percent in 1980 to 96 percent in 1989.[45] It is no wonder the anemic airlines are cannibalizing assets to stave off extinction. Figure 25.5 reveals this distressing trend.

Representative Byron Dorgan aptly noted: "I'm not so alarmed if they load up a lipstick company with debt and it fails. But if you do that to an airline, it's a real blow to the public interest."[46] Indeed it is. A collapsing infrastructure industry sends shock waves throughout the economy.

Figure 25.5
Debt as a Percentage of Total Capitalization, 1980–1989

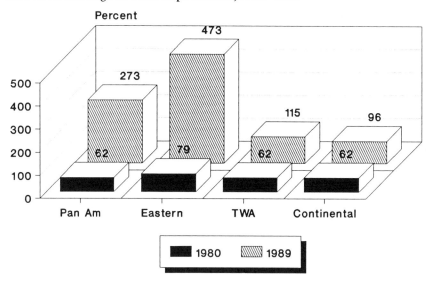

The Department of Transportation has long held jurisdiction to investigate the "fitness" of airlines plagued with debt. Here, as with so many of its other statutory responsibilities, DOT has shown no enthusiasm for protecting the public interest.

The enormous debt assumed by Pan Am and Eastern (to shore up declining revenues) and Continental and TWA (to pay off exorbitant debt accumulated by corporate raiders) appears to be dragging these airlines down a black hole.

Unfortunately, low debt has subjected some airlines to leveraged buyouts. Low debt suggests there are lots of assets owned that can be sold to pay off the debt assumed during the acquisition. For example, Northwest had one of the lowest percentage of aircraft leased (4%) in the industry before its leveraged buyout.[47] To thwart potential LBOs, some airlines have sold aircraft and leased them back, a strategy that reduces the inventory of aircraft that could finance an LBO but nonetheless increases the long-term costs of doing business, whether the debt shows up on the books of the airline or not.

Low Wages/Flexible Work Rules

Some airlines have broken unions and thereby reduced costs. Continental and TWA are prime examples. Although Continental has lower labor costs than any other major airline, not even that has kept it out of bankruptcy. Labor acrimony, perhaps enhanced by the tactics of its former chairman, Frank Lorenzo, cost it dearly in the 1980s.

The airline industry is a service industry. Happy employees can give passengers a lovely trip and lure them back for another, and another. Angry, embittered employees can do the opposite.

Other airlines have convinced unions to settle for two-tier wage rates, with the B scale at entry grade. American, United, and Delta are examples. More than half of the present pilots and flight attendants at American, for example, are on the B scale. Some of the flight attendants at the two-tier airlines, earning between $950 and $1,220 a month,[48] qualify for food stamps.

In most service industries, salaries account for a disproportionate share of operating costs. But low wages do not guarantee survival. People Express collapsed despite its rock-bottom wages. Continental, America West and Midway, also with relatively low wages, have struggled in the contemporary environment.[49]

As a percentage of operating expenses, Delta has among the highest labor costs and Continental the lowest.[50] Yet Delta has thrived under deregulation, and most analysts predict it to be one of the few surviving airlines. There seems to be a rather poor correlation between low wages and survival, despite former Secretary Skinner's allegations to the contrary.

Superior Service

Airline service has degenerated universally under deregulation, so consumers have been taught not to expect much. Consumer polls reveal that travelers rate foreign airlines higher than our domestic ones. It is no wonder. When USAir consumed Piedmont, the latter's loyal customers were most concerned with whether USAir would continue Piedmont's practice of giving passengers the full can of soda, rather than just a cup. That one example reflects how far consumer expectations have fallen.

To pose an analogy, before deregulation we enjoyed chicken-fried steak. Now we are relegated to a diet of ground horsemeat. Consumers save billions of dollars eating horsemeat, but it just doesn't taste the same.

The point is that today it doesn't take a lot of service to stand out as being better. Consumers can be, and too often are, turned off by late arrivals and departures, dirty planes, inedible food, and embittered employees. The three largest airlines—Delta, United, and American—typically are rated higher than other domestic airlines in terms of service.

International Routes

The global air-transport market is growing, and many international markets are quite lucrative. Although traffic was down on the North Atlantic in the early 1990s, airlines that serve the North Pacific market enjoy the most attractive yields. Both Northwest and United earn a disproportionate share of their total income from international markets. Between 1987 and 1989, Northwest earned between 68 percent and 91 percent of its total operating profit from international markets while United earned between 24 percent and 34 percent.[51] Many industry analysts predict international markets will grow faster than domestic markets during this decade.

CONCENTRATION IN THE TRANSPORTATION INDUSTRY

Collapsing airlines mean more concentration. Already, the eight largest airlines account for more than 90 percent of the domestic market (up from 80 percent before deregulation). Sadly, additional concentration will send ticket prices soaring into the ionosphere.

The Brookings Institution alleges that consumers save $6 billion a year because of airline deregulation. Not so. Fuel-adjusted real airfares fell at a significantly faster rate during the decade *before* deregulation than in the decade after. Except for a brief spate of sharply lower fares in the 1977–79 period, post-deregulation fuel- and inflation-adjusted fares fell at a 30 percent slower rate per mile than in the pre-deregulation period.[52] The Brookings studies wholly ignore the pre-deregulation trend of falling ticket prices (which for four decades was driven by technological improvements) and attribute all price savings since promulgation of the Airline Deregulation Act of 1978 to deregulation.

Paradoxically, whereas deregulation was supposed to produce more competition, lower prices, and better service, it has instead produced more concentration, higher prices, and miserable service. Every major prediction made by the textbook economists has proven wrong.

The airline story could itself be considered a curious aberration if the concentration epidemic was not also plaguing every other mode of transport. But under deregulation, the number of major railroads dwindled from 12 to 7, with no significant new entry. Two-thirds of the general-freight trucking companies collapsed, with no significant new entry. And with the merger of Greyhound and Trailways, the bus duopoly became a monopoly and is now in bankruptcy; here too, there has been no significant new entry.[53]

FOREIGN OWNERSHIP

Now that deregulation has failed to produce the near perfect model of textbook competition the laissez-faire economists predicted, the deregulationists are proposing to sell our domestic industry off to foreign airlines. Already Northwest, Delta, Continental, America West, and Hawaiian Airlines have significant foreign equity. The DOT has suggested that, insofar as foreign ownership is concerned, the sky is the limit.

In 1989, Secretary of Transportation Samuel Skinner expressed legitimate concern over the Checchi group's acquisition of Northwest Airlines, not only because the LBO would increase Northwest's debt fourfold but also because the $400-million equity participation by KLM Royal Dutch Airlines would give it about 57 percent of total equity.[54] Secretary Skinner appeared to interpret Section 101(16) of the Federal Aviation Act to limit foreign equity to 25 percent. As Skinner said:

While KLM's voting share technically fell within the statute's numerical limits [which requires that the airline's president and two-thirds of its board and other managing officers be U.S. citizens and that not less than 75 percent of voting interest be owned and controlled by U.S. citizens], we concluded that KLM's ownership of 57 percent of NWA Inc.'s total equity, together with the existence of other links between the carriers and KLM's position as a competitor, could create the potential for the exercise of influence and control over the carrier's decisions. This would be inconsistent with the law.[55]

Remarkably, what Secretary Skinner then declared would be, in his words, "inconsistent with the law" he now proclaims to be well within the law.

The statute has not been amended since then-Secretary Skinner found that KLM's gargantuan ownership was inconsistent with the law. The U.S. Department of Transportation continues to hold jurisdiction, under Section 401 of the Federal Aviation Act, to scrutinize the fitness of airlines (which includes safety and compliance fitness) and, under Section 101(16), to review foreign ownership. Under present law, foreign ownership is limited to 25 percent of the voting stock of U.S. airlines, and no foreign airline can ply the domestic trade.

In a radical departure from precedent and a tortuous interpretation of law, in 1991 the DOT announced that it will allow foreign equity ownership of up to 50 percent. Former Secretary Skinner also proposed that statutory limits on voting ownership be increased to 49 percent.[56] The DOT has even proposed to put the exchange of cabotage rights (the opportunity for foreign airlines to serve domestic routes) on the table in negotiations with the government of Canada, despite the legislative prohibition. Actually, foreign airlines don't need cabotage rights if they can buy access to the U.S. market.

Foreign alliances with U.S. airlines began in the 1980s with shared frequent-flyer programs, then entered computer reservations systems, and now have turned to outright equity ownership. Table 25.1 reveals the alliances of the two dominant European computer reservations systems.

International airline alliances have been stimulated by the prospect of liberalizing European transport in 1992.[57] Having witnessed the intense

Table 25.1
European Computer Reservations Systems Partners

Covia	Amadeus
United	Texas Air
British Airways	Air France
KLM	Lufthansa
Swissair	Iberia
Alitalia	SAS
USAir	

shakeout produced by deregulation in America, foreign management believes that the liberalization of competition rules will result in extreme concentration. The conventional wisdom is that when the dust settles from U.S. deregulation and international aviation liberalization, only a handful of global megacarriers will dominate air transport. Several industry experts predict that the world's air-transport system will eventually be dominated by just eight to ten global megacarriers.

Wanting to be among the survivors motivated the contemporary surge in international combinations and alliances. Moreover, with Europe's aviation infrastructure even more saturated than America's, opportunities for growth are largely limited to acquiring or affiliating with existing airlines.

Foreign airlines are deeply interested in penetrating the U.S. passenger market—a market larger than that of the rest of the world combined. In the last few years, KLM bought a huge piece of Northwest, SAS purchased a chunk of Continental, Singapore Airlines and Swissair each acquired a slice of Delta, and British Airways (which gobbled up British Caledonian) sought a share of United Airlines. Table 25.2 depicts the substantial foreign airline ownership in U.S. flag carriers.

The equity interests by Scandinavian Airline System (SAS) in Continental Airline Holdings was inspired by the American carriers' need for a substantial infusion of new capital. From SAS's perspective, the Texas Air alliance gave it new feed into its transatlantic routes; SAS moved its international hub from New York Kennedy Airport to Newark, where Texas Air's Continental and Eastern could provide domestic feed.[58] (However, SAS may have overextended itself; it is now retrenching.) Swissair's and Singapore Airlines' interest in Delta appears to have been inspired by a different reason—the desire of Delta to have friendly partners poised to fend off LBOs.

But most are motivated by foreign airlines' interests in creating operat-

Table 25.2
Foreign Airline Ownership of U.S. Airlines

Foreign Airline	Percentage Ownership	U.S. Airline
SAS	18.4%	Continental
Swissair	5%	Delta
Singapore Airlines	5%	Delta
Ansett Airlines	17%	America West
Japan Air Lines	20%	Hawaiian Airlines
KLM	49%	Northwest
British Air	15%*	United

* proposed; later withdrawn

ing and market alliances. Thus, foreign airlines invest "dumb equity," accepting suboptimal returns because they anticipate synergistic revenue on the passenger feed that U.S. airlines promise them and the diminution of competition thereby created.

Not only U.S.–foreign but also other international aviation alliances are emerging, including British Airway's acquisition of British Caledonian and Air France's purchase of UTA. Table 25.3 reveals the major ownership interests of foreign airlines. Here's a college board exam question: if Delta owns 5 percent of Swissair, and Swissair owns 5 percent of SAS, and SAS owns 18.4 percent of Continental, how much of Continental does Delta control?

Foreign ownership raises serious anticompetition concerns. Many international markets are already among the highest priced, fastest growing, most lucrative and least competitive. As noted above, United and Northwest both earn a disproportionate share of their profits from the transpacific market. How vigorous a competitor would they be if Japan Air Lines (or, for that matter, Korean Air Lines or Cathay Pacific) owned a significant chunk of either?

KLM now owns 49 percent of Northwest. Both airlines serve Amsterdam and Minneapolis (their respective hubs), as well as interior European and U.S. cities. How can we expect vigorous competition between an airline (Northwest) and its owner (KLM)? We didn't see it between Continental and Eastern once Frank Lorenzo's Texas Air subdued both.

Further, most foreign airlines are owned, in whole or part, by their governments. Monopoly is not the antithesis of competition; socialism is. A government-owned or -subsidized airline need not make a profit to stay

Table 25.3
Cross-Ownership Agreements between Foreign Airlines[59]

Purchaser	Percentage Ownership	Target
Air France	1.5%	Austrian Airlines
ANA	5%	Austrian Airlines
Cathay Pacific	35%	Dragonair
Iberia	85%	AerolineasArgentinas
SAS	5%	Swissair
SAS	35%	Lan Chile
SAS	25%	Airlines of Britain
SAS	16%	CTA
Swissair	10%	Austrian Airlines
Swissair	5%	SAS

alive and therefore lacks a proper competitive discipline. Its presence in a free market creates an unlevel playing field. Government treasuries have financial resources beyond the wildest dreams of privately owned companies. Foreign governments can subsidize losses or underwrite the capital requirements necessary to develop monopoly positions.

At the outset of deregulation, some predicted that ultimately only a handful of airlines would survive and that these would be nationalized as wards of the state. Never could they have imagined that the few surviving airlines would be wards of foreign governments.

Today, about 8 percent of Northwest is owned by the government of the Netherlands. About 8 percent of Continental and Eastern are owned by the Scandanavian governments. We have now embarked on a regime of partial nationalization, not by our government but by foreign governments.

Foreign ownership restrictions have long existed for many of our essential infrastructure industries—airlines, intercoastal and inland shipping, telecommunications, broadcasting, electric power production, and nuclear energy. These restrictions were added to our law not because of blind xenophobia but because of legitimate national security considerations.

Aviation is essential to national security. As Operation Desert Storm confirmed, the nation depends on the aircraft of our domestic airlines committed to the Civil Reserve Aviation Fleet (CRAF) as the essential logistical means to ferry troops and supplies to distant battlefields. We need the

CRAF fleet for airlift capacity in time of war. Foreign ownership may jeopardize access to CRAF aircraft. The air force simply doesn't have enough C-5As to do the job.

On August 2, 1990, Iraq invaded Kuwait. Two weeks later, the CRAF fleet was activated—the first time since its creation in 1951. Calling up the CRAF fleet was essential to meet the demands of the most massive airlift since the Berlin Airlift in 1948. During the first two months of activation, CRAF planes flew more than 500 missions and carried 66,000 passengers (mostly soldiers) and 22,000 tons of cargo. In the recent Persian Gulf crisis, we relied on our domestic Civil Reserve Aviation Fleet to ferry 60 percent of the soldiers and 23 percent of the supplies to the battlefield. Yet Skinner would have foreign governments sit on the boards of directors of U.S. airlines.

Similarly, we maintain a federally subsidized U.S.-flag fleet of ocean carriers because of the lesson we learned in World War I—when we looked around for essential ships to ferry troops and supplies across the Atlantic, there were nearly none. Not that long ago, the federal government bailed out a collapsing Conrail and Lockheed in part because of their importance to national security. Transportation is essential to our national defense.

Of course, we could commandeer the aircraft of foreign airlines if we needed them—seize the property of foreign companies as other nations have done to American firms. But acquisition of capacity is not the only problem.

Those who argue for foreign ownership of domestic airlines forget that most of the technological breakthroughs of aviation were inspired by military applications—delivering troops and bombs. Imagine a world where we had never prohibited foreign ownership or foreign airline competition. How many Pearl Harbors would we have suffered if the dominant domestic airlines in 1940 had been Lufthansa and Japan Air Lines?

Although we fought wars with Britain in two centuries, British Airways doesn't look like much of a national security threat these days. But our alliances are constantly shifting, so that an Aeroflot looks more or less threatening depending on the point in history at which you ask the question. We embraced Stalin to fight Hitler, and Syria's Assad to destroy Saddam Hussein. Today, would we want Donald Trump to sell the Trump Shuttle to Iraqi Airways?

In 1974, the Shah of Iran proposed to buy Pan American World Airways. Had former Secretary of Transportation Skinner been calling the shots then, he might well have allowed it. After all, Iran was then our closest ally in that part of the world.

We all know the tragic events that transpired in Iran after the fall of the Shah. If the foreign ownership rules adopted by DOT in 1991 had been in effect in 1974, would Iranian President Rafsanjani today be chairman of

Pan Am's Board, and would Pan Am's CRAF 747s be parked on Iranian military airfields next to Iraqi jet fighters?

We need to keep our essential infrastructure industries out of foreign hands so that we don't wake up one day in the midst of a global crisis wondering why we were so shortsighted as to allow our industries to be crippled by our adversaries. We don't want foreign owners sabotaging, disrupting, or delaying the free movement of commerce, or communications, or electric power or, indeed, putting their grubby hands on nuclear fuel rods. We need a healthy domestic infrastructure capable of loyally serving the nation in times of crisis.

Moreover, foreign ownership jeopardizes the integrity of bilateral air-transport negotiations between the United States and foreign governments. International routes are traded by nations on a bilateral basis, usually with candid input from their carriers.[60] Multiple allegiances may well jeopardize the integrity of that process.

CONCLUSIONS

Foreign ownership restrictions didn't cause the disintegration of our domestic airline industry. Neither did the fuel crisis of 1991–92.

Look around the world. No foreign airline is in as sorry shape as ours— none are liquidating operating assets, none are in bankruptcy, and none have died—despite the fact that international aviation fuel costs more than domestic fuel and the entire world is feeling the pangs of recession.

Surely, we need to alleviate the economic crisis plaguing the airline industry and threatening healthy competition. To do that, we had best start tackling the true cause rather than hastily grasping for radical alternatives that might endanger our national security.

There are more than two temperatures at which to cook a pot of stew. In the 1970s, the competitive dial was set on LOW. The stew wasn't warm enough so Congress turned the dial up to HIGH with the Airline Deregulation Act of 1978. The competitive bubbles began to boil, causing stew to splatter over the side of the pot. The aroma was sweet for a short while, until it turned foul with smoke. Before the stew burns a charcoal black, Congress should turn the dial down to MEDIUM so that we can have stew the public can eat.

The public owns the trillion-dollar airport and airway infrastructure. Common sense suggests that the public ought to have some say in how the airlines use that system. Consider that all the stock of all the airlines could be purchased on Wall Street for less than $15 billion, or a mere 1.5 percent of the value of the public investment.

Unlike the highways, where people have direct access in their privately owned automobiles, the airport and airway system can be accessed by the great majority of citizens only via the commercial airlines. Yet the destinations, the terms, the conditions, and the prices of services are all dictated by private monopolists and oligopolists, with no input from the public, which owns 98.5 percent of the system.

Deregulation gave away the public system to private monopolists. It replaced the U.S. Civil Aeronautics Board, which protected the public interest, with the chief executive officers of a handful of airlines, who treat the public system as their private Monopoly board, buying and selling properties while charging the public exorbitant rents. They are allowed to turn a profit by selling assets owned by the nation—landing slots and international routes. Deregulation transformed the air transport system from a public utility into segmented and shared regional and city-pair monopolies and a national oligopoly. The equivalent would be deregulating the trucking industry and giving the interstate highways to the trucking companies—letting them set the rates and service conditions of public access and allowing the trucking companies to sell these monopoly rights to the Dutch government.

The tyranny of monopoly gave birth to economic and antitrust regulation in the nineteenth century. (Congress regulated the monopoly railroads in 1887 and passed the Sherman Antitrust Act just three years later.) A nation that fails to learn from history is doomed to repeat it.

The *Wall Street Journal* recently asked Americans to identify the industries in which they have the most, or the least, confidence. The largest number by far, 43 percent, said they had no confidence in the airline industry. The disapproval ratings for the industries that followed—insurance (27%), banking (23%), oil and gas (22%), and stockbrokers (22%)—were not nearly as high.[61]

Note the common denominator of these five industries. Insurance has never been regulated by the federal government, and airlines, banks, oil and gas companies, and securities have all undergone significant deregulation during the last decade.

Before deregulation, our transportation system was universally acclaimed to be the world's finest. Since then, the deterioration in our transportation infrastructure, public and private, would embarrass a Third World nation. The potholes we dodge on the highways and the aging jets in which we fly are symptoms of a malignant illness.

The failure of deregulation disproves the implicit thesis of the theology of laissez-faire—that unconstrained human greed will produce a better society. It is time for a spoonful of regulatory medicine, while there is still some modicum of competition to preserve. It is time to roll back deregulation, not to the strict regime of the early 1970s but to an enlightened regime of responsible government oversight. It is time for regulatory reform.

NOTES

1. Dempsey, *The Rise and Fall of the Civil Aeronautics Board,* 11 TRANSP. L. J. 91 (1979).

2. *See generally* P. DEMPSEY, FLYING BLIND: THE FAILURE OF AIRLINE DEREGULATION (1990).

3. Passell, *Why Only a Few Big Airlines Prosper in a Deregulated Sky,* N. Y. TIMES, at A1, C8 col. 1.

4. *See* Dempsey, *The Rise and Fall of the Civil Aeronautics Board: Opening Wide the Floodgates of Entry,* 11 TRANSP. L. J. 91 (1979).

5. *Eastern Asset Distribution February 5, 1991,* AVIATION DAILY, Feb. 7, 1991, at 258. However, the U.S. Department of Justice has objected to United's acquisition of Eastern's landing slots and gates at Washington National Airport. United is already the dominant airline at Washington Dulles Airport.

6. *Losses Color 1990 Red for U.S. Airlines,* AVIATION DAILY, Jan. 2, 1991, at 5.

7. *Id.*

8. *Id.; Midway-USAir Deal Anti-Trust Implications under Review,* AVIATION DAILY, Oct. 22, 1990, at 143.

9. *DOT Approves Continental Tokyo Route Transfer to American,* AVIATION DAILY, Jan. 10, 1991, at 60.

10. *USAir Furloughs More Employees, Slates Several Facilities for Closing,* AVIATION DAILY, Feb. 12, 1991, at 280.

11. American, United, and Delta account for 47 percent of the revenue passenger miles flown by U.S. carriers in 1990. AVIATION DAILY, Jan. 29, 1991, at 189. The five disintegrating airlines—Continental, TWA, Pan Am, Eastern, and Midway—together accounted for 28 percent of the revenue passenger miles in 1990. *Id.*

12. P. DEMPSEY, *supra* note 2.

13. *Five Major Airlines Enough for Competition, Secretary Says,* AVIATION DAILY, Feb. 6, 1991, at 241.

14. *DOT Secretary, Labor Differ on Blame for Industry's Ills,* AVIATION DAILY, Feb. 11, 1991, at 273.

15. *DOT Secretary Opens Door for Increased Foreign Ownership of U.S. Airlines,* AVIATION DAILY, Jan. 24, 1991, at 151; McGinley, *Transport Aide Backs Raising Limit On Foreign Holdings in U.S. Airlines,* WALL ST. J., Feb. 20, 1991, at A8.

16. Shugrue, Jr., *More Airlines Will Share Eastern's Fate Unless We Act Now to Save Them,* USA TODAY, Jan. 31, 1991.

17. Flint, *Don't Blame It All on Fuel,* AIR TRANSPORT WORLD, Feb. 1991, at 32.

18. *See Testimony of Philip Baggaley* (vice president, Standard & Poor's) *before the Aviation Subcomm. of the House Comm. on Public Works and Transportation* (Feb. 6, 1991), at 3. Baggaley says the industry's profit margin was 1 percent during this period. As we shall see below, other sources suggest the profit margin was only .6 percent.

19. *See* Dempsey, *Antitrust Law and Policy in Transportation: Monopoly I$ the Name of the Game,* 21 GA. L. REV. 505 (1987).

20. "The decline in fuel prices [of 31 percent between 1985 and 1986] encouraged airlines to continue to operate fuel-inefficient aircraft beyond the point at which they would have been retired." Flint, *supra* note 17.

21. Several of these characteristics, or derivations of them, have been identified by other sources, including work done on the subject by Airline Economics, Inc.

22. Of the gates at the nation's 66 largest airports, 88 percent are leased to airlines, and 85 percent of the leases are for exclusive use. *Intelligence,* AVIATION DAILY, Aug. 20, 1990, at 323.

23. GENERAL ACCOUNTING OFFICE, AIRLINE COMPETITION: HIGHER FARES AND REDUCED COMPETITION AT CONCENTRATED AIRPORTS (1990).

24. GENERAL ACCOUNTING OFFICE, AIRLINE COMPETITION: INDUSTRY OPERATING AND MARKETING PRACTICES LIMIT MARKET ENTRY 4 (1990).

25. General Accounting Office, *Testimony of Kenneth Mead before the Aviation Subcomm. of the U.S. Senate Commerce Comm.* 6 (Apr. 5, 1990).

26. *Id. at* 6.

27. GENERAL ACCOUNTING OFFICE, *supra* note 24.

28. GENERAL ACCOUNTING OFFICE, *supra* note 23, at 27. Airlines attempt to induce travel agents to book flights with them by offering commission overrides, which offer economic inducements for exceeding quotas. A poll of travel agents reveals that more than half "usually" or "sometimes" select a carrier in order to obtain override commissions. *Id.* at 29.

29. GENERAL ACCOUNTING OFFICE, AIRLINE COMPETITION: IMPACT OF COMPUTERIZED RESERVATIONS SYSTEMS (1986).

30. *See* Nomani, *Fare Warning: How Airlines Trade Price Plans,* WALL ST. J., Oct. 9, 1990, at B1.

31. *See* Cowan & Gargan, *Mirage of Discount Air Fares Is Frustrating to Many Fliers,* N.Y. TIMES, Apr. 22, 1991, at 1.

32. General Accounting Office, *Testimony of Kenneth Mead before the Subcomm. on Aviation of the House Comm. on Public Works and Transportation: Meeting the Aging Aircraft Challenge* (Oct. 10, 1989).

33. UAL CORPORATION, ANNUAL REPORT 7 (1990).

34. Memorandum from Samuel K. Skinner to Congressman James Oberstar, Oct. 25, 1990.

35. AMR CORPORATION, ANNUAL REPORT 27 (1990).

36. GENERAL ACCOUNTING OFFICE, *supra* note 23, at 24.

37. *US Airline Deregulation a Financial Disaster, AFN Study Shows,* COMMUTER REGIONAL AIRLINE NEWS (Apr. 8, 1991), at 8.

38. Smith, *Pan Am Stock Soars As Icahn Makes New Bid,* WALL ST. J., Dec. 18, 1990, at A4.

39. Mahoney, *Financial Fog Still Dogs Continental Airlines,* DENVER POST, Aug. 12, 1990, at G-1.

40. AVIATION DAILY, July 30, 1990, at 192; AVIATION DAILY, Feb. 19, 1991, at 326.

41. AVIATION DAILY, Feb. 13, 1991, at 297.

42. Smith, *supra* note 38.

43. AVIATION DAILY, Feb. 13, 1991, at 297.

44. *Id.*

45. *Id.*

46. Smith, *Trump Bid $7.54 Billion to Acquire American Air*, WALL ST. J., Oct. 6, 1989, at A3.

47. AVIATION DAILY, November 6, 1986.

48. *Flight Attendant Work Force Grows 10 Percent, Salaries Mostly Unchanged*, AVIATION DAILY, Feb. 12, 1991, at 285.

49. Continental has the lowest labor costs, as a percentage of operating expenses, of all major U.S. airlines. AVIATION DAILY, Feb. 11, 1991, at 276.

50. *Id.*

51. M. Jedel, Post Deregulation Strategic Employment Relations Response of the Successful, Surviving Major Domestic Airlines: A Story Not Fully Told 42 (unpublished monograph).

52. P. DEMPSEY, *supra* note 2.

53. P. DEMPSEY, THE SOCIAL AND ECONOMIC CONSEQUENCES OF DEREGULATION (1989).

54. *Statement of Samuel Skinner before the Aviation Subcomm. of the House Comm. on Public Works and Transportation* (Oct. 4, 1989), at 4. Had the management/pilot deal for United not fallen through, British Airways was prepared to supply $570 million, or 78%, of the total $965-million equity. Valente & McGinley, *UAL Machinists Refuse to Back Buy-Out Plan*, WALL ST. J., Oct. 5, 1989, at A6.

55. *Statement of Samuel Skinner, supra* note 54, at 4–5. In September 1989, Skinner jawboned Checchi and Northwest into agreeing, inter alia, to limit KLM's voting stock to 25 percent and to limit KLM's representation on Northwest's board of directors to "matters relevant to KLM's pecuniary interest, recusing himself or herself when the board is dealing with certain matters, such as bilateral negotiations and competitive issues." *Id.* at 6.

56. McGinley, *Transport Aide Backs Raising Limit on Foreign Holdings in U.S. Airlines*, WALL ST. J., Feb. 20, 1991, at A8.

57. Dempsey, *Aerial Dogfights over Europe: The Liberalization of EEC Air Transport*, 53 J. AIR L. & COM. 615 (1988); P. DEMPSEY, *supra* note 2, at 93–108, 241–56.

58. *Repeating Mistakes*, JOURNAL OF COMMERCE, Aug. 30, 1989, at 8A.

59. *Testimony of Helane Becker* (vice president, Lehman Brothers) *before the Subcomm. on Aviation of the House Comm. on Public Works and Transportation* (Feb. 6, 1991), at 5. *Going Steady*, ECONOMIST, July 22, 1989, at 39.

60. *See generally* P. DEMPSEY, *supra* note 2.

61. Winans & Dahl, *Airlines Skid on Bad Moves, Bad News*, WALL ST. J., Sept. 2, 1989, at B1.

Part IV

PROPOSED SOLUTIONS: THE PROPER RELATIONSHIP BETWEEN GOVERNMENT AND THE MARKET

26

REREGULATION: DARE WE SPEAK IT?

Before deregulation, the United States enjoyed what was universally acclaimed to be the "world's finest system of transportation." Our service was excellent, our fleet was young and technologically efficient, labor enjoyed stability of working conditions and decent wages, and inflation-adjusted airfares had been falling for four decades. But Alfred Kahn thought he could do it better.

For a short while, he did. In the late 1970s, as chairman of the Civil Aeronautics Board, Kahn proved that regulatory reform was a good idea. He proved that airlines could, by lowering fares, tap the elasticities of demand and thereby fill seats that otherwise would have flown empty. It was a win-win situation. Consumers enjoyed lower fares, and airlines enjoyed unprecedented profitability.

Intoxicated with success, Kahn thought that if a little regulatory reform was good, then wholesale deregulation would be better. As a textbook economist, Kahn made three critical assumptions: (1) there were no economies of scale of significance in the airline industry (he described airlines as "marginal costs with wings"); (2) there were no significant barriers to entry, except licensing requirements imposed by regulators; and (3) even if some airlines enjoyed market power here and there, they could not reap monopoly profits because the industry was essentially "contestable"—supracompetitive profits would attract new entrants like sharks to the smell of blood. Thus, Kahn predicted, the airline industry would not become highly concentrated, as many critics of deregulation feared, and the public would enjoy near perfect levels of competition.

After a decade of deregulation, two things are clear. First, many of the essential assumptions advanced by free market economists regarding the inherently competitive nature of the transportation industry (e.g., the likelihood that new competitive entry would emerge because of perceived low barriers to entry and few economies of scale, as well as "contestability" of markets) were, simply, specious. The excessive optimism of how competitive the market would be stemmed from hostility to government regulation and euphoria over textbook economics.[1] Because the essential foundations of this theory were specious, the predictions have proven wrong.

Second, and as a consequence of this erroneous reasoning, the predictions that rested on these assumptions have not been sustained by the empirical evidence. Textbook economics produces one result under deregulation—near perfect competition. But the world in which we live has produced quite another—an anemic industry of megacarriers providing poor service and highly discriminatory pricing. Where airlines compete head to head, they have a tendency to engage in below-cost pricing, leading to debt, bankruptcies, and mergers—concentration by any path. In markets where they share monopolies, they charge radically higher prices to make up for their losses in competitive markets—rampant discrimination. Of course, the anemic nature of the industry is being supplanted by emerging oligopolies and monopolies, which enable carriers to exert market power. Raising prices and cutting service will improve the health of the industry while consumers forgo those deep discounts of which deregulation's proponents have been so proud.

But rather than acknowledge that the theory of deregulation is a failure, deregulation proponents tenaciously insist that the product of deregulation is a profound success—lower fares and more people flying than ever before (although they dodge the pre-deregulation trends on both, which were superior).

If there are problems in the airline industry, deregulation proponents blame everything but deregulation—the failure of the government to enforce the antitrust laws or its failure to build additional infrastructure. And any problems with deregulation can be cured by still more deregulation—repeal the cabotage laws, and let the foreign airlines in, some demand.

To his credit, Kahn has reluctantly admitted that many of the fundamental assumptions on which deregulation was based (including the existence of economies of scale and scope, as well as the theory of contestability) were either overstated or erroneous.[2] He has also admitted that many of the predictions of how deregulation would affect the transportation industry, labor, and the public were overly optimistic.[3] Nonetheless, he still clings to the position that, on balance, airline deregulation has been a success.[4]

But Kahn's dogged insistence that deregulation is a success produces a tragic result. Many in Congress and most in the White House believe he is right. So long as they do, despite the crisis of disintegrating airlines (Eastern, Pan Am, Midway, TWA, Continental, America West, and possibly USAir), we will have nothing more from Congress than a few rather ineffective Band-Aids at most.

If we want to preserve the level of competition that now exists, a more comprehensive solution to the problems of deregulation must be enacted—not reregulation of the kind we had in the early 1970s but something between that and the current environment of laissez-faire—something perhaps akin to the regulatory reform we enjoyed in the late 1970s. Yet as long as policymakers inside the Washington, D.C., beltway praise the Emperor Kahn's new clothes, a meaningful legislative solution is unlikely.

Therefore, Kahn facilitates three unfortunate results: (1) several more major airlines will disintegrate; (2) where the surviving airlines enjoy market power, they will behave as rational wealth maximizers and gouge consumers (as they already do at concentrated hubs); and (3) this situation will become so intolerable that Congress will eventually be compelled to reregulate. Unfortunately, reregulation of monopolies is a much more comprehensive endeavor than regulation of competition.

By insisting deregulation is a success, Kahn inadvertently prevents the development of a more moderate solution that could preserve the level of competition that now exists, leading inevitably to increased and intolerable concentration and "public utility" reregulation.

What will we have gone through all this for? Service is poor; we fly the oldest fleet of aircraft in the developed world; the industry is more concentrated; labor has been crushed; declines in pricing have been no better than pre-deregulation trends; and with still more concentration (which is quite likely, if not inevitable), pricing will grow worse.

Pricing deserves a few more words. We will have a highly discriminatory pricing structure, one in which (1) Fortune 500 companies (indeed, any firm that does more than half a million dollars in travel business) will receive low, unrestricted fares, (2) vacation travelers, by virtue of yield management, will receive highly restricted but low-excursion fares, but (3) everybody in between (particularly small business travelers and professionals) will pay through the nose.

One day, a historian will be able to write the following words. Because of recession and fuel crisis in the early 1970s, the government (Civil Aeronautics Board) imposed stringent regulation in order to preserve the number of firms that then existed—a healthy dose of short-term regulation to preserve long-term competition. Not one airline went bankrupt, but tight regulation was politically intolerable and led to a legislative response of deregulation. For every action, it seems, there is an equal and opposite reaction.

Despite recession and a fuel crisis in the early 1990s, our government took no action to preserve the number of firms that then existed. Its hands were tied with deregulation. Its failure to regulate in the short-term produced many airline failures and ultimately long-term concentration. The resultant national oligopoly and regional monopolies, and their highly discriminatory pricing and service offerings, were politically intolerable in an infrastructure industry as important to commerce, communications, and national defense as airlines and served as a catalyst for legislative reregulation.

Kahn has, on occasion, admitted that government needs to do more, saying that the problems that have emerged "urgently cry out for at least some government remedies."[5] He has called for more stringent antitrust and safety regulation. He has acknowledged a need for more consumer protection. He has even conceded that some sort of pricing regulation may be appropriate to deal with predatory behavior by large firms and that it may be time to consider price ceilings.[6]

There are essentially four alternatives for the protection of economic and social values in an important, privately owned infrastructure industry, like transportation. They are

1. heavy-handed regulation;
2. regulatory reform ("light-handed" economic regulation);
3. economic deregulation and antitrust regulation; or
4. laissez-faire.

The first alternative can be as debilitating to the infrastructure and the public it serves as the last.[7] Neither rigid governmental control (like that which existed at the CAB in the early 1970s) nor anarchy (like that which exists today) is a desirable alternative. The responsible choice is between alternatives two and three. It is our belief that alternative two, enlightened regulation, is the better approach.[8] Kahn prefers alternative three.[9]

Kahn has suggested, in a number of forums, that antitrust laws are an adequate substitute for economic regulation in protecting the public interest. They are not. As we have seen, under deregulation, the railroads, airlines, and bus and motor carriers have become more highly concentrated than at any other time in their history. The imperative to merge stems from the destructive competitive environment of deregulation and from the economic anemia created by traffic dilution. Carriers hemorrhaging dollars and facing the alternative of a merger or eventual bankruptcy quite logically choose the former.

Antitrust laws have not effectively been used to deter such consolidations. Take airlines. Although not a single merger has been given antitrust immunity under Section 414 of the Federal Aviation Act, no one has filed a private antitrust action in opposition. Neither have civil or criminal antitrust opportunities been employed, more than incidentally, to challenge predatory behavior by larger transportation firms. Contemporary case law on predation generally does not favor the plaintiff.[10]

In addition to the lack of political will exhibited by the U.S. Department of Justice during the last decade to pursue antitrust violations other than price-fixing, the disincentive for private parties to use antitrust as a civil means of correcting market failure has a number of reasons, including the high cost and consumption of time in pursuing an antitrust action, the significant evidentiary hurdles, and the fact that contemporary case law is not particularly sympathetic to plaintiffs alleging predation. An aggrieved party stands a better chance of prevailing if he or she follows on the coat-tails of a successful government civil or criminal action, in part because the complex evidentiary record has been assembled. But the lack of contemporary Justice Department enthusiasm for areas of antitrust, other than price-fixing, makes that less feasible. Building such a record from scratch can be extremely expensive.

Kahn blames the "complaisant" DOT antitrust policy in approving every merger submitted to it for the high level of concentration that now exists in the industry. True, national levels of concentration have been advanced by the huge mergers permitted by an irresponsible DOT. But as we saw in Chapter 20 only three hubs (Detroit, Minneapolis/St. Paul, and St. Louis) owe their single-carrier dominance to mergers. The overwhelming majority of hub monopolies owe their existence to the entry and exit opportunities, which are the very heart of deregulation. Today, 85 percent of America's city-pairs are monopolies or duopolies. Moreover, whether our government pursues a lax antitrust approach and approves all mergers, as it did in the 1980s, or pursues a get-tough strategy and insists on liquidations, as it appears to be doing in the 1990s, the inevitable result under either scenario is the same—increased concentration.

Neither do the antitrust laws provide any protection against pricing or service discrimination. The Robinson-Patman Act prohibits discrimination in the sale of goods, not services.

Hence, antitrust is an inadequate substitute for responsible economic regulation in protecting public-interest values of assuring a healthy competitive environment and advancing social objectives that do not find a high priority in a regime of laissez-faire. Kahn has been critical of what he has referred to as the "ideologues of laissez faire."[11] But because the alternative he proposes is, quite simply, not pragmatically available at this point in our legal history, stripping away economic regulation inevitably subjects the industry and the public it serves to alternative four, laissez-faire.

The net result of deregulation is that the five-member Civil Aeronautics Board has, in effect, been replaced by the chief executive officers of the largest five or six airlines. If we learned nothing else from the era of the railroad robber barons, we should have learned that the transportation industry has too many social and economic externalities to allow it to be manipulated by a handful of unconstrained monopolists. The quasi–public utility nature of the transportation industry suggests the need for enlightened regulation in the public interest.

Much of the blame for such undesirable social and economic consequences must, of course, be placed on the shoulders of the governmental officials who implemented deregulation. Both Presidents Carter and Reagan appointed free market economists and deregulation ideologues to the transportation agencies—the Civil Aeronautics Board, the Interstate Commerce Commission (ICC), and the Department of Transportation. As a consequence, these agencies became highly political institutions, taking their marching orders from a decidedly ideological White House. It was the political winds blowing down Pennsylvania Avenue that created the turbulence in the airline industry.

In true free market zealotry, those who led the CAB, the ICC, and the DOT during the past decade embraced deregulation as an end in itself. Deregulation should have been the means, not the end; entry and pricing—if gradually, carefully, and responsibly liberalized—could have produced a more competitive environment than what has emerged.[12] Had a more responsible and practical approach been employed, the turmoil would have been much less onerous. The inherent tendency of the industry to engage in destructive competition could have been ameliorated by responsible governmental oversight.

But zealotry demands immediacy, and there was little evidence of the gradual transition that was mandated by the deregulation legislation. As we saw in Part II of this book, precious little attention was paid to the statutes that Congress passed to implement deregulation. Economists and zealots tend to see truth clearly and view legislation as a nuisance when it conflicts with their vision of nirvana. Unfortunately, Congress has been incapable or unwilling to exert its Constitutional power to regulate interstate and foreign commerce and to reign the agencies in.

The time has come to contemplate rolling back deregulation, reestablishing the appropriate role of government in leveling the playing field, correcting market failure, and protecting those economic and social interests that do not find a high priority in a regime of laissez-faire.

A rising chorus of experts is calling for more regulation. The Washington attorney Robert Reed Gray noted, "There has to be some reregulation to make the industry more tolerable for the people of this country."[13] The *Washington Post* columnist Hobart Rowen put it this way:

The public is entitled to more attention to air safety; to the vigorous use of antitrust laws to break up airline monopolies; to an improvement in the quality of service; to the creation of an air travelers' lobby; and (as Kahn now suggests) to assurance that the roster of experienced air traffic controllers will be brought up to snuff.

If Congress does these things, it doesn't have to use the name "reregulation." A rose by any other name . . .[14]

His colleague Carl Rowan said it even more strongly:

Deregulation has forced all the once-great airlines to become service cheapos, union busters, fare slashers and con artists. . . . We have air travel bedlam.

Re-regulation is what America needs. It would bring back to the air travel business companies who know costs and responsibilities, and the tragedy of economic shortcuts. It would enable passengers to know that you get what you pay for—thorough inspections, first-rate pilots, decent food, [and] competent mechanics.[15]

Regulation is by no means a new concept. It has traditionally been employed to facilitate a number of public-policy objectives that either might not find a high priority in the free market or are necessary to avoid the problems surrounding the existence of imperfect competition. As was said by Vermont Royster, editor emeritus of the *Wall Street Journal:*

Regulation to protect consumers is almost as old as civilization itself. Tourists to the ruins of Pompeii see an early version of the bureau of weights and measures, a place where the townsfolk could go to be sure they weren't cheated by the local tradesmen. Unfortunately a little larceny is too common in the human species.

So regulation in some form or other is one of the prices we pay for our complex civilization. And the more complicated society becomes, the more need for some watching over its many parts. We shouldn't forget that a great deal of regulation we encounter today in business or in our personal lives arose from a recognized need in the past.[16]

In the United States, private ownership of the means of production has been deemed to provide the optimum incentives for efficiency in our economy. Nonetheless, the need for government to facilitate the market's ability to accomplish desirable social and economic objectives has long been recognized.[17]

For us to achieve societal ends other than those resulting from humankind's pursuit of wealth, the regulatory mechanism provides broad perimeters for production and pricing of privately owned firms. Regulation provides an equitable balance of public-interest objectives and market imperatives.[18]

NOTES

1. Many economists have been honest enough to admit that many of the fundamental assumptions on which deregulation theory rested were specious. See Levine, *The Legacy of Airline Deregulation: Public Benefits, But New Problems,* Av. WEEK & SPACE TECH., Nov. 9, 1987, at 161.

2. *Testimony of Alfred Kahn before the California Public Utilities Commission* (Jan. 31, 1989).

3. *Id.*

4. Kahn, *Airline Deregulation—A Mixed Bag, But a Clear Success Nevertheless,* 16 TRANSP. L.J. (1988).

5. *Air Travel Altered by Deregulation,* DENVER POST, Oct. 31, 1988, at 9C.

6. *Ex-Official Suggests Lid on Air Fares*, Rocky Mountain News, Nov. 5, 1987, at 100. *Interview with Alfred Kahn*, Antitrust (Fall 1988), at 6–7.

7. Dempsey, *The Rise and Fall of the Civil Aeronautics Board—Opening Wide the Floodgates of Entry*, 11 Transp. L. J. 91 (1979).

8. Dempsey, *Market Failure and Regulatory Failure as Catalysts for Political Change: The Choice between Imperfect Regulation and Imperfect Competition*, 45 Wash. & Lee L. Rev. (1989).

9. Kahn appears to have a particular aversion to economic regulation, describing regulators as "typically very anal." Testimony of Alfred Kahn, *supra* note 1, at 6246.

10. *See* Matsushita Electric Industrial Co. v. Zenith Radio Corporation, 475 U.S. 574 (1986).

11. *Airline Deregulation, Hearings before the Subcomm. on Antitrust of the Senate Judiciary Comm.*, 100th Cong., 1st Sess., 64 (1987).

12. *The Rise & Fall of the CAB*, *supra* note 7, at 176.

13. Knox, *Policy Shift Silent Factor in Crash?*, Rocky Mountain News, Oct. 4, 1988, at 2B.

14. Rowen, *Airline Deregulation at 10: Did the Theory Fail?*, Washington Post, Oct. 16, 1988, at H1, H10.

15. Rowan, *Airline Deregulation Bigger Threat Than Bomb*, Arkansas Gazette, Jan. 1, 1989.

16. Royster, *'Regulation' Isn't a Dirty Word*, Wall St. J., Sept. 9, 1987, at 36.

17.

America's economic system is based on the belief that a competitive, free enterprise system is the best means of achieving national economic goals. Among these goals are minimum unemployment, a low rate of inflation, adequate supplies of goods and services, and an increasing standard of living.

In some industries, the operation of the competitive, free enterprise system does not result in attaining these economic goals. This is because these goals sometimes conflict with the principal goal of private business, which is to maximize profits. For example, it may be more profitable for businesses to limit the supply of a product, thereby raising its price, than to produce a large enough supply to satisfy demand for the product. Limiting supply, however, may reduce the number of jobs in the industry, cause inflation, and negatively impact the standard of living. . . . [T]o prevent this from occurring, government regulation may be used as a means of altering the existing market (i.e. economic environment) to achieve economic goals.

Government regulation is also used to achieve political and social goals when the economic system is unable to achieve these goals. These include such goals as national defense, regional development, and social equity. Like economic goals, political and social goals sometimes cannot be achieved through the economic system because they conflict with businesses' goal to maximize profits.

Colorado State Auditor, Performance Audit of the Public Utilities Commission 14–15 (1988).

18. Dempsey, *supra* note 8.

27

PUTTING THE AIRLINES BACK ON COURSE: A MODEST LEGISLATIVE AGENDA

To suggest a need for reform of deregulation is not to say that we need to return to the tight-fisted regulatory regime of the early 1970s.[1] Nor could we, even if we wanted to. The structural changes have been so profound that we cannot restore what was lost when Pandora's box was opened. CAB Chairman Kahn was true to his promise: "We will so scramble the eggs that no one will be able to put them back into their shells again."[2] But we do need enlightened governmental oversight to correct for market failure and to achieve desirable social benefits.

Among the issues that should be considered are entry, pricing, antitrust, small community access, consumer protection, safety, and regulatory reorganization.

ENTRY

Let us address the most difficult question first—whether entry should be regulated. A good argument could be made that thin air-transport markets capable of supporting only a single carrier are in the nature of natural monopolies and should, like local electric, telephone, and gas distribution markets, be limited to but a single regulated firm. Since only one firm can survive, it would be wasteful of society's resources to have two fight it out to the death.

If entry regulation is imposed, monopoly pricing must, of course, be constrained. Hence, rate regulation is essential. But limiting entry can induce lethargy over the long term. To prevent this, the regulatory agency might issue a certificate for a specific term of years and be willing to replace the incumbent with a more vigorous firm at the end of the term if the incumbent appears not to be as efficient and economical as it might.

For reasons discussed above, spokes between rival carrier hubs may be, oddly enough, natural duopolies. Since only carriers with beyond-segment feed into the city-pair market can ordinarily survive, those without a hub in one of the end points will likely fail.

The more difficult question is whether entry should be limited in other markets, and here it is difficult to say. Enhanced competition is undoubtedly good for consumers, at least in the short run as carriers enter into a competitive war of price discounting. But as we saw in earlier chapters, because carriers competing vigorously tend to price below cost and engage in destructive competition, they hemorrhage dollars unduly, slash service, defer maintenance and replacement of aged equipment, and spiral downward into bankruptcy or, as an alternative, merge into larger and larger firms.

Alfred Kahn has suggested that the cabotage laws be repealed so that foreign airlines can compete in domestic markets. That not only would reintroduce the problems of destructive competition from which the industry is only now escaping but also would create national security concerns.

In the same way that local distribution electric power, gas, and telephone companies are efficient monopolies, airline hubs provide some system distribution efficiencies and economies of scale. But megacarrier domination of multiple hubs reduces the likelihood of new entry and pricing and service innovations.

One means of enhancing national and city-pair competition might be to impose a limit on the number of hubs a carrier may dominate. Assume, for example, that Congress passed a law prohibiting an airline from dominating more than 60 percent of the gates, landings, takeoffs, and passengers at more than a single airport. In other words, an airline could maintain a monopoly at only one airport. Let us further assume that an airline with a hub monopoly would be prohibited from having more than 25 percent of the gates, landings, takeoffs, and passengers at any other airport.

Several beneficial results would be realized. Carriers would be forced to divest themselves of all hubs but one. Thus, for example, Northwest Airlines (which today dominates the hubs of Minneapolis/St. Paul, Detroit, and Memphis) might be split into three carriers: Northwest, hubbed in Minneapolis; Air Michigan, hubbed in Detroit; and Air Memphis, hubbed in Memphis. Similarly, the other megacarriers would likely split or spin off lesser hubs. No longer would the national system be dominated by a handful of gargantuan airlines. City-pair competition would improve.

Moreover, our Air Memphis might eventually saturate its growth opportunities on spokes radiating from Memphis. This might encourage expansion into other nonhub markets, thereby restoring some of the nonstop service that deregulation eradicated.

Hubbing-and-spoking, the dominant megatrend on the deregulation landscape, is choking the air transport system. New nonstop service overflying hubs might be inaugurated if airlines could receive a protected franchise for a term of years. A franchise to serve any city-pair not receiving nonstop service ought to be available to an airline promising to provide at least one round-trip a day. The airline would receive an exclusive franchise to serve the market for, say, 3 to 5 years. If necessary, designated carriers would receive access to congested airport gates and slots, perhaps through the use of federal eminent domain power, to condemn the necessary property at fair market value and sell it to the franchisee. Preference would be given to weak airlines and new entrants.

Of course, for the same reasons that price ceilings are imposed on electric, gas, and telephone monopolies, price ceilings would have to be imposed on airline monopolies as well—to prohibit the extraction of monopoly rents. To protect consumers, the government could set average yields in the market no higher than industry-average yields for similar stage lengths.

Although in earlier periods of American history, direct subsidies were given to bail out transportation firms such as Conrail, Chrysler, and Lockheed (even Amtrak), direct subsidies are today beyond the power of the U.S. Treasury, with its $3-trillion deficit. Nonetheless, weaker carriers, new entrants, and carriers that can best enhance the competitive environment ought to be favored in distributing postal subsidies, international routes, and landing slots. However, these franchises ought not be allowed to be sold for profit. (They generally end up in the hands of the megacarriers when sold.) They should be issued on a limited-term basis and issued, at expiration or on surrender, to whatever carrier best fulfills public needs. The piecemeal sale of carriers (as is being done at TWA and Pan Am) only makes these carriers less attractive for acquisition as a whole property and makes them less viable in the long term.

PRICING

Free market economists predicted that pricing under deregulation would reflect carrier costs. But rates instead tend to reflect the level of competition in a given market. Many markets are so thin that they can support only a single carrier. As we have seen, under deregulation nearly two-thirds of America's city-pair markets are served by but a single airline. Many are in the nature of natural monopolies, for which economic regulation has long been recognized as a legitimate remedy.

Government regulation should be imposed to prohibit the extraction of monopoly or oligopoly rents. An industry-wide mileage-based formula could be devised as a benchmark by which to assess reasonableness of rates, bringing down those that cannot be sustained by a cost justification. Of course, shorter trips have higher per-mile costs than longer ones, so the formula would have to reflect that.

Regulation of rates should be imposed only where the airline has a market share enabling it to exert market power. Thus, rate review might be imposed only on complaint of consumers or in city-pair markets in which the offending airline has more than, say, 40 percent of the market (or, if you prefer a more scientific measure, an appropriate threshold of the Herfindahl-Hirschman Index) and where the rate in question exceeds industry-average fares plus, say, 10 percent—unless the airline can show good cause why the rates should be higher, cause usually in the form of extraordinary costs attributable to serving the market in question. The burden of proof should be placed on the airline charging the allegedly excessive rate. Tight time deadlines should be placed on the agency reviewing the rate, and the agency should be given the power to order refunds of excess fares collected and to order the rate lowered.

The range of rates ought to include not only a ceiling but also a floor, to prohibit predatory pricing and pricing below fully compensatory levels. Even Kahn seems to have acknowledged the propensity of airlines to engage in predatory behavior, saying, "The airline industry clearly demonstrates the dangers of permitting unrestricted responses by incumbents to counter competitive entry, particularly with selective, pinpointed, or targeted price reductions."[3] Pricing below costs to drive a competitor out should be circumscribed.

Regulation can protect smaller competitors from the predatory practices of larger rivals trying to drive them out of business. Judicial antitrust remedies ordinarily award economic compensation only to those injured by such anticompetitive conduct and do not restore the lost competitor to the market. For example, Sir Freddie Laker, victorious in an out-of-court settlement with predatory, competing airlines did not reenter the transatlantic market in which he had pioneered bargain-basement "no frills" service.[4] Thus, the consumer interest in a competitive environment often remains unvindicated by antitrust remedies. In contrast, economic regulation can keep the market flush with small and medium-size competitors engaged in a healthy battle, competitors that will thus discipline the costs and prices of their larger rivals.

The inherent tendency of airlines to engage in destructive competition (because of the instantly perishable nature of the product sold and the extremely low short-term marginal costs of production) also provides a legitimate rationale for economic regulation. Within this "zone of reasonableness" between the aforementioned price ceiling and floor, market forces should establish the rate charged. Carriers with lower costs or lesser service offerings ought to be able to offer their product to consumers at a relatively lower price.

Price discrimination ought also to be reined in a bit, at least between markets. The Robinson-Patman Act prohibits price discrimination in the sale of goods. When the legislation was enacted, there was little perceived need for a prohibition against price discrimination in the sale of services, for the service sector was then a relatively small segment of the American economy, and price discrimination in the infrastructure industries was circumscribed by the regulatory agencies.

But things today are quite different. The regulatory agencies that were established to prohibit discrimination no longer do. And today we have an economy dominated by the service sector. It is time to consider either amending the Robinson-Patman Act to prohibit discrimination in the sale of services or reestablishing the regulatory mechanism for its prohibition.

Although carriers should be free to manage yield to fill seats that otherwise might fly empty and to offer a range of fares to lure customers who might not otherwise fly, discrimination between markets based on the existence of competitive alternatives, rather than costs, should be circumscribed. Recently, a passenger flying from Washington to Cleveland via Detroit paid less than a passenger seated beside him flying from Washington to Detroit.[5] The first rate-regulation provisions ever promulgated by Congress in 1887 included prohibiting a railroad from charging a customer more for a shorter haul than a longer haul on the same line in the same direction. Such a provision would do much to cure the inverse relation between price and costs in the airline industry.

ANTITRUST

Related to Robinson-Patman and other pricing questions are the myriad of antitrust issues that have arisen under deregulation. As noted earlier, in the decade following the Airline Deregulation Act of 1978, there were 51 airline mergers.[6] Until Congress stripped it of its jurisdiction, the Department of Transportation approved each and every one of the 21 mergers submitted to it.

The legislation governing airline mergers and acquisitions should be amended to make them more difficult for competing carriers to consummate. Statutory criteria for mergers should be tightened to emphasize antitrust concerns. Of course, prohibitions against monopoly pricing would do much to ameliorate the problems created by concentration.

The dominance of incumbents is facilitated not only by their stranglehold over the "fortress hubs" but also by the consumer loyalty generated by the free mileage awarded under frequent-flyer programs. Congress should consider a tax on such benefits to discourage their use. As Severin Borenstein has noted, the tax-free nature of the frequent-flyer benefit tends to discourage monitoring by the principal (employer) of the agent (the employee who receives benefits). In effect, businesses pay higher fares than they otherwise would and are reimbursed by the taxpayers.[7] Divestiture of the computer reservations systems owned by the airlines should also be considered, since opportunities for anticompetitive conduct abound.

SMALL COMMUNITY ACCESS

Even if perfect competition existed in transportation (and it does not), society frequently views the achievement of social goals as warranting some sacrifice of allocative efficiency. One public-policy objective that may be enhanced by economic regulation is an equitable geographical distribution of the opportunity to participate in economic growth. Traditionally, prohibitions against rate discrimination required carriers to price their services to small communities at or below cost, facilitating economic growth in all geographic regions. Small towns and rural communities are served by fewer competitors than urban centers and in the absence of regulation are more prone to monopolistic exploitation.

Adam Smith recognized that the width and breadth of the market—the crucial engine for extending the division of labor in his vision—is determined in part by the price and availability of transportation services.[8] The transportation infrastructure is the foundation on which the rest of commerce is built. Without adequate and reasonably priced transportation services, small towns and rural communities cannot sustain economic growth. To have a healthy economy, all communities, large and small, must have nondiscriminatory access to the transportation infrastructure. If a small town does not enjoy adequate transportation service at a fair price, it will be isolated from the mainstream of commerce and will wither on the vine.

Transportation firms are the veins and arteries through which commerce flows. This gives them the leverage to facilitate or impede commerce and makes their rate and service offerings critically important to all who require access to the market for the sale of their products.

If we are to abandon any notion of entry regulation and cross-subsidization at the federal level (and perhaps we should not), then government subsidies for small community access should be not only continued but expanded to provide improved airline service. If the pragmatic political realities of budget deficits preclude sufficient subsidies for air service, then entry and exit regulation should be reconsidered. Establishing a service territory for which a carrier is responsible can be an effective mechanism for insuring adequate service to small towns and rural communities.

CONSUMER PROTECTION

Something has to be done about such practices as bait-and-switch advertising, false and misleading advertising, unrealistic scheduling, demand-based flight cancellations, and the like. Before 1985, the Civil Aeronautics Board provided comprehensive oversight of consumer-related airline policies.[9] Today, government regulations address only two areas of potential abuse: overbooking and lost or damaged baggage. In all other areas of consumer liability, the rules have unilaterally been dictated by the airlines themselves. The judiciary has been less than enthusiastic about picking up the pieces of the shattered regulatory regime of consumer protection.[10]

Deliberate overbooking is a practice that has received the federal government's seal of approval. Carriers routinely book reservations for more passengers than they have seats, assuming some will be "no shows." When there are more passengers than seats, airlines are obliged to ask for volunteers, sometimes bribing them with free flight coupons or paying them a modest penalty.[11] It seems highly unfair for the airline to sell a consumer a nonrefundable ticket when the "confirmed reservation" given the passenger turns out not to be confirmed at all.

As to lost or damaged luggage, government regulations limit liability on domestic flights to $1,250 per person and on international flights to $20 per kilogram.[12] All other liability rules of airlines are required to be set forth in their unilateral "Conditions of Contract of Carriage." Many of these rules are patently unfair to consumers.

Governmental oversight would be prudent in several other areas. For example, penalties for market-inspired flight cancellations should be increased and made mandatory. Carrier liability for missed connections resulting from flight delays should be imposed. Travel agent commission overrides, which provide an incentive for consumer fraud, should be outlawed. Width across seats, and distance between them, should be designated so that average-size people can enjoy a comfortable flight on a long trip without having their knees jammed against the seat in front.

Moreover, the government must intervene to protect consumers against false and misleading advertising. Bait and switch is a pervasive problem. An airline advertises, say, a $199 fare to Orlando; when the consumer calls, he or she is informed that those seats are sold but that there is a bargain immediate-purchase, nonrefundable, Saturday-stay-over seat available for $279. The $199 fare might have been available for only a very few seats. And the fine print often fails to explain the restrictions adequately. Consumer protection demands sensible advertising regulation. Tighter airline advertising regulation has been endorsed by the attorney generals of more than 40 states.[13] Congress could eliminate federal preemption over such questions, letting the state attorney generals loose.

SAFETY

An important public-policy objective that must be promoted by regulation is enhanced margins of safety. Regulation is superior to judicially ordained tort-damage awards for injuries in that however well money can ease the pain of injury, economic compensation for injury frequently cannot restore health and can never restore life. In contrast, regulation attempts to prevent injuries before they occur, thereby protecting the innocent from harm.

To deal with the safety problems that have arisen under deregulation, we need to do several things. As to airlines, the air traffic control system should be refurbished. The FAA needs to restaff the traffic control system beyond the pre-PATCO strike levels of 1981. FAA equipment needs to be updated and upgraded.

Congress should devote sufficient resources to building new airports and expanding existing ones. No new major airport has been built in the United States since 1974, when Dallas/Ft. Worth International Airport was constructed. Since then, national air traffic has doubled, and it will double again by the end of the century. Yet only one new major airport is now being constructed—Denver, scheduled to open in 1993. Local opposition (the not-in-my-backyard syndrome) to noise, congestion, and pollution throws a monkey wrench into new airport expansion and development. Perhaps it is time to consider federal legislation preempting local opposition to regional airport construction.

Congestion at hub airports can be reduced by regulating landings and takeoffs and by imposing peak-period landing fees.[14] This will help flatten out usage somewhat and reduce congestion. Landing fees should also reflect the opportunity costs of delay, which would suggest that a higher landing fee be imposed on small aircraft and a smaller fee imposed on larger aircraft, thereby favoring the larger number of human users of finite public resources.

Enhanced safety requires that more attention be paid to the economic health of firms, since economic anemia seems to be associated with deferred maintenance. But not only economically unhealthy carriers are suspect. Carriers purchased by corporate raiders are also of concern. For example, the Consumer Federation of America has accused Carl Icahn of using TWA's profits to finance his raids on other firms rather than plough back profits into badly needed new aircraft.[15] Frank Lorenzo also stripped Continental and Eastern of essential assets.[16]

Hence, the regulation of carrier fitness should be taken more seriously by DOT, through licensing. The FAA should keep a keener eye on aircraft and pilot qualifications. If that proves inadequate to improve the margin of safety, then more comprehensive regulation that enhances the economic health of the industry may be required. It is doubtful that safety can ever be separated from the economic health of airlines.

A NEW INDEPENDENT FEDERAL
TRANSPORTATION COMMISSION

Much of what is wrong with deregulation is the fault of the agencies that have implemented it and the zeal with which they embraced laissez-faire ideology. The statutes that ordained deregulation called for gradual entry and pricing liberalization. Yet its interpretation was irresponsible, which in large part was attributable to the dominance by and strong ideological agenda of the White House.

We should have expected White House domination of the DOT, for the DOT is, after all, an executive branch agency. Hence, Congress was asking for trouble when it transferred the remaining regulatory responsibilities of the CAB—on its "sunset" on December 31, 1984—to the DOT.[17]

Many of the critics of regulatory commissions allege that after the first decade or two of existence, the commissions tend to favor the interests of the industry they regulate (they become "captured"). After all, the industry is the one constituency regularly before the agency, year after year, pleading its case and looking to the agency for relief while other groups may come and go. The regulated industry is also the best financed of the constituencies that appear before the agency.

A related problem is that of the "revolving door," whereby former government officials are recruited by the industry to serve as executive officers. Ironically, this phenomenon appears under deregulation as well. For example, Alfred Kahn, Mike Levine, and Phil Bakes of the deregulationist CAB and Elliot Seiden of the Reagan Justice Department's Antitrust Division subsequently joined Frank Lorenzo's Texas Air empire.

In the final analysis, there are important regulatory functions to be performed by government, and we have to create a mechanism to perform them without undue political and ideological bias. To avoid the problem of "capture," we should sweep the regulatory functions pertaining to all of transportation (i.e., those functions formerly carried out by the CAB and now the DOT for airlines, by the ICC for rail and motor carriers, by the Federal Maritime Commission for ocean carriers, and by the Federal Energy Regulatory Commission for pipelines) into a new *U.S. Transportation Commission,* an independent federal agency outside the executive branch, or into a Federal Transportation Court, with original, appellate, and regulatory jurisdiction over all modes of transport. An agency with jurisdiction over airlines, motor carriers, bus companies, pipelines, railroads, and domestic and international water carriers would be difficult to capture by any single firm or transport mode.

To enhance its independence, the new Federal Transportation Commission should be composed of at least seven members appointed by the president, with the advice and consent of the Senate, to serve staggered and nonrenewable six-year terms. The members should be selected from a list of candidates prepared by a blue-ribbon panel of industry, labor, and consumer members appointed by the Senate and the president, thereby enhancing the Constitutional mandate of legislative "advice and consent." By calling on an independent body to recommend potential candidates for nomination, we can reduce the propensity of some presidents to fill commissions with political cronies.

The skills and competence of the men and women who serve will, in the final analysis, determine how well broader social needs are fulfilled. Potential commissioners should be selected on the basis of their competence, skill, and neutrality on the issues they will confront. They must have a deep and abiding respect for the law and the supremacy of the legislative branch in defining the perimeters within which they shall administer the regulatory function. Not just the substantive law, which defines the agency's jurisdictional limits, but also the procedural and evidentiary requirements of due process must command the commissioners' fidelity, for the agency will inevitably be quasi-judicial in nature. The commission must be filled with individuals who possess judicial temperament. As Joseph Eastman, Franklin Roosevelt's transportation coordinator, said: "The important qualifications [of a commissioner] are ability to grasp and comprehend facts quickly, and to consider them in their relation to the law logically and with an open mind. Zealots, evangelists, and crusaders have their value before an administrative tribunal, but not on it."[18]

It is a fact of language, politics, and pragmatics that legislation must be drawn broadly, not only because such statutes cannot be drafted with perfect precision (because of practical politics and the limitations of the English language) but also because some flexibility is desirable to enable the commission to address new challenges as they arise. Nonetheless, Congress should make a better effort to tighten the agency's discretion and identify more precisely its jurisdictional perimeters. Congressional committees should perform more rigorous oversight hearings more often, raking appointed officials over the coals when they stray beyond congressional intent. The judiciary should also take a hard look at the orders and rules emanating from regulatory agencies, striking down those that are ultra vires. Legislative and judicial checks and balances should be used to pull the agency to the center, away from political and economic extremes.

To avoid political bias, the commission should include no more than a simple majority of commissioners of a single political party. To alleviate the likelihood of White House domination of the agency's affairs, the commission should be free to elect its own chairman, and no commissioner should be eligible for reappointment. To avoid proindustry bias, commissioners should be restricted in working for the regulated industry when they leave the commission.

Improved process will vastly improve the regulatory function. In fact, had a neutral and responsible regulatory agency without a strong ideological agenda implemented deregulation during the past decade, it is quite likely that the results would have been significantly less onerous.

But suggesting that there is an appropriate role for a regulatory agency should not be construed to mean that we need to return to the rigid regulatory regime of the late 1960s and early 1970s. The period of modest regulatory reform of 1976–78 proved that both the industry and the public it serves can benefit significantly from enlightened regulation. Allowing carriers modest pricing flexibility so that they could tap the elasticities of demand and could fill capacity proved to be a win-win situation for both the airlines and consumers.

Moreover, not even the most omniscient regulatory commission can make all the decisions concerning levels of production and pricing. We leave that to individual, privately owned firms, with regulatory bodies identifying the broad parameters within which the firms may lawfully operate. Regulation at the margins, while allowing privately owned firms to satiate consumer demands, is all that is required. Government should set the perameters, not the particulars, of lawful behavior.

NOTES

1. Dempsey, *The Rise and Fall of the Civil Aeronautics Board—Opening Wide the Floodgates of Entry*, II TRANSP. L.J. (1979).

2. Having scrambled the eggs, it is a bit obscene for Kahn now to claim that if deregulation is a failed experiment, we can simply re-regulate. Said Kahn, "Way back in 1977–78, I recognized the possibility that the price competition we expected deregulation to release would prove to be only temporary, with the industry eventually settling back into non-price-competitive oligopoly; my response was that there was no reason for the government to continue systematically to suppress competition in the first place; and if indeed that eventually ever came to pass, that would be the time to consider re-regulating." Buckley, *Airlines Invite Talk of Regulation*, ROCKY MOUNTAIN NEWS, July 6, 1989, at 35.

3. *Interview with Alfred Kahn*, ANTITRUST (Fall 1988) at 7. Kahn continued, "The nature of entry is not independent of the policies of the incumbents. . . . If you know that if you enter a market you will immediately be met on the nose or even under the nose, that will affect your willingness to enter." *Id.* Kahn expounded on the problem of allowing a competitor to be driven from the market via predatory means:

As for the increasingly respectable view among economists that predation is nothing to worry about—why incur the cost of driving a rival from the market when you're unlikely to be able to sustain monopoly profits because rivals can always reenter?—my answer then was and still is: Does anybody really think that new price competitors will come to the consumer's rescue as promptly as their defunct predecessors? As I once heard Irwin Stelzer observe, a hiker might not pay much attention to a "no trespassing" sign standing alone, but if he sees the field behind it littered with bodies of previous trespassers, it's reasonable to suppose he will respect it.

Kahn, *Deregulatory Schizophrenia*, 75 CALIF. L. REV. 1063, 1067 (1987).

4. *See* Dempsey, *The International Rate and Route Revolution in North Atlantic Passenger Transportation*, 17 COLUM. J. TRANSNAT'L L. 393 (1978).

5. *See* G. BROWN, THE AIRLINE PASSENGER'S GUERILLA HANDBOOK; STRATEGIES & TACTICS FOR BEATING THE AIR TRAVEL SYSTEM (1989).

6. Carroll, *Higher Fares, Better Service Are Forecast*, USA TODAY, Oct. 24, 1988, at 2B.

7. S. Borenstein, *Hubs and High Fares: Airport Dominance and Market Power*

in the U.S. Airline Industry (University of Michigan: Institute of Public Policy Studies, March 1988), at 4.

8. A. Smith, An Inquiry into the Nature and Causes of the Wealth of Nations 19–23 (1985 ed.).

9. *Airlines Reluctant to Disclose Rules,* Wall St. J., Oct. 6, 1988, at B1.

10. P. Dempsey & W. Thoms, Law and Economic Regulation in Transportation 268–73 (1986).

11. *Airline vs Consumer: Your Rights,* Consumer Reports Travel Letter, Apr. 1989, at 40.

12. *Id.*

13. *Tightening of Ad Rules for Airlines Supported,* Rocky Mountain News, Dec. 15, 1987, at 13-B.

14. It should be noted that to the extent that landing fees do not reflect full social costs, the efficiency of hub and spoking may be partly illusory: it intensively uses a resource, peak-period airport facilities, whose private cost lies below its social cost due to subsidization.

15. Answer of the Consumer Federation of America to the Petition of the International Federation of Flight Attendants, DOT Docket No. 45792.

16. Dempsey, *Corporate Pirates Assault the Heavens: Leveraged Buy-Outs and the Airline Industry,* 2 DePaul Bus. L.J. 59 (1989).

17. P. Dempsey, Law and Foreign Policy in International Aviation 234–39 (1987).

18. Eastman, *Twelve Point Primer,* 16 Transp. L.J. 175, 176–77 (1987).

28

CONCLUSIONS

A little more than a decade ago, Congress passed the Airline Deregulation Act of 1978, abolishing the Civil Aeronautics Board, which had regulated the airline industry for 40 years. It was assumed that deregulation would create a healthy competitive environment, with lots of airlines offering a wide array of price and service options and a high level of safety.

During the 1980s, deregulation swept not only through transportation[1] but through the other infrastructure industries as well—telecommunications, broadcasting, cable television, savings and loans, banking, oil and gas, securities, and to a lesser extent, electric utilities. Fortunately the high-water mark of deregulation as a blossoming political movement seems to be behind us, having peaked late in the Carter and early in the Reagan administrations. The flower has lost its bloom. As the American people have had more experience with the grand experiment of deregulation, they have become less enamored with it. Congress has not passed a major deregulation bill in recent years and is considering various reregulation proposals for a number of industries too hastily deregulated, including banking and securities, and for those transport modes that have experienced the most comprehensive deregulation—airlines and railroads.

Our federal experiment with deregulation proves that transportation is not a purely competitive industry and that the theoretical benefits of pure competition have not emerged. The airline industry has become an oligopoly of megacarriers. During the past decade of deregulation, the shakeout of more than 200 airline bankruptcies and more than 50 mergers led to unprecedented levels of concentration.

By 1991, the eight largest airlines in the nation accounted for 95 percent of the domestic passenger market. Regionally, all major hub airports but three are dominated by a single airline. And in city-pair markets, nearly two-thirds are airline monopolies.

A decade ago, deregulation's proponents scoffed at the Cassandras who claimed that deregulation would ultimately result in an industry dominated by four to six giant airlines. Deregulators were mostly starry-eyed free market economists who believed that economic barriers to entry in the airline industry were relatively trivial and that new entry, or the threat thereof, could restore the competitive equilibrium if markets became concentrated. This was the theory of contestability, which provided a major intellectual justification for deregulation. The theory essentially posits that if incumbent airlines raise rates to supracompetitive levels, new entrants will be attracted like sharks to the smell of blood.

In the short run, they were. In the early years of deregulation, new low-cost airlines emerged to rival the established carriers. But where have all the flowers gone? Where are the Donald Burrs and the Sir Freddie Lakers today, with their discount prices and spartan service? The spartan service survived, but the new entrepreneurs have fled a ruthlessly predatory economic environment, never to return, and have taken their discounts with them. With the creation of frequent-flyer programs, travel agent commission overrides, and megacarrier dominance of fortress hubs and computer reservations systems, new entry is today highly unlikely.

To the extent that some pricing competition has occurred (albeit at the expense of a sharp decline in service and safety), these benefits have been unevenly distributed in favor of large markets not dominated by a single carrier. Moreover, such benefits may be a short-term phenomenon. As the dust kicked up by deregulation begins to settle, the benefits are seriously jeopardized by an unprecedented level of industry concentration. As nearly every study of airline pricing under deregulation reveals, there is a strong correlation between less competition and higher prices. Higher prices and poorer service provided by an oligopoly of megacarriers appears to be the megatrend of deregulation.

The empirical results of deregulation also demonstrate that much is lost when the government declines to promote the public's interest in achieving broader societal benefits. Prudently administered economic regulation can accomplish the important public-policy goal of correcting imperfections in the market, such as those resulting from economies of scale and scope, barriers to entry, market power, inequality of bargaining power, insufficiency of information, and externalities.[2] It can also advance important social objectives that do not find a high priority in a regime of laissez-faire. The primordial imperative of economic man is the accumulation of wealth, and this may conflict with society's desire to accomplish other important objectives, such as stimulating economic growth in rural communities and small towns or enhancing safety.

The market for transportation services is not perfectly competitive. Significant economies of scale, scope, and density do exist. Economic barriers to new entry in several of the modes are formidable. Oligopolies and monopolies have resulted. The theory of contestable markets has not been sustained by the empirical evidence.

Another bankruptcy here or another merger there and the industry will be dominated by an even smaller handful of gargantuan airlines. The net result of deregulation is that we will have replaced the five-member Civil Aeronautics Board with the chief executive officers of the five major airlines. And we *know* they have the public interest at heart.

Without government protection, the sky is the limit for prices. The time has come for Congress to reestablish responsible oversight of this industry. After a decade of deregulation, the word *reregulation* no longer sounds so awful. Indeed, it is now *deregulation* that has the disharmonious tone. Already there are growing chants for enhanced regulation of banking, broadcasting, and securities, as well as airline safety and consumer protection.

Only regulation can promote public-interest values that do not find a high priority in a regime of laissez-faire.[3] Economic regulation, responsibly and prudently administered, can foster the following social and economic policies:

- *Avoidance of Problems of Imperfect Competition.* Regulation can avoid problems of concentrations of wealth and power—the oligopoly or monopoly power of large carriers. Market power enables a firm to maximize its profits by raising prices and/or lowering service. The transfer of wealth from consumers to producers is regressive in character and, therefore, undesirable.

- *Equality of Access.* Regulation can insure that all users of infrastructure services, large and small, enjoy equality of access to the market for the sale of their products. Prohibitions against price and service discrimination allow small communities the same opportunity to compete that large cities have.

- *Economic Growth.* Regulation can enhance the social policy of encouraging a geographic distribution of economic growth. Thus, under regulation, small towns and rural communities enjoy adequate service at a fair price, in spite of the fact that less competition for such traffic exists than in larger markets. Regulation can foster economic growth in rural areas by requiring nondiscriminatory access to infrastructure services. Fairly priced transportation services help facilitate access to the broader American economic pie by a larger number and more diverse group of participants. Opportunities for both wealth and pluralism are thereby enhanced. Adequate and reasonably priced infrastructure services are essential for economic growth.

- *Productivity.* Regulation can prevent overcapacity in the transportation industry and thereby improve carrier productivity and economic health. Under regulation, destructive competition can be avoided. It is destructive competition that has mandated the hub-and-spoke systems of air transport as carriers have sought monopoly opportunities to avoid the hemorrhaging of dollars caused by deregulation. Before deregulation, there had been significant growth in the use of widebodied aircraft flying nonstop in dense markets. Under deregulation, hub systems demand the use of relatively smaller aircraft, thereby creating a slide in efficiency. Moreover, circuitous routing via hub airports wastefully consumes fuel and time and narrows the margin of safety.

- *Safety.* By enhancing productivity, economic regulation allows efficient and well-managed carriers to earn a reasonable return on investment and thereby enables them to repair or replace aged and worn equipment.

Private ownership of the means of production inspires the efficient and economical allocation of scarce resources. Important public benefits, these all ought to be encouraged under enlightened regulation. But government oversight of some managerial decisions can protect other public-interest values, beyond allocative efficiency. Administrative agencies can temper market imperatives with protection of broader social and economic values, can insure that the economies and efficiencies of private ownership are tapped for the public good, can avoid the problems of imperfect competition, and can foster those public-interest values that do not find a high priority in an environment of laissez-faire.

Deregulation went awry because it rested on a foundation of false assumptions. A few industries resemble public utilities: the public interest is paramount, and market failure cannot be tolerated. Airlines are among them.

The debate over what should be done with an infrastructure industry as important to the nation as airlines has been cast in terms of two options: regulation, of the type that existed before the mid-1970s, and deregulation, of the kind we have today. But neither rigid, heavy-handed regulation nor the existing environment of laissez-faire is a desirable alternative.

The public debate must begin to move beyond these polar extremes and explore more moderate alternatives between. Neither governmental control nor unregulated competition is a perfect environment. The real choice is between imperfect regulation and imperfect competition. But if applied with a gentle touch, economic regulation ought to be able to yield the best of both worlds—the economies and efficiencies of private ownership and the accomplishment of social and economic policies in the highest public interest.

Regulation of some type is dearly needed, for the public interest in safe, adequate, reliable, efficient, and reasonably priced transportation service is paramount. At the very least, we need governmental oversight of monopoly and predatory pricing, consumer protection, and safety. The time has come to take a fresh look at the mess that deregulation has made and to devise an enlightened response.

Transportation has too vast a social and economic impact to leave it to the whims of a dwindling club of unconstrained monopolists. We ought to have the courage and wisdom to admit we made a mistake. We need to rectify a market that has gone sour. The time has come to roll back deregulation.

NOTES

1. Legislation deregulating transportation was not confined to aviation. The federal statutes partially deregulating various aspects of the transportation industry include the following:

The Railroad Revitalization and Regulatory Reform Act of 1976

The Air Cargo Act of 1977

The Airline Deregulation Act of 1978

The International Air Transportation Competition Act of 1979

The Motor Carrier Act of 1980

The Staggers Rail Act of 1980

The Household Goods Transportation Act of 1980

The Bus Regulatory Reform Act of 1982

The Shipping Act of 1984

The Civil Aeronautics Board Sunset Act of 1984

The Freight Forwarder Deregulation Act of 1986

2. Dempsey, *Market Failure and Regulatory Failure as Catalysts for Political Change: The Choice between Imperfect Regulation and Imperfect Competition*, 45 Wash. & Lee L. Rev. (1989).

3. *Id.*

EPILOGUE: TOWARD A NATIONAL DEREGULATION DAY

I know I am in America, land of the free, when I hear the Fat Lady sing the national anthem off key, or I savor the benefits of deregulation. Both bring tears to my eyes. Deregulation is as American as motherhood, apple pie, and P. T. Barnum. It is high time for Congress to declare a National Deregulation Day so that we Americans can come together and celebrate its profound accomplishments.

Nothing brought Americans so close as airline deregulation. By jamming seats together and flying us through constipated hub airports, we rub elbows and knees for hours. Pity the FAA won't let airlines sell standing room in the aisles.

Deregulation added excitement to what once was a dull trip. After 200 bankruptcies and 50 mergers, we now fly the oldest and most repainted fleet of aircraft in the developed world. Buying an airline ticket has become as thrilling as a trip to Las Vegas. If we buy a nonrefundable ticket three weeks ahead, promise to sleep in a strange city on Saturday night, and pledge our first-born child, we fly at a fraction of the price paid by the poor guy in the rumpled business suit sandwiched between us.

But not just airlines were deregulated. Before deregulation, we had to thumb through two thick bus schedules (one for Greyhound and one for Trailways). Now there's just one bus line, and it skirts most of those boring little towns in Buffalo Commons.

Before deregulation, television was a "vast wasteland." No longer. Now we can watch full-length half-hour commercials, or network episodes where teenagers passionately surrender their virginity. Couldn't see that before deregulation.

Cable TV prices have risen at a rate only three times inflation. We now get to watch 20 times the "Gilligan's Island" reruns at only three times the price.

Sure, before deregulation we could pick up our telephones and get a clear, crisp line to anywhere in America. But now we can pick up our car phone and dial 1-900-PORN from anywhere in town. We couldn't hear that before deregulation.

Of course, savings and loan deregulation will cost us half a trillion dollars, minimum ($5,000+ per taxpayer). You could hear them sing "Pennies from Heaven" on Wall Street as they took their one-third cut. It's a small price to pay for those cute personal checks with the misty nature scenes that once were available only from banks.

We have a lot to be thankful for. We live in a country where every American has a God-given right to play Monopoly, where the strong can exploit the weak, and a fool and his money are soon parted with junk bonds.

We should support the patriotic members of Congress who courageously wage a relentless war on the last vestiges of regulation in industries like banking, trucking, and electric utilities.

Sure, credit card interest rates have become frozen at the 18–21 percent levels to which they ascended in the late 1970s, when peacetime inflation reached its highest levels ever. More banks bankrupted in the last decade than in the Great Depression. Maybe Congress didn't deregulate enough. Full deregulation will bring the emergence of the two or three megabanks that can finally control this country.

Motor carrier deregulation bankrupted two-thirds of the general freight trucking companies. But there is still one-third to go.

Electric utility deregulation will free the corporate raiders to devour the last great bastion of public assets. Leveraged buyouts are yet another chapter in the predatory saga of Market Darwinism, a creature of the Jeffrey Dahmer school of economics. While it's hard to find an asset-stripped company now stronger, more productive, or better able to compete in a global environment, gaze with envy at the mountain of dollars looted by the corporate pirates. Although some went to prison (Club Fed, actually), most realized the American dream, walking away as zillionaires, scot free. Look while you can, before the lights go dim. The electric company is next.

The Fortune 500 lavishly finance a gaggle of Washington-based think tanks to advance the theology of laissez-faire and its implicit thesis—that unconstrained human greed will produce a better society. Their platoons of free market economists (alchemists, really) collate mountains of data to prove that stripping away layers of government produces perfect competition and saves consumers billions. So too, stripping the Earth's atmosphere of ozone will let in more light.

And it will. Before deregulation, America was floundering, without clear direction, coddling its working class while the Japanese were kicking out its teeth. The Reagan administration gave us direction. It kicked the teeth out of organized labor while coddling the Japanese.

Free trade is trade deregulated. Free trade means a working man can keep his job so long as he's willing to live in a mud hut and eat rice. Those who can't find work will be able to stretch their unemployment checks further with the billions of dollars they save from deregulation.

Let's celebrate National Deregulation Day on a day we all can remember—every Friday the 13th. Bring out the flag and potato salad and bless America, land of the free, where the rich get richer, and the best government is no government. And please, let the Fat Lady sing.

INDEX

About the Authors

PAUL STEPHEN DEMPSEY is Professor of Law and Director of the Transportation Law Program at the University of Denver College of Law. He formerly served as an attorney with the Interstate Commerce Commission and Civil Aeronautics Board. He has been a Fulbright Fellow, was designated the University of Denver's Outstanding Scholar, and has received the Transportation Lawyers Association's Distinguished Service Award.

ANDREW R. GOETZ is Assistant Professor Geography at the University of Denver. He has published research focusing on the geographic distribution of air transportation services since deregulation.